University Texts in the Mathematical Sciences

Editors-in-Chief

Raju K. George, Indian Institute of Space Science and Technology, Thiruvananthapuram, India

Shalabh, Indian Institute of Technology Kanpur, Kanpur, India

S. Kesavan, The Institute of Mathematical Sciences, Chennai, India

Sujatha Ramdorai, University of British Columbia, Vancouver, Canada

Associate Editors

Enrique Zuazua, Universidad Autónoma de Madrid, Madrid, Spain

Kaneenika Sinha, Indian Institute of Science Education and Research, Pune, India

Kapil Hari Paranjape, Indian Institute of Science Education and Research, Mohali, India

K. N. Raghavan, The Institute of Mathematical Sciences, Chennai, India

Kaushal Verma, Indian Institute of Science, Bengaluru, India

Riddhi Shah, Jawaharlal Nehru University, New Delhi, India

V. Ravichandran, National Institute of Technology, Trichy, India

Textbooks in this series cover a wide variety of courses in mathematics, statistics and computational methods. Ranging across undergraduate and graduate levels, books may focus on theoretical or applied aspects. All texts include frequent examples and exercises of varying complexity. Illustrations, projects, historical remarks, program code and real-world examples may offer additional opportunities for engagement. Texts may be used as a primary or supplemental resource for coursework and are often suitable for independent study.

Abhishek Mishra

Special Integrals

Abhishek Mishra
Department of Computer Science
and Information Systems
Birla Institute of Technology and Science
Pilani, Rajasthan, India

ISSN 2731-9318　　　　　　ISSN 2731-9326　(electronic)
University Texts in the Mathematical Sciences
ISBN 978-981-97-7513-2　　　ISBN 978-981-97-7514-9　(eBook)
https://doi.org/10.1007/978-981-97-7514-9

Mathematics Subject Classification: 26A06, 26A09, 33B10, 33E20, 40C10, 40C15

© The Editor(s) (if applicable) and The Author(s), under exclusive license to Springer Nature Singapore Pte Ltd. 2025

This work is protected by the Indian Copyright Act, 1957. Its unauthorized and commercial use is prohibited without obtaining permission from the author. It can be used for the purpose of research or private study.

This work is subject to copyright. All rights are solely and exclusively licensed by the Publisher, whether the whole or part of the material is concerned, specifically the rights of translation, reprinting, reuse of illustrations, recitation, broadcasting, reproduction on microfilms or in any other physical way, and transmission or information storage and retrieval, electronic adaptation, computer software, or by similar or dissimilar methodology now known or hereafter developed.
The use of general descriptive names, registered names, trademarks, service marks, etc. in this publication does not imply, even in the absence of a specific statement, that such names are exempt from the relevant protective laws and regulations and therefore free for general use.
The publisher, the authors and the editors are safe to assume that the advice and information in this book are believed to be true and accurate at the date of publication. Neither the publisher nor the authors or the editors give a warranty, expressed or implied, with respect to the material contained herein or for any errors or omissions that may have been made. The publisher remains neutral with regard to jurisdictional claims in published maps and institutional affiliations.

This Springer imprint is published by the registered company Springer Nature Singapore Pte Ltd.
The registered company address is: 152 Beach Road, #21-01/04 Gateway East, Singapore 189721, Singapore

If disposing of this product, please recycle the paper.

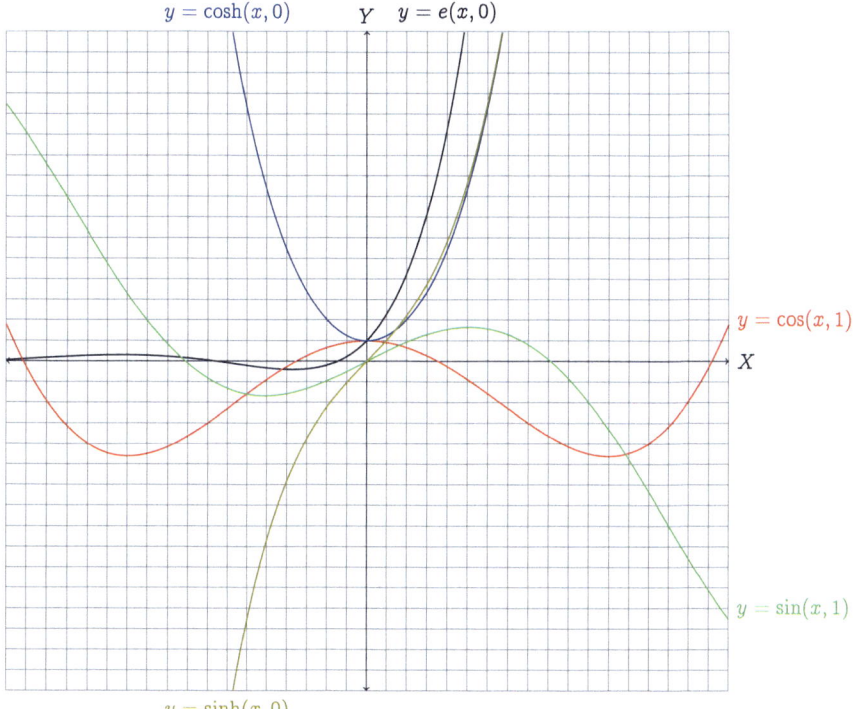

Pseudo-Exponential Functions in $[-18; 18] \times [-16; 16]$

Dedicated to my philosopher father, from whom I Inherited the habit of reading books.

Preface

Having extensive experience in teaching theoretical computer science courses, I have written many lecture notes and designed many original problems for my students to solve. I was thinking of writing a book using these lecture notes and problems. I had to decide on the topic for my first book. I developed my interest in theoretical computer science during my Ph.D. However, even before that, calculus was my first love during my schooldays. Having already done some original work in calculus during my schooldays, I decided to write this book on it.

My motivation for writing this book dates back to the summer of 1997. After taking my class 10 board exam, I was trying to find the sum of the following infinite series.

$$\sum_{j=0}^{\infty} \frac{1}{(j!)^2} = \frac{1}{(0!)^2} + \frac{1}{(1!)^2} + \frac{1}{(2!)^2} + \frac{1}{(3!)^2} + \ldots = ?$$

I succeeded in finding an integral representation for the series as [4.22]

$$\frac{1}{\pi} \int_0^\pi e^{2\cos\theta} \, d\theta = \sum_{j=0}^{\infty} \frac{1}{(j!)^2} = \frac{1}{(0!)^2} + \frac{1}{(1!)^2} + \frac{1}{(2!)^2} + \frac{1}{(3!)^2} + \ldots$$

This motivated me to explore further integral representations of series. This resulted in about a 150 page-long manuscript listing about 380 identities (Mishra, 1997).

The present book is based on the above manuscript. The manuscript contains several theorems without proof that are used in the derivation of the identities. I have proved all the theorems in the chapters. I have given the identities as problems to solve. I found very few of the identities in the manuscript to be incorrect, which I have not included in the book. Some of the identities had minor corrections. I have included the corrected versions of these identities in the book.

Later, I was able to find 32 of the identities in the existing literature. [1.1–1.6], [1.15], and [2.9–2.10] are Lehmer's identities (Lehmer, 1985). [1.50] is a special case of Eq. (2.2.2) on page 37 in Sect. 2.2 of the book by Borwein and Borwein

(1987). Seventy-two similar identities, mainly involving the binomial coefficients, are given in Chap. 1.

[1.73] and [1.83] are Ramanujan's identities. They are given as entry 24 on p. 396 and entry 29 on p. 399, respectively, in Chap. 31 of the book by Berndt (1994). Twenty similar identities involving powers of 2 are given in Chap. 1.

[2.5] and [2.20] are given as Examples XXI-8 and XXI-7, p. 158, respectively; and [2.21] is given as Eq. 3 in Sect. 129 on p. 156 in the book by Loney (1928). Fifty-two similar identities involving the circular and hyperbolic trigonometrical functions are given in Chap. 2.

[2.8–2.9] are given as Theorems 6.6.1–6.6.2 on p. 119, [1.37] is given as Eq. (6.6.10) on p. 120, and [2.10] is given as Eq. (6.6.21) on p. 122 in Sect. 6.6; [4.143] is given as Theorem 6.4.1 on p. 113 in Sect. 6.4; and [4.147] is given as Proposition 12.5.1 on p. 246 in Sect. 12.5 in the book by Boros and Moll (2006).

[4.2] is given as Example 3 on p. 113 in Sect. 6.21 of the book by Whittaker and Watson (2016). [4.83] and [5.18] are Eq. (6) on p. 20, [4.91] and [5.19] are Eq. (7) on p. 21, [4.77] and [5.12] are Eq. (8) on p. 21, [4.85] and [5.13] are Eq. (9) on p. 21 in Sect. 2.2; and [4.22] and [4.24] are special cases of Eq. (9) on p. 79 in Sect. 3.71 of the book by Watson (2011). Three hundred eighteen similar identities involving special integrals and series sums are given in the three chapters (Chaps. 4–6). This is the unique contribution of the present book. To the best of my knowledge, this is the first book with similar topics covered.

The three prerequisites for reading this book are knowledge of *Calculus* (Thomas et al., 2018; Apostol, 2017; Hardy, 1921; Piskunov, 1996a,b; Maron, 1973; Edwards, 1896; Bromwich, 2018), *Complex Analysis* (Ahlfors, 2017; Brown and Churchill, 2017a; Whittaker and Watson, 2016; Lang, 2013; Convay, 1995), and *Fourier Series* (Brown and Churchill, 2017b; Tolstov, 1977; Carslaw, 2018). In *Calculus*, you should be able to differentiate and integrate functions of a single variable. Knowledge of double integrals and differential equations is desirable but not essential in reading major portions of the book. In *Complex Analysis*, you should be familiar with the complex elementary functions and Taylor's and Maclaurin's Series. In *Fourier Series*, you should be able to compute the coefficients of the Fourier series.

Part I of the book has six chapters. In Chap. 1 ("Binomial Series"), you will learn the differentiation and integration methods for summing a series. You will learn how to apply these techniques to find the sum of some well-known series. You will also learn the technique of summing some interesting series involving integral and fractional powers of 2. The problems listed in this chapter mainly involve the binomial coefficients.

In Chap. 2 ("Trigonometrical Series"), you will learn the use of complex variables and the techniques of Chap. 1 to derive some interesting trigonometrical series. The problems listed in this chapter are series sum identities involving the circular and hyperbolic trigonometrical functions.

In Chap. 3 ("Bessel Functions"), first you will learn about the Bessel coefficients $J_n(z)$ and $I_n(z)$. After that, you will learn about their generalizations: the Bessel functions $J_\nu(z)$ and $I_\nu(z)$. The problems listed in this chapter are some of the properties of the Bessel functions.

In Chap. 4 ("Special Integrals"), you will learn the use of complex variables in the integrals and how to use them to get some integral representations of some series sums. The problems listed in this chapter are some special integrals derived using the theorems listed in the chapter.

In Chap. 5 ("Generalized Bessel Functions"), first you learn about the Mittag-Leffler functions $E_\alpha(z)$ (Mittag-Leffler, 1903; Erdélyi, 1955) and their various generalizations: $E_{\alpha,\beta}(z)$ by Wiman (1905), $E_{\alpha,\beta}^{\gamma}(z)$ by Prabhakar (1971), and $E_{\alpha,\beta}^{\gamma,q}(z)$ by Shukla and Prajapati (2007). Then you will learn about the generalized Bessel functions $\phi(\rho, \beta; z)$ as defined by Wright (1933). Using the ideas of Chap. 3, as a special case of $\phi(\rho, \beta; z)$, the pseudo-exponential functions are defined. You will learn how to get their integral representations in the problems listed in this chapter. You will learn that these functions exhibit similar differentiation and integration properties as compared to the exponential function and the circular and hyperbolic trigonometrical functions. The problems listed in this chapter involve the pseudo-exponential functions and the properties of the above-listed functions.

Chapter 6 ("Series Sums Using Special Integrals"), is a continuation of Chap. 4, in which you will learn some more interesting properties of the pseudo-exponential functions. You will also learn a generalized theorem for series sums. As a special case of this theorem, you can derive some of the theorems of Chap. 4. The problems listed in this chapter are some series sums obtained by making use of special integrals.

Part II of the book has solutions to the problems in the six chapters in Part I. You should attempt the problems before referring to the solutions. If you can find a better solution, I will be very grateful to you for knowing the solution. The Appendix ("Mapping from the Manuscript to the Book") tells where to find the corresponding entries of the manuscript in the book.

I am highly grateful to my cousin, Vishal Mishra. He came from Delhi to our place at Banaras Hindu University in Varanasi to complete his Bachelor of Technology in Computer Science and Engineering at the Institute of Technology during the period 1992–1996. During his four years of stay with us, I studied in the school: classes 6–9. I got inspired by him and started asking him mathematics questions. Class 6 was not fruitful for me because I was frightened of algebra. But in class 7, I got deeply interested in mathematics. Under his expert guidance, within six months, I was studying class 12 mathematics. His method of explaining the mathematical concepts was excellent. Taking full advantage of his mentorship, in class 9, I started studying his *Calculus* (Thomas and Finney, 1995). As a result, by class 10, I became proficient in calculus, complex analysis, and Fourier series, and was able to write the manuscript (Mishra, 1997).

I am highly grateful to the anonymous referee for giving me more insights and references that helped to improve the quality of this book. I am also highly grateful to Shamim Ahmad from Springer Nature for the book review process.

I am thankful to my colleagues Sundaresan Raman, Kamlesh Tiwari, Ashutosh Bhatia, and Sudeept Mohan. They helped me in my application for sabbatical leave for writing this and other books. Unfortunately, it was unsuccessful, and I had to drop the idea of writing other books.

Last but not least, I would like to thank my family for their support and encouragement. My mother was happy to know that I was writing a book. My sons Ishitva and Ivaan took away my file for keeping the drafts of the book for playing! My wife Gitika constantly enquired about the status of my book. She was eagerly waiting for my book to be published. This motivated me to make extra efforts to complete the book.

I have reviewed the book several times and found some errors in each revision. If you can find some more errors in the book and inform me, I will be very grateful to you.

Pilani, India   Abhishek Mishra

Contents

Part I Special Integrals

1 Binomial Series ... 3
   1.1 The Differentiation Method for Series Sum ... 3
   1.2 The Integration Method for Series Sum ... 10
   1.3 A Recursive Formula for π ... 12
   Problems ... 15

2 Trigonometrical Series ... 39
   2.1 Trigonometrical Series ... 39
   2.2 Trigonometrical Products ... 50
   Problems ... 55

3 Bessel Functions ... 69
   3.1 Euler's Constant γ ... 69
   3.2 The Gamma Function $\Gamma(z)$ ... 71
   3.3 The Bessel Coefficients $J_n(z)$ ... 72
   3.4 The Bessel Coefficients $I_n(z)$ ... 75
   3.5 The Bessel Functions $J_\nu(z)$ ... 78
   3.6 The Bessel Functions $I_\nu(z)$ ... 79
   Problems ... 80

4 Special Integrals ... 87
   4.1 Special Integrals 1 ... 87
   4.2 Special Integrals 2 ... 111
   Problems ... 126

5 Generalized Bessel Functions ... 163
   5.1 The Mittag-Leffler Functions $E_\alpha(z)$ ... 165
   5.2 Wiman's Generalized Mittag-Leffler Functions $E_{\alpha,\beta}(z)$ ... 165
   5.3 Prabhakar's Generalized Mittag-Leffler Functions $E_{\alpha,\beta}^{\gamma}(z)$ ... 166

	5.4 The Shukla-Prajapati Generalized Mittag-Leffler Functions $E_{\alpha,\beta}^{\gamma,q}(z)$	167
	5.5 Wright's Generalized Bessel Functions $\phi(\alpha, \beta; z)$	167
	5.6 The Pseudo-Exponential Functions	168
	5.7 Derivatives of the Pseudo-Exponential Functions	174
	5.8 Integrals of the Pseudo-Exponential Functions	177
	Problems	178
6	**Series Sums Using Special Integrals**	**191**
	6.1 Polynomial Dependence in the Pseudo-Exponential Functions	191
	6.2 Differential Equations Satisfied by the Pseudo-Exponential Functions	195
	6.3 Series Transformations	199
	Problems	203

Part II Solutions

7	**Solutions to Binomial Series**	**223**
8	**Solutions to Trigonometrical Series**	**257**
9	**Solutions to Bessel Functions**	**271**
10	**Solutions to Special Integrals**	**281**
11	**Solutions to Generalized Bessel Functions**	**333**
12	**Solutions to Series Sums Using Special Integrals**	**355**

Appendix A: Mapping from the Manuscript to the Book 371

References 381

Notations

$[m.n]$	Problem n in Chapter m
$(m.n)$	Equation n in Chapter m
$\mathbb{N}$	The set of all natural numbers: $\{1, 2, 3, 4, \ldots\}$
$\mathbb{N} - 1$	The set of all non-negative integers: $\{n - 1 \mid n \in \mathbb{N}\}$
$\mathbb{N} + 1$	The set of all natural numbers greater than 1: $\{n + 1 \mid n \in \mathbb{N}\}$
$-\mathbb{N}$	The set of all negative integers: $\{-n \mid n \in \mathbb{N}\}$
$2\mathbb{N}$	The set of all even natural numbers: $\{2n \mid n \in \mathbb{N}\}$
$2\mathbb{N} - 2$	The set of all non-negative even integers: $\{2n - 2 \mid n \in \mathbb{N}\}$
$2\mathbb{N} - 1$	The set of all odd natural numbers: $\{2n - 1 \mid n \in \mathbb{N}\}$
$2\mathbb{N} + 1$	The set of all odd natural numbers greater than 1: $\{2n + 1 \mid n \in \mathbb{N}\}$
$-(2\mathbb{N} - 1)$	The set of all negative odd integers: $\{-(2n - 1) \mid n \in \mathbb{N}\}$
$-2\mathbb{N}$	The set of all negative even integers: $\{-2n \mid n \in \mathbb{N}\}$
$\mathbb{Z}$	The set of all integers: $\{sn \mid n \in \mathbb{N}, s \in \{0, 1, -1\}\}$
$2\mathbb{Z} + 1$	The set of all odd integers: $\{2z + 1 \mid z \in \mathbb{Z}\}$
$2\mathbb{Z}$	The set of all even integers: $\{2z \mid z \in \mathbb{Z}\}$
$\mathbb{R}$	The set of all real numbers
$\mathbb{R}^+$	The set of all positive real numbers: $\{x \mid x \in \mathbb{R}, x > 0\}$
$\mathbb{R}^-$	The set of all negative real numbers: $\{x \mid x \in \mathbb{R}, x < 0\}$
$\mathbb{C}$	The set of all complex numbers
LHS	Left-Hand Side
RHS	Right-Hand Side
$(f(j))_{j=1}^{n}$	The ordered n-tuple $(f(1), f(2), \ldots, f(n-1), f(n))$
S^2	The 2-dimensional Cartesian product $S \times S$
xS or Sx	x multiplied by each element of the set S: $\{xs \mid s \in S\}$
$[m..n]$	The closed integer interval $\{p \mid m \leq p \leq n, p \in \mathbb{Z}\}$
$[x, y]$	The closed real interval $\{z \mid x \leq z \leq y, z \in \mathbb{R}\}$
(x, y)	The open real interval $\{z \mid x < z < y, z \in \mathbb{R}\}$

$[x, y)$	The semi-closed real interval $\{z \mid x \leq z < y, z \in \mathbb{R}\}$
$\forall$	For all
$\exists$	There exists
$\forall (n, r) \in$ $(\mathbb{N} - 1) \times [0..n]$	$= \{(n, r) \mid n \in \mathbb{N} - 1, r \in [0..n]\}$
$\forall (z, \theta) \in \mathbb{Z} \times$ $\left((4z-1)\frac{\pi}{n}, (4z+1)\frac{\pi}{n}\right)$	$= \left\{(z, \theta) \mid z \in \mathbb{Z}, \theta \in \left((4z-1)\frac{\pi}{n}, (4z+1)\frac{\pi}{n}\right)\right\}$
$\lfloor x \rfloor$	The greatest integer $\leq x$
$\lceil x \rceil$	The smallest integer $\geq x$
0^0	$= 1$
$[n \geq 0]$	$= 1$ if $n \geq 0$; $= 0$, otherwise
$\sum_{j=m}^{n} a_j \sum_{k=p}^{q} b_k$	$= \left(\sum_{j=m}^{n} a_j\right)\left(\sum_{k=p}^{q} b_k\right)$
$\Re(z)$	Real part of the complex number z
$\Re(\mathbb{C})^+$	The set of all complex numbers with positive real part: $\{w \mid w \in \mathbb{C}, \Re(w) > 0\}$
$\zeta(x)$	The Riemann's zeta function defined by $\zeta(x) = \sum_{j=1}^{\infty} \frac{1}{j^x}$ for $x > 1$

Part I
Special Integrals

Chapter 1
Binomial Series

In this chapter, we will learn about differentiation and integration methods for series sums. We will apply these techniques to get some well-known series sums. The problems listed in this chapter mainly involve the binomial coefficients. Many of the problems are special cases of problems in Chap. 2. The remaining problems can be solved by the techniques developed in this chapter.

We will also learn about summing series involving powers of 2. All these series have one thing in common. They involve two types of terms: several similar terms and one dissimilar term. The dissimilar term consumes the similar terms successively, finally yielding a dissimilar term, which is the sum of the series.

1.1 The Differentiation Method for Series Sum

The idea of using derivatives for finding series sums is straightforward. Given an identity involving a variable x, simply differentiate both sides of the identity with respect to x to get a new identity.

Definition 1.1 (*The Differentiation Method for Series Sum*)
Let

$$f(x) = \sum_{j=0}^{n} a_j x^j, \tag{1.1}$$

be a function. In the differentiation method for series sum, we differentiate (1.1) with respect to x to get

$$f'(x) = \sum_{j=1}^{n} j a_j x^{j-1}.$$

© The Author(s), under exclusive license to Springer Nature Singapore Pte Ltd. 2025
A. Mishra, *Special Integrals*, University Texts in the Mathematical Sciences,
https://doi.org/10.1007/978-981-97-7514-9_1

The above method can also be applied when the series in (1.1) is infinite. We can get an alternative method for the sum of *Arithmetic Progressions (AP)* using this method.

Example 1.2 (*Finding the Sum of First n Natural Numbers*)
Prove that $\forall n \in \mathbb{N}$,

$$\sum_{j=1}^{n} j = 1 + 2 + 3 + \cdots + n = \frac{n(n+1)}{2}.$$

Proof The sum of the *Geometric Progression (GP)* $\left(x^j\right)_{j=0}^{n}$ is given by

$$\sum_{j=0}^{n} x^j = \frac{x^{n+1} - 1}{x - 1}. \tag{1.2}$$

Differentiating (1.2) with respect to x, we get

$$\sum_{j=1}^{n} j x^{j-1} = \frac{(n+1)x^n(x-1) - \left(x^{n+1} - 1\right)}{(x-1)^2}$$

$$= \frac{nx^{n+1} - (n+1)x^n + 1}{(x-1)^2}. \tag{1.3}$$

Taking the limit as $x \to 1$ in (1.3), we get

$$\lim_{x \to 1} \sum_{j=1}^{n} j x^{j-1} = \lim_{x \to 1} \frac{nx^{n+1} - (n+1)x^n + 1}{(x-1)^2}$$

$$\iff \sum_{j=1}^{n} j = \lim_{y \to 0} \frac{n(1+y)^{n+1} - (n+1)(1+y)^n + 1}{y^2}$$

$$= \lim_{y \to 0} \left(n \left(\frac{1}{y^2} + (n+1)\frac{1}{y} + \frac{(n+1)n}{2} + y \sum_{j=3}^{n+1} \binom{n+1}{j} y^{j-3} \right) \right.$$

$$\left. -(n+1) \left(\frac{1}{y^2} + \frac{n}{y} + \frac{n(n-1)}{2} + y \sum_{j=3}^{n} \binom{n}{j} y^{j-3} \right) + \frac{1}{y^2} \right)$$

$$= \lim_{y \to 0} \frac{n - (n+1) + 1}{y^2} + \lim_{y \to 0} \frac{n(n+1) - (n+1)n}{y}$$

$$+ \lim_{y \to 0} \frac{n^2(n+1) - n(n+1)(n-1)}{2}$$

$$+ \lim_{y \to 0} y \left(\sum_{j=3}^{n+1} n \binom{n+1}{j} y^{j-3} - \sum_{j=3}^{n} (n+1) \binom{n}{j} y^{j-3} \right)$$

1.1 The Differentiation Method for Series Sum

$$= \frac{n(n+1)(n-(n-1))}{2}$$
$$= \frac{n(n+1)}{2}.$$

□

Example 1.3 (*Finding the Sum of First n Odd Natural Numbers*)
Prove that $\forall n \in \mathbb{N}$,

$$\sum_{j=1}^{n}(2j-1) = 1 + 3 + 5 + \cdots + 2n - 1 = n^2.$$

Proof The sum of the GP
$\left(x^{2j-1}\right)_{j=1}^{n}$ is given by

$$\sum_{j=1}^{n} x^{2j-1} = \frac{x^{2n+1} - x}{x^2 - 1}. \tag{1.4}$$

Differentiating (1.4) with respect to x, we get

$$\sum_{j=1}^{n}(2j-1)x^{2j-2} = \frac{\left((2n+1)x^{2n} - 1\right)\left(x^2 - 1\right) - 2x\left(x^{2n+1} - x\right)}{\left(x^2 - 1\right)^2}$$
$$= \frac{(2n-1)x^{2n+2} - (2n+1)x^{2n} + x^2 + 1}{\left(x^2 - 1\right)^2}. \tag{1.5}$$

Taking the limit as $x \to 1$ in (1.5), we get

$$\lim_{x \to 1} \sum_{j=1}^{n}(2j-1)x^{2j-2} = \lim_{x \to 1} \frac{(2n-1)x^{2n+2} - (2n+1)x^{2n} + x^2 + 1}{\left(x^2 - 1\right)^2}$$

$$\iff \sum_{j=1}^{n}(2j-1)$$

$$= \lim_{y \to 0} \frac{(2n-1)(1+y)^{2n+2} - (2n+1)(1+y)^{2n} + (1+y)^2 + 1}{((1+y)^2 - 1)^2}$$

$$= \lim_{y \to 0} \frac{1}{y^2(2+y)^2}\left((2n-1)\left(1 + (2n+2)y + \frac{(2n+2)(2n+1)}{2}y^2\right.\right.$$
$$\left.+ y^3 \sum_{j=3}^{2n+2}\binom{2n+2}{j}y^{j-3}\right)$$
$$- (2n+1)\left(1 + 2ny + \frac{2n(2n-1)}{2}y^2 + y^3 \sum_{j=3}^{2n}\binom{2n}{j}y^{j-3}\right)$$
$$+ (1 + 2y + y^2) + 1\Big)$$

$$= \lim_{y \to 0} \frac{(2n-1) - (2n+1) + 1 + 1}{y^2(2+y)^2} + \lim_{y \to 0} \frac{(2n-1)(2n+2) - 2n(2n+1) + 2}{y(2+y)^2}$$
$$+ \lim_{y \to 0} \frac{(2n-1)(n+1)(2n+1) - n(2n+1)(2n-1) + 1}{(2+y)^2}$$
$$+ \lim_{y \to 0} \frac{y}{(2+y)^2} \left(\sum_{j=3}^{2n+2} (2n-1) \binom{2n+2}{j} y^{j-3} - \sum_{j=3}^{2n} (2n+1) \binom{2n}{j} y^{j-3} \right)$$
$$= \frac{(2n+1)(2n-1)((n+1) - n) + 1}{4}$$
$$= n^2.$$

□

With more lengthy calculations, we can compute the sum of squares and cubes of consecutive natural numbers.

Example 1.4 (*Finding the Sum of Squares of First n Natural Numbers*)
Prove that $\forall n \in \mathbb{N}$,

$$\sum_{j=1}^{n} j^2 = 1^2 + 2^2 + 3^2 + \cdots + n^2 = \frac{n(n+1)(2n+1)}{6}.$$

Proof Multiplying (1.3) by x, we get

$$\sum_{j=1}^{n} jx^j = \frac{nx^{n+2} - (n+1)x^{n+1} + x}{(x-1)^2}. \tag{1.6}$$

Differentiating (1.6) with respect to x, we get

$$\sum_{j=1}^{n} j^2 x^{j-1} = \frac{1}{(x-1)^4} \left(\left(n(n+2)x^{n+1} - (n+1)^2 x^n + 1 \right)(x-1)^2 \right.$$
$$\left. - 2(x-1) \left(nx^{n+2} - (n+1)x^{n+1} + x \right) \right)$$
$$= \frac{1}{(x-1)^3} \left(\left(n(n+2)x^{n+1} - (n+1)^2 x^n + 1 \right)(x-1) \right.$$
$$\left. - 2 \left(nx^{n+2} - (n+1)x^{n+1} + x \right) \right)$$
$$= \frac{1}{(x-1)^3} \left((n(n+2) - 2n)x^{n+2} \right.$$
$$\left. + \left(2(n+1) - n(n+2) - (n+1)^2 \right) x^{n+1} + (n+1)^2 x^n - x - 1 \right)$$
$$= \frac{n^2 x^{n+2} - \left(2n^2 + 2n - 1 \right) x^{n+1} + (n+1)^2 x^n - x - 1}{(x-1)^3}.$$
$$\tag{1.7}$$

1.1 The Differentiation Method for Series Sum

Taking the limit as $x \to 1$ in (1.7), we get

$$\lim_{x \to 1} \sum_{j=1}^{n} j^2 x^{j-1} = \lim_{x \to 1} \frac{n^2 x^{n+2} - (2n^2 + 2n - 1) x^{n+1} + (n+1)^2 x^n - x - 1}{(x-1)^3}$$

$$\iff \sum_{j=1}^{n} j^2 = \lim_{y \to 0} \frac{1}{y^3} \left(n^2 (1+y)^{n+2} - (2n^2 + 2n - 1)(1+y)^{n+1} \right.$$
$$\left. + (n+1)^2 (1+y)^n - (1+y) - 1 \right)$$

$$= \lim_{y \to 0} \frac{1}{y^3} \left(n^2 \left(1 + (n+2)y + \frac{(n+2)(n+1)}{2} y^2 + \frac{(n+2)(n+1)n}{6} y^3 \right.\right.$$
$$\left. + y^4 \sum_{j=4}^{n+2} \binom{n+2}{j} y^{j-4} \right)$$
$$- (2n^2 + 2n - 1) \left(1 + (n+1)y + \frac{(n+1)n}{2} y^2 + \frac{(n+1)n(n-1)}{6} y^3 \right.$$
$$\left. + y^4 \sum_{j=4}^{n+1} \binom{n+1}{j} y^{j-4} \right)$$
$$+ (n+1)^2 \left(1 + ny + \frac{n(n-1)}{2} y^2 + \frac{n(n-1)(n-2)}{6} y^3 \right.$$
$$\left.\left. + y^4 \sum_{j=4}^{n} \binom{n}{j} y^{j-4} \right) - (1+y) - 1 \right)$$

$$= \lim_{y \to 0} \frac{n^2 - (2n^2 + 2n - 1) + (n+1)^2 - 1 - 1}{y^3}$$
$$+ \lim_{y \to 0} \frac{n^2(n+2) - (2n^2 + 2n - 1)(n+1) + n(n+1)^2 - 1}{y^2}$$
$$+ \lim_{y \to 0} \frac{n^2(n+2)(n+1) - n(2n^2 + 2n - 1)(n+1) + n(n+1)^2(n-1)}{2y}$$
$$+ \lim_{y \to 0} \left(\frac{n^3(n+2)(n+1)}{6} - \frac{n(2n^2 + 2n - 1)(n+1)(n-1)}{6} \right.$$
$$+ \frac{n(n+1)^2(n-1)(n-2)}{6} \right) + \lim_{y \to 0} y \left(n^2 \sum_{j=4}^{n+2} \binom{n+2}{j} y^{j-4} \right.$$
$$\left. - (2n^2 + 2n - 1) \sum_{j=4}^{n+1} \binom{n+1}{j} y^{j-4} + (n+1)^2 \sum_{j=4}^{n} \binom{n}{j} y^{j-4} \right)$$

$$= \frac{n(n+1)\left(n^2(n+2) - (2n^2+2n-1)(n-1) + (n+1)(n-1)(n-2)\right)}{6}$$

$$= \frac{n(n+1)(2n+1)}{6}.$$

□

Example 1.5 (*Finding the Sum of Cubes of First n Natural Numbers*)
Prove that $\forall n \in \mathbb{N}$,

$$\sum_{j=1}^{n} j^3 = 1^3 + 2^3 + 3^3 + \cdots + n^3 = \left(\frac{n(n+1)}{2}\right)^2.$$

Proof Multiplying (1.7) by x, we get

$$\sum_{j=1}^{n} j^2 x^j = \frac{n^2 x^{n+3} - (2n^2+2n-1) x^{n+2} + (n+1)^2 x^{n+1} - x^2 - x}{(x-1)^3}. \quad (1.8)$$

Differentiating (1.8) with respect to x, we get

$$\sum_{j=1}^{n} j^3 x^{j-1} = \frac{1}{(x-1)^6} \left((n^2(n+3)x^{n+2} - (2n^2+2n-1)(n+2)x^{n+1} \right.$$
$$+ (n+1)^2(n+1)x^n - 2x - 1)(x-1)^3$$
$$\left. - 3(x-1)^2 \left(n^2 x^{n+3} - (2n^2+2n-1)x^{n+2} + (n+1)^2 x^{n+1} - x^2 - x \right) \right)$$
$$= \frac{1}{(x-1)^4} \left((n^2(n+3)x^{n+2} - (2n^2+2n-1)(n+2)x^{n+1} \right.$$
$$+ (n+1)^3 x^n - 2x - 1)(x-1)$$
$$\left. - 3 \left(n^2 x^{n+3} - (2n^2+2n-1)x^{n+2} + (n+1)^2 x^{n+1} - x^2 - x \right) \right) \quad (1.9)$$
$$= \frac{1}{(x-1)^4} \left((n^2(n+3) - 3n^2) x^{n+3} + (3(2n^2+2n-1) \right.$$
$$- n^2(n+3) - (2n^2+2n-1)(n+2)) x^{n+2}$$
$$+ ((n+1)^3 + (2n^2+2n-1)(n+2) - 3(n+1)^2) x^{n+1}$$
$$\left. - (n+1)^3 x^n + (3-2)x^2 + (3+2-1)x + 1 \right)$$
$$= \frac{1}{(x-1)^4} \left(n^3 x^{n+3} - (3n^3 + 3n^2 - 3n + 1) x^{n+2} \right.$$
$$\left. + (3n^3 + 6n^2 - 4) x^{n+1} - (n^3 + 3n^2 + 3n + 1) x^n + x^2 + 4x + 1 \right).$$

Taking the limit as $x \to 1$ in (1.9), we get

1.1 The Differentiation Method for Series Sum

$$\lim_{x \to 1} \sum_{j=1}^{n} j^3 x^{j-1} = \lim_{x \to 1} \frac{1}{(x-1)^4} \left(n^3 x^{n+3} - \left(3n^3 + 3n^2 - 3n + 1\right) x^{n+2} \right.$$
$$\left. + \left(3n^3 + 6n^2 - 4\right) x^{n+1} - \left(n^3 + 3n^2 + 3n + 1\right) x^n + x^2 + 4x + 1 \right)$$

$$\iff \sum_{j=1}^{n} j^3 = \lim_{y \to 0} \frac{1}{y^4} \left(n^3 (1+y)^{n+3} - \left(3n^3 + 3n^2 - 3n + 1\right)(1+y)^{n+2} \right.$$
$$+ \left(3n^3 + 6n^2 - 4\right)(1+y)^{n+1}$$
$$\left. - \left(n^3 + 3n^2 + 3n + 1\right)(1+y)^n + (1+y)^2 + 4(1+y) + 1 \right)$$

$$= \lim_{y \to 0} \frac{1}{y^4} \left(n^3 \left(1 + (n+3)y + \frac{(n+3)(n+2)}{2} y^2 + \frac{(n+3)(n+2)(n+1)}{6} y^3 \right. \right.$$
$$\left. + \frac{(n+3)(n+2)(n+1)n}{24} y^4 + y^5 \sum_{j=5}^{n+3} \binom{n+3}{j} y^{j-5} \right)$$
$$- \left(3n^3 + 3n^2 - 3n + 1\right) \left(1 + (n+2)y + \frac{(n+2)(n+1)}{2} y^2 \right.$$
$$\left. + \frac{(n+2)(n+1)n}{6} y^3 + \frac{(n+2)(n+1)n(n-1)}{24} y^4 + y^5 \sum_{j=5}^{n+2} \binom{n+2}{j} y^{j-5} \right)$$
$$+ \left(3n^3 + 6n^2 - 4\right) \left(1 + (n+1)y + \frac{(n+1)n}{2} y^2 + \frac{(n+1)n(n-1)}{6} y^3 \right.$$
$$\left. + \frac{(n+1)n(n-1)(n-2)}{24} y^4 + y^5 \sum_{j=5}^{n+1} \binom{n+1}{j} y^{j-5} \right)$$
$$- \left(n^3 + 3n^2 + 3n + 1\right) \left(1 + ny + \frac{n(n-1)}{2} y^2 + \frac{n(n-1)(n-2)}{6} y^3 \right.$$
$$\left. \left. + \frac{n(n-1)(n-2)(n-3)}{24} y^4 + y^5 \sum_{j=5}^{n} \binom{n}{j} y^{j-5} \right) + (1+y)^2 + 4(1+y) + 1 \right)$$

$$= \lim_{y \to 0} \frac{1}{y^4} \left(n^3 - \left(3n^3 + 3n^2 - 3n + 1\right) + \left(3n^3 + 6n^2 - 4\right) \right.$$
$$\left. - \left(n^3 + 3n^2 + 3n + 1\right) + 1 + 4 + 1 \right)$$
$$+ \lim_{y \to 0} \frac{1}{y^3} \left(n^3 (n+3) - \left(3n^3 + 3n^2 - 3n + 1\right)(n+2) \right.$$
$$\left. + \left(3n^3 + 6n^2 - 4\right)(n+1) - n\left(n^3 + 3n^2 + 3n + 1\right) + 2 + 4 \right)$$
$$+ \lim_{y \to 0} \frac{1}{2y^2} \left(n^3 (n+3)(n+2) - \left(3n^3 + 3n^2 - 3n + 1\right)(n+2)(n+1) \right.$$

9

$$+ n\left(3n^3 + 6n^2 - 4\right)(n+1) - n\left(n^3 + 3n^2 + 3n + 1\right)(n-1) + 2\Big)$$
$$+ \lim_{y \to 0} \frac{n(n+1)}{6y}\left(n^2(n+3)(n+2) - \left(3n^3 + 3n^2 - 3n + 1\right)(n+2)\right.$$
$$+ \left(3n^3 + 6n^2 - 4\right)(n-1) - (n+1)^2(n-1)(n-2)\Big)$$
$$+ \lim_{y \to 0} \frac{n(n+1)}{24}\left(n^3(n+3)(n+2) - \left(3n^3 + 3n^2 - 3n + 1\right)(n+2)(n-1)\right.$$
$$+ \left(3n^3 + 6n^2 - 4\right)(n-1)(n-2) - (n+1)^2(n-1)(n-2)(n-3)\Big)$$
$$+ \lim_{y \to 0} y \left(n^3 \sum_{j=5}^{n+3}\binom{n+3}{j}y^{j-5} - \left(3n^3 + 3n^2 - 3n + 1\right)\sum_{j=5}^{n+2}\binom{n+2}{j}y^{j-5}\right.$$
$$+ \left(3n^3 + 6n^2 - 4\right)\sum_{j=5}^{n+1}\binom{n+1}{j}y^{j-5} - \left(n^3 + 3n^2 + 3n + 1\right)\sum_{j=5}^{n}\binom{n}{j}y^{j-5}\Big)$$
$$= \frac{n(n+1)}{24}\left(n^5 + 5n^4 + 6n^3 - \left(3n^3 + 3n^2 - 3n + 1\right)\left(n^2 + n - 2\right)\right.$$
$$+ \left(3n^3 + 6n^2 - 4\right)\left(n^2 - 3n + 2\right) - \left(n^2 - 1\right)\left(n^3 - 4n^2 + n + 6\right)\Big)$$
$$= \left(\frac{n(n+1)}{2}\right)^2.$$

$\square$

1.2 The Integration Method for Series Sum

Similar to the differentiation method of series sum, given an identity in variable x, we can integrate both sides of the identity with respect to x to get a new identity.

Definition 1.6 (*The Integration Method for Series Sum*)
Let
$$f(x) = \sum_{j=0}^{n} a_j x^j \tag{1.10}$$

be a function. In the integration method for series sum, we integrate (1.10) with respect to x from 0 to x to get

$$\int_0^x f(x)\,dx = \sum_{j=0}^{n} \frac{a_j}{j+1} x^{j+1}.$$

1.2 The Integration Method for Series Sum

We can also apply the above method to infinite series sums. A simple application of this method to GP gives an identity involving a *Harmonic Progression (HP)* and a series for $\tan^{-1} x$.

Example 1.7 (*Finding the Sum of Reciprocals of First n Natural Numbers*)
Prove that $\forall n \in \mathbb{N}$,

$$\sum_{j=1}^{n} \frac{1}{j} = 1 + \frac{1}{2} + \frac{1}{3} + \cdots + \frac{1}{n}$$

$$= \sum_{j=1}^{n} \frac{(-1)^{j-1}}{j} \binom{n}{j} \qquad (1.11)$$

$$= \frac{\binom{n}{1}}{1} - \frac{\binom{n}{2}}{2} + \frac{\binom{n}{3}}{3} + \cdots + (-1)^{n-1} \frac{\binom{n}{n}}{n}.$$

Proof The sum of the GP $(x^j)_{j=0}^{n-1}$ is given by

$$\sum_{j=0}^{n-1} x^j = \frac{x^n - 1}{x - 1}. \qquad (1.12)$$

Integrating (1.12) with respect to x from 0 to 1, we get

$$\sum_{j=0}^{n-1} \frac{x^{j+1}}{j+1} \bigg|_0^1 = \int_0^1 \frac{x^n - 1}{x - 1} dx$$

$$\iff \sum_{j=1}^{n} \frac{1}{j} = \int_{-1}^{0} \frac{(1+y)^n - 1}{y} dy$$

$$= \int_{-1}^{0} \frac{1}{y} \left(\sum_{j=0}^{n} \binom{n}{j} y^j - 1 \right) dy$$

$$= \sum_{j=1}^{n} \binom{n}{j} \int_{-1}^{0} y^{j-1} dy$$

$$= \sum_{j=1}^{n} \binom{n}{j} \frac{y^j}{j} \bigg|_{-1}^{0}$$

$$= \sum_{j=1}^{n} \frac{(-1)^{j-1}}{j} \binom{n}{j}.$$

$\square$

Example 1.8 (*Infinite Series for* $\tan^{-1} x$)
Prove that $\forall x \in [-1, 1]$,

$$\tan^{-1} x = \sum_{j=0}^{\infty} (-1)^j \frac{x^{2j+1}}{2j+1} = x - \frac{x^3}{3} + \frac{x^5}{5} - \frac{x^7}{7} + \cdots \quad (1.13)$$

Proof The sum of the GP $\left((-1)^j x^{2j}\right)_{j=0}^{\infty}$ is given by

$$\sum_{j=0}^{\infty} (-1)^j x^{2j} = \frac{1}{1+x^2}. \quad (1.14)$$

Integrating (1.14) with respect to x from 0 to x, we get

$$\sum_{j=0}^{\infty} (-1)^j \frac{x^{2j+1}}{2j+1} \bigg|_0^x = \tan^{-1} x \bigg|_0^x$$

$$\iff \sum_{j=0}^{\infty} (-1)^j \frac{x^{2j+1}}{2j+1} = \tan^{-1} x.$$

$\square$

Example 1.9 (*Infinite Series for* $\tanh^{-1} x$)
Prove that $\forall x \in (-1, 1)$,

$$\tanh^{-1} x = \sum_{j=0}^{\infty} \frac{x^{2j+1}}{2j+1} = x + \frac{x^3}{3} + \frac{x^5}{5} + \frac{x^7}{7} + \cdots \quad (1.15)$$

Proof Replace x with ix in (1.13). $\square$

1.3 A Recursive Formula for π

As an example of terms involving powers of 2, we will derive a recursive formula for π.

Example 1.10 (*A Recursive Formula for* π)
Let the sequence $(y_j)_{j=1}^{\infty}$ be defined as

$$y_j = \begin{cases} 2, & \text{if } j = 1; \\ \sqrt{2^{2j-1} - \sqrt{2^{4j-2} - 2^{2j} y_{j-1}^2}}, & \text{if } j \geq 2. \end{cases} \quad (1.16)$$

1.3 A Recursive Formula for π

Prove that
$$\lim_{n\to\infty} y_n = \pi.$$

Proof We will prove by induction that $\forall n \in \mathbb{N}$,
$$y_n = 2^n \sin \frac{\pi}{2^n}. \tag{1.17}$$

Let the induction hypothesis $P(n)$ be (1.17).

Basis Step: $P(1)$ states that
$$y_1 = 2 = 2\sin \frac{\pi}{2},$$

which is true. Assuming $P(n)$ to be true, from (1.16) and (1.17), we have

$$y_{n+1} = \sqrt{2^{2n+1} - \sqrt{2^{4n+2} - 2^{2n+2} y_n^2}}$$

$$= \sqrt{2^{2n+1} - \sqrt{2^{4n+2} - 2^{2n+2} \cdot 2^{2n} \sin^2 \frac{\pi}{2^n}}}$$

$$= \sqrt{2^{2n+1} - \sqrt{2^{4n+2} - 2^{4n+2} \sin^2 \frac{\pi}{2^n}}}$$

$$= \sqrt{2^{2n+1} - 2^{2n+1}\sqrt{1 - \sin^2 \frac{\pi}{2^n}}}$$

$$= \sqrt{2^{2n+1} - 2^{2n+1} \cos \frac{\pi}{2^n}}$$

$$= 2^n \sqrt{2\left(1 - \cos \frac{\pi}{2^n}\right)}$$

$$= 2^n \sqrt{4 \sin^2 \frac{\pi}{2^{n+1}}}$$

$$= 2^{n+1} \sin \frac{\pi}{2^{n+1}},$$

proving $P(n+1)$ to be true. From (1.17), we have

$$\lim_{n\to\infty} y_n = \lim_{n\to\infty} 2^n \sin \frac{\pi}{2^n}$$

$$= \lim_{n\to\infty} \pi \left(\frac{\sin \frac{\pi}{2^n}}{\frac{\pi}{2^n}}\right)$$

$$= \pi \cdot 1$$

$$= \pi.$$

□

Definition 1.11 (*Repeated Exponential Square Root Functions*)
$\forall n \in \mathbb{N}$, we define the function $R_n : \mathbb{N}^n \to \mathbb{R}^+$ as follows:

$$R_n\left((m_j)_{j=1}^n\right) = \sqrt{2^{m_1} + \sqrt{2^{m_2} + \sqrt{\cdots + \sqrt{2^{m_{n-1}} + \sqrt{2^{m_n}}}}}}.$$

Example 1.12 (*Solving the Recursive Formula for π*)
Prove that the solution of (1.16) is given by

$$y_n = \begin{cases} 2, & \text{if } n = 1; \\ R_1(3), & \text{if } n = 2; \\ \sqrt{2^{2n-1} - R_{n-1}\left((2^j n - 2^j + 1)_{j=2}^n\right)}, & \text{if } n \geq 3. \end{cases} \quad (1.18)$$

Proof We will prove (1.18) by induction. Let the induction hypothesis $P(n)$ be (1.18).
Basis Step: $P(1)$ states that $y_1 = 2$, which is already given in (1.16). $P(2)$ states that $y_2 = R_1(3) = \sqrt{2^3}$, which can be proved using (1.17) as follows:

$$y_2 = 2^2 \sin\frac{\pi}{2^2} = \frac{2^2}{\sqrt{2}} = \sqrt{2^3}.$$

Induction Step: Assuming $P(n)$ to be true, using (1.16) and (1.18), we can prove $P(n+1)$ to be true as follows:

$$y_{n+1} = \sqrt{2^{2n+1} - \sqrt{2^{4n+2} - 2^{2n+2} y_n^2}}$$

$$= \sqrt{2^{2n+1} - \sqrt{2^{4n+2} - 2^{2n+2}\left(2^{2n-1} - R_{n-1}\left((2^j n - 2^j + 1)_{j=2}^n\right)\right)}}$$

$$= \sqrt{2^{2n+1} - \sqrt{2^{4n+1} + 2^{2n+2} R_{n-1}\left((2^j n - 2^j + 1)_{j=2}^n\right)}}$$

$$= \sqrt{2^{2n+1} - \sqrt{2^{4n+1} + R_{n-1}\left((2^j n + 2^j + 2^j n - 2^j + 1)_{j=2}^n\right)}}$$

$$= \sqrt{2^{2n+1} - \sqrt{2^{4n+1} + R_{n-1}\left((2^{j+1} n + 1)_{j=2}^n\right)}}$$

$$= \sqrt{2^{2n+1} - R_n\left((2^{j+1} n + 1)_{j=1}^n\right)}$$

$$= \sqrt{2^{2n+1} - R_n\left((2^j n + 1)_{j=2}^{n+1}\right)}.$$

$\square$

As an example of a limit involving powers of 2, we will derive an alternative definition of $\log x$.

Example 1.13 (*An Alternative Definition of* $\log x$)
Prove that $\forall x \in \mathbb{R}^+$,
$$\log x = \lim_{n \to \infty} 2^n \left(x^{1/2^n} - 1 \right). \tag{1.19}$$

Proof We have
$$\lim_{n \to \infty} 2^n \left(x^{1/2^n} - 1 \right) = \lim_{n \to \infty} 2^n \left(e^{\log(x)/2^n} - 1 \right)$$
$$= \lim_{n \to \infty} 2^n \sum_{j=1}^{\infty} \frac{\log^j x}{j! 2^{nj}}$$
$$= \log x + \lim_{n \to \infty} \sum_{j=2}^{\infty} \frac{\log^j x}{j! 2^{n(j-1)}}$$
$$= \log x.$$

$\square$

Problems

[1.1] (Lehmer, 1985) Prove that
$$\frac{1}{3} + \frac{2\pi\sqrt{3}}{27} = \sum_{j=1}^{\infty} \binom{2j}{j}^{-1}$$
$$= \frac{1}{\binom{2}{1}} + \frac{1}{\binom{4}{2}} + \frac{1}{\binom{6}{3}} + \frac{1}{\binom{8}{4}} + \cdots$$

[1.2] (Lehmer, 1985) Prove that
$$\frac{1}{5} + \frac{4\sqrt{5}}{25} \log \left(\frac{1+\sqrt{5}}{2} \right) = \sum_{j=1}^{\infty} (-1)^{j-1} \binom{2j}{j}^{-1}$$
$$= \frac{1}{\binom{2}{1}} - \frac{1}{\binom{4}{2}} + \frac{1}{\binom{6}{3}} - \frac{1}{\binom{8}{4}} + \cdots$$

[1.3] (Lehmer, 1985) Prove that

$$\frac{\pi\sqrt{3}}{9} = \sum_{j=1}^{\infty} \frac{1}{j}\binom{2j}{j}^{-1}$$

$$= \frac{1}{1\binom{2}{1}} + \frac{1}{2\binom{4}{2}} + \frac{1}{3\binom{6}{3}} + \frac{1}{4\binom{8}{4}} + \cdots$$

[1.4] (Lehmer, 1985) Prove that

$$\frac{2\sqrt{5}}{5}\log\left(\frac{1+\sqrt{5}}{2}\right) = \sum_{j=1}^{\infty} \frac{(-1)^{j-1}}{j}\binom{2j}{j}^{-1}$$

$$= \frac{1}{1\binom{2}{1}} - \frac{1}{2\binom{4}{2}} + \frac{1}{3\binom{6}{3}} - \frac{1}{4\binom{8}{4}} + \cdots$$

[1.5] (Lehmer, 1985) Prove that

$$\frac{\pi^2}{18} = \sum_{j=1}^{\infty} \frac{1}{j^2}\binom{2j}{j}^{-1}$$

$$= \frac{1}{1^2\binom{2}{1}} + \frac{1}{2^2\binom{4}{2}} + \frac{1}{3^2\binom{6}{3}} + \frac{1}{4^2\binom{8}{4}} + \cdots$$

[1.6] (Lehmer, 1985) Prove that

$$2\log^2\left(\frac{1+\sqrt{5}}{2}\right) = \sum_{j=1}^{\infty} \frac{(-1)^{j-1}}{j^2}\binom{2j}{j}^{-1}$$

$$= \frac{1}{1^2\binom{2}{1}} - \frac{1}{2^2\binom{4}{2}} + \frac{1}{3^2\binom{6}{3}} - \frac{1}{4^2\binom{8}{4}} + \cdots$$

[1.7] Prove that

$$2 - \frac{\pi\sqrt{3}}{3} = \sum_{j=1}^{\infty} \frac{1}{j(2j+1)} \binom{2j}{j}^{-1}$$

$$= \frac{1}{1\cdot 3 \cdot \binom{2}{1}} + \frac{1}{2\cdot 5 \cdot \binom{4}{2}} + \frac{1}{3\cdot 7 \cdot \binom{6}{3}} + \frac{1}{4\cdot 9 \cdot \binom{8}{4}} + \cdots$$

[1.8] Prove that

$$2\sqrt{5}\log\left(\frac{1+\sqrt{5}}{2}\right) - 2 = \sum_{j=1}^{\infty} \frac{(-1)^{j-1}}{j(2j+1)} \binom{2j}{j}^{-1}$$

$$= \frac{1}{1\cdot 3 \cdot \binom{2}{1}} - \frac{1}{2\cdot 5 \cdot \binom{4}{2}} + \frac{1}{3\cdot 7 \cdot \binom{6}{3}} - \frac{1}{4\cdot 9 \cdot \binom{8}{4}} + \cdots$$

[1.9] Prove that

$$\frac{\pi}{3} - \frac{1}{4} + 4\sqrt{3} - 4\log\left(\frac{e^2}{2}\left(1+\frac{\sqrt{3}}{2}\right)\right) = \sum_{j=1}^{\infty} \frac{1}{(2j+1)(2j+2)2^{4j}} \binom{2j}{j}$$

$$= \frac{\binom{2}{1}}{3\cdot 4^2 \cdot 2^4} + \frac{\binom{4}{2}}{5\cdot 6^2 \cdot 2^8} + \frac{\binom{6}{3}}{7\cdot 8^2 \cdot 2^{12}} + \frac{\binom{8}{4}}{9\cdot 10^2 \cdot 2^{16}} + \cdots$$

[1.10] Prove that

$$4\sqrt{5} + \frac{1}{4} - 2\log\left(\frac{1+\sqrt{5}}{2}\right) - 4\log\left(\frac{e^2}{2}\left(1+\frac{\sqrt{5}}{2}\right)\right)$$

$$= \sum_{j=1}^{\infty} \frac{(-1)^{j-1}}{(2j+1)(2j+2)2^{4j}} \binom{2j}{j}$$

$$= \frac{\binom{2}{1}}{3\cdot 4^2 \cdot 2^4} - \frac{\binom{4}{2}}{5\cdot 6^2 \cdot 2^8} + \frac{\binom{6}{3}}{7\cdot 8^2 \cdot 2^{12}} - \frac{\binom{8}{4}}{9\cdot 10^2 \cdot 2^{16}} + \cdots$$

[1.11] Prove that

$$\frac{\pi^2}{18} + \frac{2\pi\sqrt{3}}{3} - 4 = \sum_{j=1}^{\infty} \frac{1}{j^2(2j+1)} \binom{2j}{j}^{-1}$$

$$= \frac{1}{1^2 \cdot 3 \cdot \binom{2}{1}} + \frac{1}{2^2 \cdot 5 \cdot \binom{4}{2}} + \frac{1}{3^2 \cdot 7 \cdot \binom{6}{3}} + \frac{1}{4^2 \cdot 9 \cdot \binom{8}{4}} + \cdots$$

[1.12] Prove that

$$4 + 2\log^2\left(\frac{1+\sqrt{5}}{2}\right) - 4\sqrt{5}\log\left(\frac{1+\sqrt{5}}{2}\right) = \sum_{j=1}^{\infty} \frac{(-1)^{j-1}}{j^2(2j+1)} \binom{2j}{j}^{-1}$$

$$= \frac{1}{1^2 \cdot 3 \cdot \binom{2}{1}} - \frac{1}{2^2 \cdot 5 \cdot \binom{4}{2}} + \frac{1}{3^2 \cdot 7 \cdot \binom{6}{3}} - \frac{1}{4^2 \cdot 9 \cdot \binom{8}{4}} + \cdots$$

[1.13] Prove that

$$\frac{3}{2} - \frac{\pi\sqrt{3}}{6} - \frac{\pi^2}{18} = \sum_{j=1}^{\infty} \frac{1}{j(2j+1)(2j+2)} \binom{2j}{j}^{-1}$$

$$= \frac{1}{1 \cdot 3 \cdot 4 \cdot \binom{2}{1}} + \frac{1}{2 \cdot 5 \cdot 6 \cdot \binom{4}{2}} + \frac{1}{3 \cdot 7 \cdot 8 \cdot \binom{6}{3}} + \frac{1}{4 \cdot 9 \cdot 10 \cdot \binom{8}{4}} + \cdots$$

[1.14] Prove that

$$2\log^2\left(\frac{1+\sqrt{5}}{2}\right) + \sqrt{5}\log\left(\frac{1+\sqrt{5}}{2}\right) - \frac{3}{2}$$

$$= \sum_{j=1}^{\infty} \frac{(-1)^{j-1}}{j(2j+1)(2j+2)} \binom{2j}{j}^{-1}$$

$$= \frac{1}{1 \cdot 3 \cdot 4 \cdot \binom{2}{1}} - \frac{1}{2 \cdot 5 \cdot 6 \cdot \binom{4}{2}} + \frac{1}{3 \cdot 7 \cdot 8 \cdot \binom{6}{3}} - \frac{1}{4 \cdot 9 \cdot 10 \cdot \binom{8}{4}} + \cdots$$

[1.15] (Lehmer, 1985) Prove that

$$\frac{\pi}{3} = \sum_{j=0}^{\infty} \frac{1}{(2j+1)2^{4j}} \binom{2j}{j}$$

$$= 1 + \frac{\binom{2}{1}}{3 \cdot 2^4} + \frac{\binom{4}{2}}{5 \cdot 2^8} + \frac{\binom{6}{3}}{7 \cdot 2^{12}} + \cdots$$

[1.16] Prove that

$$2 \log\left(\frac{1+\sqrt{5}}{2}\right) = \sum_{j=0}^{\infty} \frac{(-1)^j}{(2j+1)2^{4j}} \binom{2j}{j}$$

$$= 1 - \frac{\binom{2}{1}}{3 \cdot 2^4} + \frac{\binom{4}{2}}{5 \cdot 2^8} - \frac{\binom{6}{3}}{7 \cdot 2^{12}} + \cdots$$

[1.17] Prove that

$$\frac{\pi}{3} + 2\sqrt{3} - \frac{9}{2} = \sum_{j=1}^{\infty} \frac{1}{(2j+1)(2j+2)2^{4j}} \binom{2j}{j}$$

$$= \frac{\binom{2}{1}}{3 \cdot 4 \cdot 2^4} + \frac{\binom{4}{2}}{5 \cdot 6 \cdot 2^8} + \frac{\binom{6}{3}}{7 \cdot 8 \cdot 2^{12}} + \frac{\binom{8}{4}}{9 \cdot 10 \cdot 2^{16}} + \cdots$$

[1.18] Prove that

$$2\sqrt{5} - \frac{7}{2} - 2\log\left(\frac{1+\sqrt{5}}{2}\right) = \sum_{j=1}^{\infty} \frac{(-1)^{j-1}}{(2j+1)(2j+2)2^{4j}} \binom{2j}{j}$$

$$= \frac{\binom{2}{1}}{3 \cdot 4 \cdot 2^4} - \frac{\binom{4}{2}}{5 \cdot 6 \cdot 2^8} + \frac{\binom{6}{3}}{7 \cdot 8 \cdot 2^{12}} - \frac{\binom{8}{4}}{9 \cdot 10 \cdot 2^{16}} + \cdots$$

[1.19] Prove that

$$\frac{2\sqrt{3}}{3} = \sum_{j=0}^{\infty} \frac{1}{2^{4j}} \binom{2j}{j}$$

$$= 1 + \frac{\binom{2}{1}}{2^4} + \frac{\binom{4}{2}}{2^8} + \frac{\binom{6}{3}}{2^{12}} + \cdots$$

[1.20] Prove that

$$\frac{2\sqrt{5}}{5} = \sum_{j=0}^{\infty} \frac{(-1)^j}{2^{4j}} \binom{2j}{j}$$

$$= 1 - \frac{\binom{2}{1}}{2^4} + \frac{\binom{4}{2}}{2^8} - \frac{\binom{6}{3}}{2^{12}} + \cdots$$

[1.21] Prove that

$$\frac{8\sqrt{3}}{9} = \sum_{j=0}^{\infty} \frac{(2j+1)}{2^{4j}} \binom{2j}{j}$$

$$= 1 + \frac{3\binom{2}{1}}{2^4} + \frac{5\binom{4}{2}}{2^8} + \frac{7\binom{6}{3}}{2^{12}} + \cdots$$

[1.22] Prove that

$$\frac{8\sqrt{5}}{25} = \sum_{j=0}^{\infty} (-1)^j \frac{(2j+1)}{2^{4j}} \binom{2j}{j}$$

$$= 1 - \frac{3\binom{2}{1}}{2^4} + \frac{5\binom{4}{2}}{2^8} - \frac{7\binom{6}{3}}{2^{12}} + \cdots$$

[1.23] Prove that

$$\frac{16\sqrt{3}}{9} = \sum_{j=0}^{\infty} \frac{(2j+1)^2}{2^{4j}} \binom{2j}{j}$$

$$= 1^2 + \frac{3^2 \binom{2}{1}}{2^4} + \frac{5^2 \binom{4}{2}}{2^8} + \frac{7^2 \binom{6}{3}}{2^{12}} + \cdots$$

[1.24] Prove that

$$\frac{16\sqrt{5}}{125} = \sum_{j=0}^{\infty} (-1)^j \frac{(2j+1)^2}{2^{4j}} \binom{2j}{j}$$

$$= 1^2 - \frac{3^2 \binom{2}{1}}{2^4} + \frac{5^2 \binom{4}{2}}{2^8} - \frac{7^2 \binom{6}{3}}{2^{12}} + \cdots$$

[1.25] Prove that

$$\frac{160\sqrt{3}}{27} = \sum_{j=0}^{\infty} \frac{(2j+1)^3}{2^{4j}} \binom{2j}{j}$$

$$= 1^3 + \frac{3^3 \binom{2}{1}}{2^4} + \frac{5^3 \binom{4}{2}}{2^8} + \frac{7^3 \binom{6}{3}}{2^{12}} + \cdots$$

[1.26] Prove that

$$-\frac{32\sqrt{5}}{125} = \sum_{j=0}^{\infty} (-1)^j \frac{(2j+1)^3}{2^{4j}} \binom{2j}{j}$$

$$= 1^3 - \frac{3^3 \binom{2}{1}}{2^4} + \frac{5^3 \binom{4}{2}}{2^8} - \frac{7^3 \binom{6}{3}}{2^{12}} + \cdots$$

[1.27] Prove that

$$\frac{2368\sqrt{3}}{81} = \sum_{j=0}^{\infty} \frac{(2j+1)^4}{2^{4j}} \binom{2j}{j}$$

$$= 1^4 + \frac{3^4 \binom{2}{1}}{2^4} + \frac{5^4 \binom{4}{2}}{2^8} + \frac{7^4 \binom{6}{3}}{2^{12}} + \cdots$$

[1.28] Prove that

$$-\frac{448\sqrt{5}}{625} = \sum_{j=0}^{\infty} (-1)^j \frac{(2j+1)^4}{2^{4j}} \binom{2j}{j}$$

$$= 1^4 - \frac{3^4 \binom{2}{1}}{2^4} + \frac{5^4 \binom{4}{2}}{2^8} - \frac{7^4 \binom{6}{3}}{2^{12}} + \cdots$$

[1.29] Prove that

$$\pi = 3 + 6 \sum_{j=1}^{\infty} \frac{(-1)^{j-1}}{(6j)^2 - 1}$$

$$= 3 + 6 \left(\frac{1}{6^2 - 1} - \frac{1}{12^2 - 1} + \frac{1}{18^2 - 1} - \cdots \right).$$

[1.30] Prove that

$$\frac{\pi}{e^\pi - e^{-\pi}} = \sum_{j=2}^{\infty} \frac{(-1)^j}{j^2 + 1}$$

$$= \frac{1}{2^2 + 1} - \frac{1}{3^2 + 1} + \frac{1}{4^2 + 1} - \frac{1}{5^2 + 1} + \cdots$$

[1.31] Prove that

$$\frac{e^2 + 2e - 1}{2(e^2 - 1)} = \sum_{j=0}^{\infty} \frac{(-1)^j}{j^2 \pi^2 + 1}$$

$$= 1 - \frac{1}{\pi^2 + 1} + \frac{1}{2^2 \pi^2 + 1} - \frac{1}{3^2 \pi^2 + 1} + \cdots$$

Problems

[1.32] Prove that

$$\frac{3 - \csc 1}{2} = \sum_{j=0}^{\infty} \frac{(-1)^j}{j^2\pi^2 - 1}$$

$$= 1 - \frac{1}{\pi^2 - 1} + \frac{1}{2^2\pi^2 - 1} - \frac{1}{3^2\pi^2 - 1} + \cdots$$

[1.33] Prove that

$$6 = \sum_{j=1}^{\infty} (4j+1) \binom{2j+2}{4}^{-1}$$

$$= \frac{5}{\binom{4}{4}} + \frac{9}{\binom{6}{4}} + \frac{13}{\binom{8}{4}} + \frac{17}{\binom{10}{4}} + \cdots$$

[1.34] Prove that $\forall n \in \mathbb{N}$,

$$\frac{1}{n+1} = \sum_{j=0}^{n} \frac{(-1)^j}{j+1} \binom{n}{j}$$

$$= \frac{\binom{n}{0}}{1} - \frac{\binom{n}{1}}{2} + \frac{\binom{n}{2}}{3} - \cdots + (-1)^n \frac{\binom{n}{n}}{n+1}.$$

[1.35] Prove that $\forall n \in \mathbb{N}$,

$$\frac{1}{n+1} \sum_{j=1}^{n} \frac{1}{j+1} = \frac{1}{n+1} \left(\frac{1}{2} + \frac{1}{3} + \frac{1}{4} + \cdots \frac{1}{n+1} \right)$$

$$= \sum_{j=1}^{n} (-1)^{j-1} \frac{j}{(j+1)^2} \binom{n}{j}$$

$$= \frac{1\binom{n}{1}}{2^2} - \frac{2\binom{n}{2}}{3^2} + \frac{3\binom{n}{3}}{4^2} - \cdots + (-1)^{n-1} \frac{n\binom{n}{n}}{(n+1)^2}.$$

[1.36] Prove that $\forall n \in \mathbb{N}+1$,

$$0 = \sum_{j=1}^{n}(-1)^{j-1} j \binom{n}{j}$$
$$= 1\binom{n}{1} - 2\binom{n}{2} + 3\binom{n}{3} - \cdots + (-1)^{n-1} n \binom{n}{n}.$$

[1.37] (Boros and Moll, 2006) Prove that $\forall n \in \mathbb{N}$,

$$\frac{2^{2n}}{2n+1}\binom{2n}{n}^{-1} = \sum_{j=0}^{n} \frac{(-1)^j}{2j+1}\binom{n}{j}$$
$$= \frac{\binom{n}{0}}{1} - \frac{\binom{n}{1}}{3} + \frac{\binom{n}{2}}{5} - \cdots + (-1)^n \frac{\binom{n}{n}}{2n+1}.$$

[1.38] Prove that $\forall n \in 2\mathbb{N}-2$,

$$(-1)^{n/2} = \sum_{j=0}^{n/2}(-1)^j \frac{\prod_{k=0}^{j-1}\left(n^2-(2k)^2\right)}{(2j)!}$$
$$= 1 - \frac{n^2}{2!} + \frac{n^2\left(n^2-2^2\right)}{4!} - \cdots + (-1)^{n/2}$$
$$\frac{n^2\left(n^2-2^2\right)\left(n^2-4^2\right)\cdots\left(n^2-(n-2)^2\right)}{n!}.$$

[1.39] Prove that $\forall n \in 2\mathbb{N}$,

$$(-1)^{n/2+1} = \sum_{j=0}^{n/2-1}(-1)^j \frac{\prod_{k=1}^{j}\left(n^2-(2k)^2\right)}{(2j+1)!}$$
$$= 1 - \frac{n^2-2^2}{3!} + \frac{\left(n^2-2^2\right)\left(n^2-4^2\right)}{5!} - \cdots$$
$$+ (-1)^{n/2-1} \frac{\left(n^2-2^2\right)\left(n^2-4^2\right)\left(n^2-6^2\right)\cdots\left(n^2-(n-2)^2\right)}{(n-1)!}.$$

[1.40] Prove that $\forall n \in 2\mathbb{N}$,

$$(-1)^{n/2}\frac{(1-n^2)}{3} = \sum_{j=0}^{n/2-1}(-1)^j\frac{\prod_{k=1}^{j}\left(n^2-(2k)^2\right)}{(2j)!}$$

$$= 1 - \frac{n^2-2^2}{2!} + \frac{\left(n^2-2^2\right)\left(n^2-4^2\right)}{4!} - \cdots$$

$$+ (-1)^{n/2-1}\frac{\left(n^2-2^2\right)\left(n^2-4^2\right)\left(n^2-6^2\right)\cdots\left(n^2-(n-2)^2\right)}{(n-2)!}.$$

[1.41] Prove that $\forall n \in 2\mathbb{N} - 1$,

$$\frac{(-1)^{(n-1)/2}}{n} = \sum_{j=0}^{(n-1)/2}(-1)^j\frac{\prod_{k=0}^{j-1}\left(n^2-(2k+1)^2\right)}{(2j+1)!}$$

$$= 1 - \frac{\left(n^2-1^2\right)}{3!} + \frac{\left(n^2-1^2\right)\left(n^2-3^2\right)}{5!} - \cdots$$

$$+ (-1)^{(n-1)/2}\frac{\left(n^2-1^2\right)\left(n^2-3^2\right)\left(n^2-5^2\right)\cdots\left(n^2-(n-2)^2\right)}{n!}.$$

[1.42] Prove that $\forall n \in 2\mathbb{N} - 1$,

$$(-1)^{(n-1)/2}n = \sum_{j=0}^{(n-1)/2}(-1)^j\frac{\prod_{k=0}^{j-1}\left(n^2-(2k+1)^2\right)}{(2j)!}$$

$$= 1 - \frac{\left(n^2-1^2\right)}{2!} + \frac{\left(n^2-1^2\right)\left(n^2-3^2\right)}{4!} - \cdots$$

$$+ (-1)^{(n-1)/2}\frac{\left(n^2-1^2\right)\left(n^2-3^2\right)\left(n^2-5^2\right)\cdots\left(n^2-(n-2)^2\right)}{(n-1)!}.$$

[1.43] Prove that $\forall n \in 2\mathbb{N}+1$,

$$(-1)^{(n+1)/2}\frac{n}{3} = \sum_{j=0}^{(n-3)/2} (-1)^j \frac{\prod_{k=1}^{j}\left(n^2 - (2k+1)^2\right)}{(2j+1)!}$$

$$= 1 - \frac{\left(n^2 - 3^2\right)}{3!} + \frac{\left(n^2 - 3^2\right)\left(n^2 - 5^2\right)}{5!} - \cdots$$

$$+ (-1)^{(n-3)/2} \frac{\left(n^2 - 3^2\right)\left(n^2 - 5^2\right)\left(n^2 - 7^2\right)\cdots\left(n^2 - (n-2)^2\right)}{(n-2)!}.$$

[1.44] Prove that $\forall x \in \mathbb{R} - [-1, 1]$,

$$\frac{x(x+1)}{(x-1)^3} = \sum_{j=1}^{\infty} \frac{j^2}{x^j}$$

$$= \frac{1^2}{x} + \frac{2^2}{x^2} + \frac{3^2}{x^3} + \frac{4^2}{x^4} + \cdots$$

[1.45] Prove that $\forall x \in \mathbb{R} - [-1, 1]$,

$$\frac{x(x-1)}{(x+1)^3} = \sum_{j=1}^{\infty} (-1)^{j-1} \frac{i^2}{x^j}$$

$$= \frac{1^2}{x} - \frac{2^2}{x^2} + \frac{3^2}{x^3} - \frac{4^2}{x^4} + \cdots$$

[1.46] Prove that $\forall x \in \mathbb{R} - [-1, 1]$,

$$\frac{x(x^2 + 4x + 1)}{(x-1)^4} = \sum_{j=1}^{\infty} \frac{j^3}{x^j}$$

$$= \frac{1^3}{x} + \frac{2^3}{x^2} + \frac{3^3}{x^3} + \frac{4^3}{x^4} + \cdots$$

[1.47] Prove that $\forall x \in \mathbb{R} - [-1, 1]$,

$$\frac{x(x^2 - 4x + 1)}{(x+1)^4} = \sum_{j=1}^{\infty} (-1)^{j-1} \frac{j^3}{x^j}$$

$$= \frac{1^3}{x} - \frac{2^3}{x^2} + \frac{3^3}{x^3} - \frac{4^3}{x^4} + \cdots$$

[1.48] Prove that $\forall x \in \mathbb{R} - [-1, 1]$,

$$\frac{x(x+1)(x^2 + 10x + 1)}{(x-1)^5} = \sum_{j=1}^{\infty} \frac{j^4}{x^j}$$

$$= \frac{1^4}{x} + \frac{2^4}{x^2} + \frac{3^4}{x^3} + \frac{4^4}{x^4} + \cdots$$

[1.49] Prove that $\forall x \in \mathbb{R} - [-1, 1]$,

$$\frac{x(x-1)(x^2 - 10x + 1)}{(x+1)^5} = \sum_{j=1}^{\infty} (-1)^{j-1} \frac{j^4}{x^j}$$

$$= \frac{1^4}{x} - \frac{2^4}{x^2} + \frac{3^4}{x^3} - \frac{4^4}{x^4} + \cdots$$

[1.50] (Borwein and Borwein, 1987) Prove that

$$\frac{\pi}{2} \coth \pi - \frac{1}{2} = \sum_{j=1}^{\infty} \frac{1}{j^2 + 1}$$

$$= \frac{1}{1^2 + 1} + \frac{1}{2^2 + 1} + \frac{1}{3^2 + 1} + \frac{1}{4^2 + 1} + \cdots$$

[1.51] Prove that

$$\frac{\pi}{4} \tanh \frac{\pi}{2} = \sum_{j=0}^{\infty} \frac{1}{(2j+1)^2 + 1}$$

$$= \frac{1}{1^2 + 1} + \frac{1}{3^2 + 1} + \frac{1}{5^2 + 1} + \frac{1}{7^2 + 1} + \cdots$$

[1.52] Prove that
$$\frac{2\pi^2 + \pi \sinh 2\pi - 4\sinh^2 \pi}{8\sinh^2 \pi} = \sum_{j=1}^{\infty} \frac{1}{(j^2+1)^2}$$
$$= \frac{1}{(1^2+1)^2} + \frac{1}{(2^2+1)^2} + \frac{1}{(3^2+1)^2} + \frac{1}{(4^2+1)^2} + \cdots$$

[1.53] Prove that
$$\frac{\pi(\sinh \pi - \pi)}{16\cosh^2 \frac{\pi}{2}} = \sum_{j=0}^{\infty} \frac{1}{((2j+1)^2+1)^2}$$
$$= \frac{1}{(1^2+1)^2} + \frac{1}{(3^2+1)^2} + \frac{1}{(5^2+1)^2} + \frac{1}{(7^2+1)^2} + \cdots$$

[1.54] Prove that
$$\frac{1}{2} - \frac{\pi\sqrt{3}}{12} = \sum_{j=1}^{\infty} \frac{1}{(6j)^2 - 1}$$
$$= \frac{1}{6^2-1} + \frac{1}{12^2-1} + \frac{1}{18^2-1} + \frac{1}{24^2-1} + \cdots$$

[1.55] Prove that
$$\frac{\pi\sqrt{3}}{36} = \sum_{j=0}^{\infty} \frac{1}{(6j+3)^2 - 1}$$
$$= \frac{1}{3^2-1} + \frac{1}{9^2-1} + \frac{1}{15^2-1} + \frac{1}{21^2-1} + \cdots$$

[1.56] Prove that
$$\frac{\pi^2}{36} + \frac{\pi\sqrt{3}}{24} - \frac{1}{2} = \sum_{j=1}^{\infty} \frac{1}{((6j)^2-1)^2}$$
$$= \frac{1}{(6^2-1)^2} + \frac{1}{(12^2-1)^2} + \frac{1}{(18^2-1)^2} + \frac{1}{(24^2-1)^2} + \cdots$$

[1.57] Prove that

$$\frac{\pi(2\pi - 3\sqrt{3})}{216} = \sum_{j=0}^{\infty} \frac{1}{\left((6j+3)^2 - 1\right)^2}$$

$$= \frac{1}{\left(3^2 - 1\right)^2} + \frac{1}{\left(9^2 - 1\right)^2} + \frac{1}{\left(15^2 - 1\right)^2} + \frac{1}{\left(21^2 - 1\right)^2} + \cdots$$

[1.58] Prove that

$$\frac{1}{e^2 - 1} = \sum_{j=1}^{\infty} \frac{1}{j^2\pi^2 + 1}$$

$$= \frac{1}{\pi^2 + 1} + \frac{1}{2^2\pi^2 + 1} + \frac{1}{3^2\pi^2 + 1} + \frac{1}{4^2\pi^2 + 1} + \cdots$$

[1.59] Prove that

$$\frac{e-1}{4(e+1)} = \sum_{j=0}^{\infty} \frac{1}{(2j+1)^2\pi^2 + 1}$$

$$= \frac{1}{\pi^2 + 1} + \frac{1}{3^2\pi^2 + 1} + \frac{1}{5^2\pi^2 + 1} + \frac{1}{7^2\pi^2 + 1} + \cdots$$

[1.60] Prove that

$$\frac{2 + \sinh 2 - 4\sinh^2 1}{8\sinh^2 1} = \sum_{j=1}^{\infty} \frac{1}{\left(j^2\pi^2 + 1\right)^2}$$

$$= \frac{1}{\left(\pi^2 + 1\right)^2} + \frac{1}{\left(2^2\pi^2 + 1\right)^2} + \frac{1}{\left(3^2\pi^2 + 1\right)^2} + \frac{1}{\left(4^2\pi^2 + 1\right)^2} + \cdots$$

[1.61] Prove that

$$\frac{\sinh 1 - 1}{16\cosh^2 \frac{1}{2}} = \sum_{j=0}^{\infty} \frac{1}{\left((2j+1)^2\pi^2 + 1\right)^2}$$

$$= \frac{1}{\left(\pi^2 + 1\right)^2} + \frac{1}{\left(3^2\pi^2 + 1\right)^2} + \frac{1}{\left(5^2\pi^2 + 1\right)^2} + \frac{1}{\left(7^2\pi^2 + 1\right)^2} + \cdots$$

[1.62] Prove that

$$\frac{1-\cot 1}{2} = \sum_{j=1}^{\infty} \frac{1}{j^2\pi^2 - 1}$$
$$= \frac{1}{\pi^2 - 1} + \frac{1}{2^2\pi^2 - 1} + \frac{1}{3^2\pi^2 - 1} + \frac{1}{4^2\pi^2 - 1} + \cdots$$

[1.63] Prove that

$$\frac{1}{4}\tan\frac{1}{2} = \sum_{j=0}^{\infty} \frac{1}{(2j+1)^2\pi^2 - 1}$$
$$= \frac{1}{\pi^2 - 1} + \frac{1}{3^2\pi^2 - 1} + \frac{1}{5^2\pi^2 - 1} + \frac{1}{7^2\pi^2 - 1} + \cdots$$

[1.64] Prove that

$$\frac{2 + \sin 2 - 4\sin^2 1}{8\sin^2 1} = \sum_{j=1}^{\infty} \frac{1}{\left(j^2\pi^2 - 1\right)^2}$$
$$= \frac{1}{\left(\pi^2 - 1\right)^2} + \frac{1}{\left(2^2\pi^2 - 1\right)^2} + \frac{1}{\left(3^2\pi^2 - 1\right)^2} + \frac{1}{\left(4^2\pi^2 - 1\right)^2} + \cdots$$

[1.65] Prove that

$$\frac{1-\sin 1}{16\cos^2\frac{1}{2}} = \sum_{j=0}^{\infty} \frac{1}{\left((2j+1)^2\pi^2 - 1\right)^2}$$
$$= \frac{1}{\left(\pi^2 - 1\right)^2} + \frac{1}{\left(3^2\pi^2 - 1\right)^2} + \frac{1}{\left(5^2\pi^2 - 1\right)^2} + \frac{1}{\left(7^2\pi^2 - 1\right)^2} + \cdots$$

[1.66] Prove that $\forall n \in \mathbb{N}$,

$$\frac{\lfloor n!e \rfloor}{n!} = \sum_{j=0}^{n} \frac{1}{j!}$$
$$= \frac{1}{0!} + \frac{1}{1!} + \frac{1}{2!} + \cdots + \frac{1}{n!}.$$

[1.67] Prove that $\forall n \in 2\mathbb{N} - 1$,

$$\frac{1}{n!}\left\lfloor \frac{n!}{e} \right\rfloor = \frac{1}{(n+1)!}\left\lfloor \frac{(n+1)!}{e} \right\rfloor = \sum_{j=0}^{n} \frac{(-1)^j}{j!}$$

$$= \frac{1}{0!} - \frac{1}{1!} + \frac{1}{2!} - \cdots - \frac{1}{n!}.$$

[1.68] Prove that

$$\frac{10100}{970299} = \sum_{j=1}^{\infty} \frac{j^2}{100^j}$$

$$= 0.010409162536496482\ldots$$

[1.69] Prove that

$$\frac{\pi^2 \coth \pi^2 - 1}{2\pi^2} = \sum_{j=1}^{\infty} \frac{1}{j^2 + \pi^2}$$

$$= \frac{1}{1^2 + \pi^2} + \frac{1}{2^2 + \pi^2} + \frac{1}{3^2 + \pi^2} + \frac{1}{4^2 + \pi^2} + \cdots$$

[1.70] Prove that $\forall (n, x) \in (\mathbb{N} - 1) \times \left(\mathbb{R} - \left\{ -\frac{1}{j} \,\middle|\, j \in [1..n] \right\} \right)$,

$$\sum_{j=0}^{n} \frac{(-1)^j}{1 + jx}\binom{n}{j} = 1 - \frac{\binom{n}{1}}{1+x} + \frac{\binom{n}{2}}{1+2x} - \cdots + (-1)^n \frac{\binom{n}{n}}{1+nx}$$

$$= \frac{n! x^n}{\prod_{j=0}^{n}(1+jx)}$$

$$= \frac{n! x^n}{(1+x)(1+2x)(1+3x)\cdots(1+nx)}.$$

[1.71] Prove that $\forall (m,n) \in (\mathbb{N}-1)^2$, if $m=n$, then

$$\sum_{j=0}^{m-n}(-1)^j \binom{n+j}{n}\binom{m}{n+j}$$
$$= \binom{n}{n}\binom{m}{n} - \binom{n+1}{n}\binom{m}{n+1} + \binom{n+2}{n}\binom{m}{n+2} - \cdots$$
$$+ (-1)^{m-n}\binom{m}{n}\binom{m}{m}$$
$$= 1.$$

[1.72] Prove that $\forall (m,n) \in (\mathbb{N}-1)^2$, if $m \neq n$, then

$$\sum_{j=0}^{m-n}(-1)^j \binom{n+j}{n}\binom{m}{n+j}$$
$$= \binom{n}{n}\binom{m}{n} - \binom{n+1}{n}\binom{m}{n+1} + \binom{n+2}{n}\binom{m}{n+2} - \cdots$$
$$+ (-1)^{m-n}\binom{m}{n}\binom{m}{m}$$
$$= 0.$$

[1.73] (Berndt, 1994) Prove that

$$\forall \theta \in \mathbb{R} - \{0\} - \left\{\frac{1}{n\pi} \mid n \in \mathbb{Z} - \{0\}\right\} - \left\{\frac{1}{2^{m-1}(2n+1)\pi} \mid n \in \mathbb{Z}, m \in \mathbb{N}\right\},$$

$$\theta = \cot\frac{1}{\theta} + \sum_{j=1}^{\infty} \frac{1}{2^j} \tan\frac{1}{2^j\theta}$$
$$= \cot\frac{1}{\theta} + \frac{1}{2}\tan\frac{1}{2\theta} + \frac{1}{4}\tan\frac{1}{4\theta} + \frac{1}{8}\tan\frac{1}{8\theta} + \cdots$$

[1.74] Prove that $\forall (x,n) \in (\mathbb{R} - \{-1,1\}) \times \mathbb{N}$,

$$\frac{1}{x-1} = \sum_{j=0}^{n-1} \frac{2^j}{x^{2^j}+1} + \frac{2^n}{x^{2^n}-1}$$
$$= \frac{1}{x+1} + \frac{2}{x^2+1} + \cdots + \frac{2^{n-1}}{x^{2^{n-1}}+1} + \frac{2^n}{x^{2^n}-1}.$$

[1.75] Prove that $\forall (x, n) \in (\mathbb{R} - \{-1, 1\}) \times \mathbb{N}$,

$$\frac{x}{(x-1)^2} = \sum_{j=0}^{n-1} \frac{2^{2j} x^{2^j}}{(x^{2^j} + 1)^2} + \frac{2^{2n} x^{2^n}}{(x^{2^n} - 1)^2}$$

$$= \frac{x}{(x+1)^2} + \frac{2^2 x^2}{(x^2+1)^2} + \cdots + \frac{2^{2n-2} x^{2^{n-1}}}{(x^{2^{n-1}} + 1)^2} + \frac{2^{2n} x^{2^n}}{(x^{2^n} - 1)^2}.$$

[1.76] Prove that $\forall (x, y, n) \in \mathbb{R}^2 \times \mathbb{N}$, $x \neq \pm y$, $xy \neq \pm 1$,

$$\frac{x(y^2 - 1)}{(x-y)(xy-1)} = \sum_{j=0}^{n-1} \frac{2^j x^{2^j} (y^{2^{j+1}} - 1)}{(x^{2^j} + y^{2^j})(x^{2^j} y^{2^j} + 1)} + \frac{2^n x^{2^n} (y^{2^{n+1}} - 1)}{(x^{2^n} - y^{2^n})(x^{2^n} y^{2^n} - 1)}$$

$$= \frac{x(y^2 - 1)}{(x+y)(xy+1)} + \frac{2x^2 (y^4 - 1)}{(x^2 + y^2)(x^2 y^2 + 1)} + \cdots$$

$$+ \frac{2^{n-1} x^{2^{n-1}} (y^{2^n} - 1)}{(x^{2^{n-1}} + y^{2^{n-1}})(x^{2^{n-1}} y^{2^{n-1}} + 1)} + \frac{2^n x^{2^n} (y^{2^{n+1}} - 1)}{(x^{2^n} - y^{2^n})(x^{2^n} y^{2^n} - 1)}.$$

[1.77] Prove that $\forall (x, y, n) \in \mathbb{R}^2 \times \mathbb{N}$, $x \neq \pm y$,

$$\frac{y}{x-y} = \sum_{j=0}^{n-1} \frac{2^j y^{2^j}}{x^{2^j} + y^{2^j}} + \frac{2^n y^{2^n}}{x^{2^n} - y^{2^n}}$$

$$= \frac{y}{x+y} + \frac{2y^2}{x^2 + y^2} + \cdots + \frac{2^{n-1} y^{2^{n-1}}}{x^{2^{n-1}} + y^{2^{n-1}}} + \frac{2^n y^{2^n}}{x^{2^n} - y^{2^n}}.$$

[1.78] Prove that $\forall (x, y, n) \in \mathbb{R}^2 \times \mathbb{N}$, $x \neq \pm y$, $xy \neq \pm 1$,

$$\frac{y(xy-1) + x - y}{(x-y)(xy-1)}$$

$$= \sum_{j=0}^{n-1} 2^j \frac{y^{2^j}(x^{2^j} y^{2^j} + 1) + x^{2^j} + y^{2^j}}{(x^{2^j} + y^{2^j})(x^{2^j} y^{2^j} + 1)} + 2^n \frac{y^{2^n}(x^{2^n} y^{2^n} - 1) + x^{2^n} - y^{2^n}}{(x^{2^n} - y^{2^n})(x^{2^n} y^{2^n} - 1)}$$

$$= \frac{y(xy+1) + x + y}{(x+y)(xy+1)} + 2\frac{y^2(x^2 y^2 + 1) + x^2 + y^2}{(x^2 + y^2)(x^2 y^2 + 1)} + \cdots$$

$$+ 2^{n-1} \frac{y^{2^{n-1}}(x^{2^{n-1}} y^{2^{n-1}} + 1) + x^{2^{n-1}} + y^{2^{n-1}}}{(x^{2^{n-1}} + y^{2^{n-1}})(x^{2^{n-1}} y^{2^{n-1}} + 1)} + 2^n \frac{y^{2^n}(x^{2^n} y^{2^n} - 1) + x^{2^n} - y^{2^n}}{(x^{2^n} - y^{2^n})(x^{2^n} y^{2^n} - 1)}.$$

[1.79] Prove that $\forall (x, \theta, n) \in \mathbb{R}^2 \times \mathbb{N}$,

$$\frac{x\cos\theta - 1}{x^2 - 2x\cos\theta + 1} = \sum_{j=0}^{n-1} \frac{2^j\left(x^{2^j}\cos 2^j\theta + 1\right)}{x^{2^{j+1}} + 2x^{2^j}\cos 2^j\theta + 1} + \frac{2^n\left(x^{2^n}\cos 2^n\theta - 1\right)}{x^{2^{n+1}} - 2x^{2^n}\cos 2^n\theta + 1}$$

$$= \frac{x\cos\theta + 1}{x^2 + 2x\cos\theta + 1} + \frac{2\left(x^2\cos 2\theta + 1\right)}{x^4 + 2x^2\cos 2\theta + 1} + \cdots$$

$$+ \frac{2^{n-1}\left(x^{2^{n-1}}\cos 2^{n-1}\theta + 1\right)}{x^{2^n} + 2x^{2^{n-1}}\cos 2^{n-1}\theta + 1} + \frac{2^n\left(x^{2^n}\cos 2^n\theta - 1\right)}{x^{2^{n+1}} - 2x^{2^n}\cos 2^n\theta + 1}.$$

[1.80] Prove that $\forall (x, \theta, n) \in \mathbb{R}^2 \times \mathbb{N}$,

$$\frac{x\sin\theta}{x^2 - 2x\cos\theta + 1} = \sum_{j=0}^{n-1} \frac{2^j x^{2^j}\sin 2^j\theta}{x^{2^{j+1}} + 2x^{2^j}\cos 2^j\theta + 1} + \frac{2^n x^{2^n}\sin 2^n\theta}{x^{2^{n+1}} - 2x^{2^n}\cos 2^n\theta + 1}$$

$$= \frac{x\sin\theta}{x^2 + 2x\cos\theta + 1} + \frac{2x^2\sin 2\theta}{x^4 + 2x^2\cos 2\theta + 1} + \cdots$$

$$+ \frac{2^{n-1}x^{2^{n-1}}\sin 2^{n-1}\theta}{x^{2^n} + 2x^{2^{n-1}}\cos 2^{n-1}\theta + 1} + \frac{2^n x^{2^n}\sin 2^n\theta}{x^{2^{n+1}} - 2x^{2^n}\cos 2^n\theta + 1}.$$

[1.81] Prove that $\forall (x, \theta, n) \in \mathbb{R}^2 \times (\mathbb{N} - 1)$,

$$\frac{x^{2^{n+2}} - 2x^{2^{n+1}}\cos 2^{n+1}\theta + 1}{x^2 - 2x\cos\theta + 1} = \prod_{j=0}^{n}\left(x^{2^{j+1}} + 2x^{2^j}\cos 2^j\theta + 1\right)$$

$$= \left(x^2 + 2x\cos\theta + 1\right)\left(x^4 + 2x^2\cos 2\theta + 1\right)\left(x^8 + 2x^4\cos 4\theta + 1\right)$$

$$\cdots \left(x^{2^{n+1}} + 2x^{2^n}\cos 2^n\theta + 1\right).$$

[1.82] Let

$$T(x,\theta,n) = \sum_{j=0}^{n-1} \tan^{-1} \frac{x^{2^j} + \cos 2^j \theta}{\sin 2^j \theta} + \tan^{-1} \frac{x^{2^n} - \cos 2^n \theta}{\sin 2^n \theta}$$

$$= \tan^{-1} \frac{x + \cos \theta}{\sin \theta} + \tan^{-1} \frac{x^2 + \cos 2\theta}{\sin 2\theta} + \cdots$$

$$+ \tan^{-1} \frac{x^{2^{n-1}} + \cos 2^{n-1} \theta}{\sin 2^{n-1} \theta} + \tan^{-1} \frac{x^{2^n} - \cos 2^n \theta}{\sin 2^n \theta}.$$

Prove that $\forall (x, \theta, n) \in \mathbb{R} \times \left(\mathbb{R} - \left\{ \frac{m\pi}{2^j} \mid j \in [0..n], m \in \mathbb{Z} \right\} \right) \times (\mathbb{N} - 1)$,

$$T(x, \theta, n) - \tan^{-1} \frac{x - \cos \theta}{\sin \theta} \in \frac{\pi}{2} \mathbb{Z}.$$

[1.83] (Berndt, 1994) Prove that $\forall x \in \mathbb{R}^+ - \{1\}$,

$$\frac{1}{\log x} = \frac{1}{x - 1} + \sum_{j=1}^{\infty} \frac{1}{2^j \left(1 + x^{1/2^j}\right)}$$

$$= \frac{1}{x - 1} + \frac{1}{2\left(1 + x^{1/2}\right)} + \frac{1}{4\left(1 + x^{1/4}\right)} + \frac{1}{8\left(1 + x^{1/8}\right)} + \cdots$$

[1.84] Prove that $\forall x \in \mathbb{R}^+ - \{1\}$,

$$\frac{1}{\log^2 x} = \frac{x}{(x-1)^2} + \sum_{j=1}^{\infty} \frac{x^{1/2^j}}{2^{2j}\left(1 + x^{1/2^j}\right)^2}$$

$$= \frac{x}{(x-1)^2} + \frac{x^{1/2}}{2^2 \left(1 + x^{1/2}\right)^2} + \frac{x^{1/4}}{4^2 \left(1 + x^{1/4}\right)^2} + \frac{x^{1/8}}{8^2 \left(1 + x^{1/8}\right)^2} + \cdots$$

[1.85] Prove that $\forall n \in \mathbb{N}, \exists (a_k)_{k=1}^n \in \mathbb{R}^n$ and $\exists (b_k)_{k=1}^n \in \mathbb{R}^n$ such that

$$\frac{1}{\log^n x} = \sum_{k=1}^{n} \frac{a_k}{(x-1)^k} + \sum_{k=1}^{n} \sum_{j=1}^{\infty} \frac{b_k}{2^{jn} \left(1 + x^{1/2^j}\right)^k}.$$

[1.86] Prove that $\forall (x,\theta) \in \mathbb{R}^+ \times \mathbb{R}$, such that $\theta^2 + \log^2 x \neq 0$,

$$\frac{\log x}{\theta^2 + \log^2 x} = \frac{x\cos\theta - 1}{x^2 - 2x\cos\theta + 1} + \sum_{j=1}^{\infty} \frac{x^{1/2^j}\cos\frac{\theta}{2^j} + 1}{2^j \left(x^{1/2^{j-1}} + 2x^{1/2^j}\cos\frac{\theta}{2^j} + 1 \right)}$$

$$= \frac{x\cos\theta - 1}{x^2 - 2x\cos\theta + 1} + \frac{x^{1/2}\cos\frac{\theta}{2} + 1}{2\left(x + 2x^{1/2}\cos\frac{\theta}{2} + 1\right)}$$

$$+ \frac{x^{1/4}\cos\frac{\theta}{4} + 1}{4\left(x^{1/2} + 2x^{1/4}\cos\frac{\theta}{4} + 1\right)} + \frac{x^{1/8}\cos\frac{\theta}{8} + 1}{8\left(x^{1/4} + 2x^{1/8}\cos\frac{\theta}{8} + 1\right)} + \cdots$$

[1.87] Prove that $\forall (x,\theta) \in \mathbb{R}^+ \times \mathbb{R}$, such that $\theta^2 + \log^2 x \neq 0$,

$$\frac{\theta}{\theta^2 + \log^2 x} = \frac{x\sin\theta}{x^2 - 2x\cos\theta + 1} + \sum_{j=1}^{\infty} \frac{x^{1/2^j}\sin\frac{\theta}{2^j}}{2^j \left(x^{1/2^{j-1}} + 2x^{1/2^j}\cos\frac{\theta}{2^j} + 1 \right)}$$

$$= \frac{x\sin\theta}{x^2 - 2x\cos\theta + 1} + \frac{x^{1/2}\sin\frac{\theta}{2}}{2\left(x + 2x^{1/2}\cos\frac{\theta}{2} + 1\right)}$$

$$+ \frac{x^{1/4}\sin\frac{\theta}{4}}{4\left(x^{1/2} + 2x^{1/4}\cos\frac{\theta}{4} + 1\right)} + \frac{x^{1/8}\sin\frac{\theta}{8}}{8\left(x^{1/4} + 2x^{1/8}\cos\frac{\theta}{8} + 1\right)} + \cdots$$

[1.88] Prove that $\forall (x, \theta) \in \mathbb{R}^+ \times \mathbb{R}, \theta \neq \pm \log x$,

$$\frac{\log x}{\log^2 x - \theta^2} = \frac{x \cosh \theta - 1}{x^2 - 2x \cosh \theta + 1} + \sum_{j=1}^{\infty} \frac{x^{1/2^j} \cosh \frac{\theta}{2^j} + 1}{2^j \left(x^{1/2^{j-1}} + 2x^{1/2^j} \cosh \frac{\theta}{2^j} + 1 \right)}$$

$$= \frac{x \cosh \theta - 1}{x^2 - 2x \cosh \theta + 1} + \frac{x^{1/2} \cosh \frac{\theta}{2} + 1}{2 \left(x + 2x^{1/2} \cosh \frac{\theta}{2} + 1 \right)}$$

$$+ \frac{x^{1/4} \cosh \frac{\theta}{4} + 1}{4 \left(x^{1/2} + 2x^{1/4} \cosh \frac{\theta}{4} + 1 \right)} + \frac{x^{1/8} \cosh \frac{\theta}{8} + 1}{8 \left(x^{1/4} + 2x^{1/8} \cosh \frac{\theta}{8} + 1 \right)} + \cdots$$

[1.89] Prove that $\forall (x, \theta) \in \mathbb{R}^+ \times \mathbb{R}, \theta \neq \pm \log x$,

$$\frac{\theta}{\log^2 x - \theta^2} = \frac{x \sinh \theta}{x^2 - 2x \cosh \theta + 1} + \sum_{j=1}^{\infty} \frac{x^{1/2^j} \sinh \frac{\theta}{2^j}}{2^j \left(x^{1/2^{j-1}} + 2x^{1/2^j} \cosh \frac{\theta}{2^j} + 1 \right)}$$

$$= \frac{x \sinh \theta}{x^2 - 2x \cosh \theta + 1} + \frac{x^{1/2} \sinh \frac{\theta}{2}}{2 \left(x + 2x^{1/2} \cosh \frac{\theta}{2} + 1 \right)}$$

$$+ \frac{x^{1/4} \sinh \frac{\theta}{4}}{4 \left(x^{1/2} + 2x^{1/4} \cosh \frac{\theta}{4} + 1 \right)} + \frac{x^{1/8} \sinh \frac{\theta}{8}}{8 \left(x^{1/4} + 2x^{1/8} \cosh \frac{\theta}{8} + 1 \right)} + \cdots$$

[1.90] Prove that $\forall (x, y) \in \left(\mathbb{R}^+ \right)^2, x \neq y, xy \neq 1$,

$$\frac{\log x^2}{\log^2 x - \log^2 y} = \frac{y(xy - 1) + x - y}{(x - y)(xy - 1)}$$

$$+ \sum_{j=1}^{\infty} \frac{y^{1/2^j} \left(x^{1/2^j} y^{1/2^j} + 1 \right) + x^{1/2^j} + y^{1/2^j}}{2^j \left(x^{1/2^j} + y^{1/2^j} \right) \left(x^{1/2^j} y^{1/2^j} + 1 \right)}$$

$$= \frac{y(xy-1)+x-y}{(x-y)(xy-1)} + \frac{y^{1/2}\left(x^{1/2}y^{1/2}+1\right)+x^{1/2}+y^{1/2}}{2\left(x^{1/2}+y^{1/2}\right)\left(x^{1/2}y^{1/2}+1\right)}$$
$$+ \frac{y^{1/4}\left(x^{1/4}y^{1/4}+1\right)+x^{1/4}+y^{1/4}}{4\left(x^{1/4}+y^{1/4}\right)\left(x^{1/4}y^{1/4}+1\right)}$$
$$+ \frac{y^{1/8}\left(x^{1/8}y^{1/8}+1\right)+x^{1/8}+y^{1/8}}{8\left(x^{1/8}+y^{1/8}\right)\left(x^{1/8}y^{1/8}+1\right)} + \cdots$$

[1.91] Prove that $\forall (x,y) \in \left(\mathbb{R}^+\right)^2, x \neq y, xy \neq 1$,

$$\frac{\log y^2}{\log^2 x - \log^2 y} = \frac{x(y^2-1)}{(x-y)(xy-1)} + \sum_{j=1}^{\infty} \frac{x^{1/2^j}\left(y^{1/2^{j-1}}-1\right)}{2^j\left(x^{1/2^j}+y^{1/2^j}\right)\left(x^{1/2^j}y^{1/2^j}+1\right)}$$
$$= \frac{x(y^2-1)}{(x-y)(xy-1)} + \frac{x^{1/2}(y-1)}{2\left(x^{1/2}+y^{1/2}\right)\left(x^{1/2}y^{1/2}+1\right)}$$
$$+ \frac{x^{1/4}\left(y^{1/2}-1\right)}{4\left(x^{1/4}+y^{1/4}\right)\left(x^{1/4}y^{1/4}+1\right)} + \frac{x^{1/8}\left(y^{1/4}-1\right)}{8\left(x^{1/8}+y^{1/8}\right)\left(x^{1/8}y^{1/8}+1\right)} + \cdots$$

[1.92] Prove that $\forall (x,n) \in (\mathbb{R} - \{-1,1\}) \times \mathbb{N}$,

$$\frac{1+x}{1-x} = \sum_{j=0}^{n-1} 2^j \frac{x^{2^j}-1}{x^{2^j}+1} + 2^n \frac{1+x^{2^n}}{1-x^{2^n}}$$
$$= \frac{x-1}{x+1} + 2\frac{x^2-1}{x^2+1} + \cdots + 2^{n-1}\frac{x^{2^{n-1}}-1}{x^{2^{n-1}}+1} + 2^n \frac{1+x^{2^n}}{1-x^{2^n}}.$$

Chapter 2
Trigonometrical Series

In this chapter, we will learn about the application of De Moivre's theorem and Euler's formula in the derivation of trigonometrical sums and products. We will also apply the differentiation and integration methods of series sums as described in the previous chapter. Circular and hyperbolic trigonometrical identities are given as problems in this chapter.

2.1 Trigonometrical Series

De Moivre's theorem states that $\forall (\theta, m) \in \mathbb{R} \times \mathbb{Z}$,

$$(\cos\theta + i\sin\theta)^m = \cos m\theta + i\sin m\theta = e^{im\theta}.$$

By Euler's formula, $\forall \theta \in \mathbb{R}$, we have

$$e^{i\theta} = \cos\theta + i\sin\theta.$$

We can find the sum of the GP $\left(e^{ij\theta}\right)_{j=0}^{n}$, and equate its real and imaginary parts to get some trigonometrical sums.

Example 2.1 (*Trigonometrical Series 1 and 2*)
Prove that $\forall (\theta, n) \in (\mathbb{R} - 2\pi\mathbb{Z}) \times (\mathbb{N} - 1)$,

$$\sum_{j=0}^{n} \cos j\theta = 1 + \cos\theta + \cos 2\theta + \cdots + \cos n\theta = \frac{\sin\frac{(n+1)\theta}{2} \cos\frac{n\theta}{2}}{\sin\frac{\theta}{2}}, \quad (2.1)$$

and

$$\sum_{j=0}^{n} \sin j\theta = \sin\theta + \sin 2\theta + \sin 3\theta + \cdots + \sin n\theta = \frac{\sin\frac{(n+1)\theta}{2} \sin\frac{n\theta}{2}}{\sin\frac{\theta}{2}}. \quad (2.2)$$

Proof Let

$$C = \sum_{j=0}^{n} \cos j\theta, \quad (2.3)$$

and

$$S = \sum_{j=0}^{n} \sin j\theta. \quad (2.4)$$

Multiplying (2.4) by i and adding the result in (2.3), we get

$$\begin{aligned}
C + iS &= \sum_{j=0}^{n} (\cos j\theta + i \sin j\theta) \\
&= \sum_{j=0}^{n} e^{ij\theta} \\
&= \frac{e^{i(n+1)\theta} - 1}{e^{i\theta} - 1} \\
&= \frac{e^{i(n+1)\theta/2} \left(e^{i(n+1)\theta/2} - e^{-i(n+1)\theta/2}\right)}{e^{i\theta/2} \left(e^{i\theta/2} - e^{-i\theta/2}\right)} \\
&= e^{in\theta/2} \frac{\left(e^{i(n+1)\theta/2} - e^{-i(n+1)\theta/2}\right)/2i}{\left(e^{i\theta/2} - e^{-i\theta/2}\right)/2i} \\
&= \left(\cos\frac{n\theta}{2} + i \sin\frac{n\theta}{2}\right) \frac{\sin\frac{(n+1)\theta}{2}}{\sin\frac{\theta}{2}} \\
&= \frac{\sin\frac{(n+1)\theta}{2} \cos\frac{n\theta}{2}}{\sin\frac{\theta}{2}} + i \frac{\sin\frac{(n+1)\theta}{2} \sin\frac{n\theta}{2}}{\sin\frac{\theta}{2}}.
\end{aligned} \quad (2.5)$$

By equating the real and imaginary parts of (2.5), we get (2.1) and (2.2) respectively. □

Example 2.2 (*Finding the Sum of First n Natural Numbers*)
Prove that $\forall n \in \mathbb{N}$,

2.1 Trigonometrical Series

$$\sum_{j=1}^{n} j = 1 + 2 + 3 + \cdots + n = \frac{n(n+1)}{2}. \tag{2.6}$$

Proof Dividing the LHS of (2.2) by $\sin\theta$ and taking the limit as $\theta \to 0$, we get

$$\lim_{\theta \to 0} \sum_{j=0}^{n} \frac{\sin j\theta}{\sin \theta} = \sum_{j=0}^{n} \lim_{\theta \to 0} \frac{\sin j\theta}{\sin \theta}$$

$$= \sum_{j=1}^{n} \lim_{\theta \to 0} j \frac{\frac{\sin j\theta}{j\theta}}{\frac{\sin \theta}{\theta}} \tag{2.7}$$

$$= \sum_{j=1}^{n} j.$$

Dividing the RHS of (2.2) by $\sin \theta$ and taking the limit as $\theta \to 0$, we get

$$\lim_{\theta \to 0} \frac{\sin \frac{(n+1)\theta}{2} \sin \frac{n\theta}{2}}{\sin \theta \sin \frac{\theta}{2}}$$

$$= \lim_{\theta \to 0} \left(\frac{n+1}{2}\right)\left(\frac{n}{2}\right)\left(\frac{1}{2}\right)^{-1} \left(\frac{\sin \frac{(n+1)\theta}{2}}{\frac{(n+1)\theta}{2}}\right) \left(\frac{\sin \frac{n\theta}{2}}{\frac{n\theta}{2}}\right) \left(\frac{\sin \theta}{\theta}\right)^{-1} \left(\frac{\sin \frac{\theta}{2}}{\frac{\theta}{2}}\right)^{-1}$$

$$= \frac{n(n+1)}{2}.$$

(2.8)

By equating (2.7) and (2.8), we get (2.6). □

Example 2.3 (*Trigonometrical Series* 3)
Prove that $\forall (\theta, n) \in (\mathbb{R} - 2\pi\mathbb{Z}) \times (\mathbb{N} - 1)$,

$$\sum_{j=1}^{n} j \sin j\theta = \sin\theta + 2\sin 2\theta + 3\sin 3\theta + \cdots + n \sin n\theta$$

$$= \frac{n \sin \frac{(n+1)\theta}{2} \sin \frac{n\theta}{2} - (n+1)\cos \frac{(n+1)\theta}{2} \cos \frac{n\theta}{2}}{2\sin \frac{\theta}{2}}$$

$$+ \frac{\sin \frac{(n+1)\theta}{2} \cos \frac{n\theta}{2} \cos \frac{\theta}{2}}{2\sin^2 \frac{\theta}{2}}.$$

Proof Differentiating (2.1) with respect to θ, we get

$$\sum_{j=1}^{n} -j \sin j\theta$$

$$= \frac{\left(\frac{n+1}{2}\cos\frac{(n+1)\theta}{2}\cos\frac{n\theta}{2} - \frac{n}{2}\sin\frac{(n+1)\theta}{2}\sin\frac{n\theta}{2}\right)\sin\frac{\theta}{2}}{\sin^2\frac{\theta}{2}}$$

$$- \frac{\frac{1}{2}\sin\frac{(n+1)\theta}{2}\cos\frac{n\theta}{2}\cos\frac{\theta}{2}}{\sin^2\frac{\theta}{2}}$$

$$\iff \sum_{j=1}^{n} j \sin j\theta$$

$$= \frac{n\sin\frac{(n+1)\theta}{2}\sin\frac{n\theta}{2} - (n+1)\cos\frac{(n+1)\theta}{2}\cos\frac{n\theta}{2}}{2\sin\frac{\theta}{2}}$$

$$+ \frac{\sin\frac{(n+1)\theta}{2}\cos\frac{n\theta}{2}\cos\frac{\theta}{2}}{2\sin^2\frac{\theta}{2}}.$$

$\square$

Example 2.4 (*Trigonometrical Series* 4)
Prove that $\forall (\theta, n) \in (\mathbb{R} - 2\pi\mathbb{Z}) \times \mathbb{N}$,

$$\sum_{j=1}^{n} j \cos j\theta = \cos\theta + 2\cos 2\theta + 3\cos 3\theta + \cdots + n\cos n\theta$$

$$= \frac{(n+1)\cos\frac{(n+1)\theta}{2}\sin\frac{n\theta}{2} + n\sin\frac{(n+1)\theta}{2}\cos\frac{n\theta}{2}}{2\sin\frac{\theta}{2}}$$

$$- \frac{\sin\frac{(n+1)\theta}{2}\sin\frac{n\theta}{2}\cos\frac{\theta}{2}}{2\sin^2\frac{\theta}{2}}.$$

2.1 Trigonometrical Series

Proof Differentiating (2.2) with respect to θ, we get

$$\sum_{j=1}^{n} j \cos j\theta = \frac{\left(\frac{n+1}{2}\cos\frac{(n+1)\theta}{2}\sin\frac{n\theta}{2} + \frac{n}{2}\sin\frac{(n+1)\theta}{2}\cos\frac{n\theta}{2}\right)\sin\frac{\theta}{2}}{\sin^2\frac{\theta}{2}}$$

$$- \frac{\frac{1}{2}\sin\frac{(n+1)\theta}{2}\sin\frac{n\theta}{2}\cos\frac{\theta}{2}}{\sin^2\frac{\theta}{2}}$$

$$= \frac{(n+1)\cos\frac{(n+1)\theta}{2}\sin\frac{n\theta}{2} + n\sin\frac{(n+1)\theta}{2}\cos\frac{n\theta}{2}}{2\sin\frac{\theta}{2}}$$

$$- \frac{\sin\frac{(n+1)\theta}{2}\sin\frac{n\theta}{2}\cos\frac{\theta}{2}}{2\sin^2\frac{\theta}{2}}.$$

$\square$

Now, we will apply De Moivre's theorem and the binomial theorem to expand $\cos n\theta$ and $\sin n\theta$ in a series of powers of $\cos\theta$ and $\sin\theta$.

Example 2.5 (*Trigonometrical Series 5 and 6*)
Prove that $\forall (\theta, n) \in \mathbb{R} \times \mathbb{N}$,

$$\cos n\theta = \sum_{j=0}^{\lfloor n/2 \rfloor} (-1)^j \binom{n}{2j} \cos^{n-2j}\theta \sin^{2j}\theta$$

$$= \cos^n \theta - \binom{n}{2}\cos^{n-2}\theta \sin^2\theta + \binom{n}{4}\cos^{n-4}\theta \sin^4\theta \quad (2.9)$$

$$+ \cdots (-1)^{\lfloor n/2 \rfloor} \cos^{n-2\lfloor n/2 \rfloor}\theta \sin^{2\lfloor n/2 \rfloor}\theta,$$

and

$$\sin n\theta = \sum_{j=1}^{\lceil n/2 \rceil} (-1)^{j-1} \binom{n}{2j-1} \cos^{n-2j+1}\theta \sin^{2j-1}\theta$$

$$= \binom{n}{1}\cos^{n-1}\theta \sin\theta - \binom{n}{3}\cos^{n-3}\theta \sin^3\theta + \binom{n}{5}\cos^{n-5}\theta \sin^5\theta \quad (2.10)$$

$$+ \cdots (-1)^{\lceil n/2 \rceil - 1} \cos^{n-2\lceil n/2 \rceil + 1}\theta \sin^{2\lceil n/2 \rceil - 1}\theta.$$

Proof Using De Moivre's theorem and the binomial theorem, we have

$$(\cos\theta + i\sin\theta)^n = \sum_{j=0}^{n} \binom{n}{j} i^j \cos^{n-j}\theta \sin^j\theta$$

$$\iff \cos n\theta + i \sin n\theta$$

$$= \sum_{j=0}^{\lfloor n/2 \rfloor} (-1)^j \binom{n}{2j} \cos^{n-2j}\theta \sin^{2j}\theta \qquad (2.11)$$

$$+ i \sum_{j=1}^{\lceil n/2 \rceil} (-1)^{j-1} \binom{n}{2j-1} \cos^{n-2j+1}\theta \sin^{2j-1}\theta.$$

By equating the real and imaginary parts of (2.11), we get (2.9) and (2.10) respectively. □

We can also evaluate a quadratic fraction using two methods: first, by making use of the complex variables, and second, by making use of the binomial theorem. As a result, we will get expansions of $\sin n\theta / \sin \theta$ and $\cos n\theta$ in a series of powers of $\cos \theta$.

Example 2.6 (Loney, 1928) (*Trigonometrical Series 7*)
Prove that $\forall (\theta, n) \in (\mathbb{R} - \mathbb{Z}\pi) \times 2\mathbb{N}$,

$$\frac{\sin n\theta}{\sin \theta} = (-1)^{n/2+1} n \sum_{j=0}^{n/2-1} (-1)^j \frac{\prod_{k=1}^{j}(n^2 - (2k)^2)}{(2j+1)!} \cos^{2j+1}\theta$$

$$= (-1)^{n/2+1} n \left(\cos\theta - \frac{(n^2 - 2^2)}{3!} \cos^3\theta + \frac{(n^2-2^2)(n^2-4^2)}{5!} \cos^5\theta \right. \qquad (2.12)$$

$$\left. + \cdots + \frac{(n^2-2^2)(n^2-4^2)(n^2-6^2)\cdots(n^2-(n-2)^2)}{(n-1)!} \cos^{n-1}\theta \right).$$

Proof We will compute the value of $1/(1 - 2x\cos\theta + x^2)$ using two methods and equate the coefficient of x^{n-1} to get the value of $\sin n\theta/\sin\theta$ as a series in powers of $\cos\theta$.
First Method: Using Euler's formula, we have

$$\frac{1}{1 - 2x\cos\theta + x^2} = \frac{1}{1 - x(e^{i\theta} + e^{-i\theta}) + x^2}$$

$$= \frac{1}{(1 - xe^{i\theta})(1 - xe^{-i\theta})}$$

$$= \frac{1}{x(e^{i\theta} - e^{-i\theta})} \left(\frac{1}{1 - xe^{i\theta}} - \frac{1}{1 - xe^{-i\theta}} \right)$$

2.1 Trigonometrical Series

$$= \frac{1}{2ix\sin\theta}\left(\sum_{j=0}^{\infty} x^j e^{ij\theta} - \sum_{j=0}^{\infty} x^j e^{-ij\theta}\right)$$

$$= \frac{1}{2ix\sin\theta}\sum_{j=0}^{\infty} x^j \left(e^{ij\theta} - e^{-ij\theta}\right) \tag{2.13}$$

$$= \frac{1}{2ix\sin\theta}\sum_{j=0}^{\infty} 2ix^j \sin j\theta$$

$$= \sum_{j=1}^{\infty} \frac{\sin j\theta}{\sin\theta} x^{j-1}.$$

Second Method: Using the binomial theorem, we have

$$\frac{1}{1 - 2x\cos\theta + x^2} = \frac{1}{1 + x(x - 2\cos\theta)}$$

$$= \sum_{j=0}^{\infty} (-1)^j x^j (x - 2\cos\theta)^j. \tag{2.14}$$

For $(\theta, n) \in (\mathbb{R} - \mathbb{Z}\pi) \times 2\mathbb{N}$, we equate the coefficients of x^{n-1} in (2.13) and (2.14) to get

$$\frac{\sin n\theta}{\sin\theta} = \sum_{j=0}^{n/2-1} (-1)^{n/2+j} \binom{n/2+j}{2j+1} (-2\cos\theta)^{2j+1}$$

$$= (-1)^{n/2+1} \sum_{j=0}^{n/2-1} (-1)^j \frac{2^{2j+1} \prod_{k=0}^{2j}\left(\frac{n}{2}+j-k\right)}{(2j+1)!} \cos^{2j+1}\theta$$

$$= (-1)^{n/2+1} \sum_{j=0}^{n/2-1} (-1)^j \frac{\prod_{k=0}^{2j}(n+2j-2k)}{(2j+1)!} \cos^{2j+1}\theta$$

$$= (-1)^{n/2+1} n \sum_{j=0}^{n/2-1} (-1)^j \frac{\prod_{k=1}^{j}\left(n^2 - (2k)^2\right)}{(2j+1)!} \cos^{2j+1}\theta.$$

$\square$

Example 2.7 (Loney, 1928) (*Trigonometrical Series* 8)
Prove that $\forall (\theta, n) \in (\mathbb{R} - \mathbb{Z}\pi) \times (2\mathbb{N} - 1)$,

$$\frac{\sin n\theta}{\sin \theta} = (-1)^{(n-1)/2} \sum_{j=0}^{(n-1)/2} (-1)^j \frac{\prod_{k=0}^{j-1}\left(n^2 - (2k+1)^2\right)}{(2j)!} \cos^{2j}\theta$$

$$= (-1)^{(n-1)/2}\left(1 - \frac{(n^2-1^2)}{2!}\cos^2\theta + \frac{(n^2-1^2)(n^2-3^2)}{4!}\cos^4\theta \right. \tag{2.15}$$

$$\left. + \cdots + \frac{(n^2-1^2)(n^2-3^2)(n^2-5^2)\cdots(n^2-(n-2)^2)}{(n-1)!}\cos^{n-1}\theta\right).$$

Proof For $(\theta, n) \in (\mathbb{R} - \mathbb{Z}\pi) \times (2\mathbb{N} - 1)$, we equate the coefficients of x^{n-1} in (2.13) and (2.14) to get

$$\frac{\sin n\theta}{\sin \theta} = \sum_{j=0}^{(n-1)/2} (-1)^{(n-1)/2+j} \binom{(n-1)/2+j}{2j} (-2\cos\theta)^{2j}$$

$$= (-1)^{(n-1)/2} \sum_{j=0}^{(n-1)/2} (-1)^j \frac{2^{2j}\prod_{k=0}^{2j-1}\left(\frac{n-1}{2} + j - k\right)}{(2j)!} \cos^{2j}\theta$$

$$= (-1)^{(n-1)/2} \sum_{j=0}^{(n-1)/2} (-1)^j \frac{\prod_{k=0}^{2j-1}(n - 1 + 2j - 2k)}{(2j)!} \cos^{2j}\theta$$

$$= (-1)^{(n-1)/2} \sum_{j=0}^{(n-1)/2} (-1)^j \frac{\prod_{k=0}^{j-1}\left(n^2 - (2k+1)^2\right)}{(2j)!} \cos^{2j}\theta.$$

□

Example 2.8 (Loney, 1928) (*Trigonometrical Series* 9)
Prove that $\forall (\theta, n) \in (\mathbb{R} - \mathbb{Z}\pi) \times (2\mathbb{N} - 2)$,

$$\cos n\theta = (-1)^{n/2} \sum_{j=0}^{n/2} (-1)^j \frac{\prod_{k=0}^{j-1}\left(n^2 - (2k)^2\right)}{(2j)!} \cos^{2j}\theta$$

$$= (-1)^{n/2}\left(1 - \frac{n^2}{2!}\cos^2\theta + \frac{n^2(n^2-2^2)}{4!}\cos^4\theta \right. \tag{2.16}$$

$$\left. + \cdots + \frac{n^2(n^2-2^2)(n^2-4^2)\cdots(n^2-(n-2)^2)}{n!}\cos^n\theta\right).$$

2.1 Trigonometrical Series

Proof We will compute the value of $(1 - x\cos\theta)/(1 - 2x\cos\theta + x^2)$ using two methods and equate the coefficient of x^n to get the value of $\cos n\theta$ as a series in powers of $\cos\theta$.

First Method: Using Euler's formula, we have

$$\begin{aligned}
\frac{1 - x\cos\theta}{1 - 2x\cos\theta + x^2} &= \frac{1 - x\dfrac{e^{i\theta} + e^{i\theta}}{2}}{1 - x\left(e^{i\theta} + e^{-i\theta}\right) + x^2} \\
&= \frac{2 - x\left(e^{i\theta} + e^{-i\theta}\right)}{2\left(1 - xe^{i\theta}\right)\left(1 - xe^{-i\theta}\right)} \\
&= \frac{1}{2}\left(\frac{1}{1 - xe^{i\theta}} + \frac{1}{1 - xe^{-i\theta}}\right) \\
&= \frac{1}{2}\left(\sum_{j=0}^{\infty} x^j e^{ij\theta} + \sum_{j=0}^{\infty} x^j e^{-ij\theta}\right) \\
&= \sum_{j=0}^{\infty} x^j \frac{e^{ij\theta} + e^{-ij\theta}}{2} \\
&= \sum_{j=0}^{\infty} x^j \cos j\theta.
\end{aligned} \qquad (2.17)$$

Second Method: Using the binomial theorem, we have

$$\begin{aligned}
\frac{1 - x\cos\theta}{1 - 2x\cos\theta + x^2} &= \frac{1 - x\cos\theta}{1 + x(x - 2\cos\theta)} \\
&= (1 - x\cos\theta)\sum_{j=0}^{\infty}(-1)^j x^j (x - 2\cos\theta)^j.
\end{aligned} \qquad (2.18)$$

For $(\theta, n) \in (\mathbb{R} - \mathbb{Z}\pi) \times 2\mathbb{N}$, we equate the coefficients of x^n in (2.17) and (2.18) to get

$$\begin{aligned}
\cos n\theta &= \sum_{j=0}^{n/2}(-1)^{n/2+j}\binom{n/2+j}{2j}(-2\cos\theta)^{2j} \\
&\quad - \sum_{j=0}^{n/2-1}(-1)^{n/2+j}\binom{n/2+j}{2j+1}(-2\cos\theta)^{2j+1}\cos\theta \\
&= \sum_{j=0}^{n/2}(-1)^{n/2+j}\binom{n/2+j}{2j}(-2\cos\theta)^{2j} \\
&\quad - \sum_{j=1}^{n/2}(-1)^{n/2+j-1}\binom{n/2+j-1}{2j-1}(-2\cos\theta)^{2j-1}\cos\theta
\end{aligned}$$

$$= (-1)^{n/2} \left(1 + \sum_{j=1}^{n/2} (-1)^j 2^{2j-1} \left(\frac{2 \prod_{k=0}^{2j-1} \left(\frac{n}{2} + j - k \right)}{(2j)!} \right. \right.$$

$$\left. \left. - \frac{\prod_{k=0}^{2j-2} \left(\frac{n}{2} + j - 1 - k \right)}{(2j-1)!} \right) \cos^{2j} \theta \right)$$

$$= (-1)^{n/2} \left(1 + \sum_{j=1}^{n/2} (-1)^j 2^{2j-1} \left(\frac{n+2j}{2j} - 1 \right) \right.$$

$$\left. \frac{\prod_{k=0}^{2j-2} \left(\frac{n}{2} + j - 1 - k \right)}{(2j-1)!} \cos^{2j} \theta \right)$$

$$= (-1)^{n/2} \left(1 + \sum_{j=1}^{n/2} (-1)^j \frac{n \prod_{k=0}^{2j-2} (n+2j-2-2k)}{(2j)!} \cos^{2j} \theta \right)$$

$$= (-1)^{n/2} \sum_{j=0}^{n/2} (-1)^j \frac{\prod_{k=0}^{j-1} \left(n^2 - (2k)^2 \right)}{(2j)!} \cos^{2j} \theta.$$

$\square$

Example 2.9 (Loney, 1928) (*Trigonometrical Series* 10)
Prove that $\forall (\theta, n) \in (\mathbb{R} - \mathbb{Z}\pi) \times (2\mathbb{N} - 1)$,

2.1 Trigonometrical Series

$$\cos n\theta = (-1)^{(n-1)/2} n \sum_{j=0}^{(n-1)/2} (-1)^j \frac{\prod_{k=0}^{j-1}\left(n^2 - (2k+1)^2\right)}{(2j+1)!} \cos^{2j+1}\theta$$

$$= (-1)^{(n-1)/2} n \left(\cos\theta - \frac{(n^2-1^2)}{3!}\cos^3\theta + \frac{(n^2-1^2)(n^2-3^2)}{5!}\cos^5\theta \right. \quad (2.19)$$

$$\left. + \cdots + \frac{(n^2-1^2)(n^2-3^2)(n^2-5^2)\cdots(n^2-(n-2)^2)}{n!}\cos^n\theta \right).$$

Proof For $(\theta, n) \in (\mathbb{R} - \mathbb{Z}\pi) \times (2\mathbb{N} - 1)$, we equate the coefficients of x^n in (2.17) and (2.18) to get

$$\cos n\theta = \sum_{j=0}^{(n-1)/2} (-1)^{(n+1)/2+j} \binom{(n+1)/2+j}{2j+1} (-2\cos\theta)^{2j+1}$$

$$- \sum_{j=0}^{(n-1)/2} (-1)^{(n-1)/2+j} \binom{(n-1)/2+j}{2j} (-2\cos\theta)^{2j}\cos\theta$$

$$= (-1)^{(n-1)/2} \sum_{j=0}^{(n-1)/2} (-1)^j 2^{2j}$$

$$\left(2\binom{(n+1)/2+j}{2j+1} - \binom{(n-1)/2+j}{2j} \right) \cos^{2j+1}\theta$$

$$= (-1)^{(n-1)/2} \sum_{j=0}^{(n-1)/2} (-1)^j 2^{2j} \left(\frac{2\prod_{k=0}^{2j}\left(\frac{n+1}{2}+j-k\right)}{(2j+1)!} \right.$$

$$\left. - \frac{\prod_{k=0}^{2j-1}\left(\frac{n-1}{2}+j-k\right)}{(2j)!} \right) \cos^{2j+1}\theta$$

$$= (-1)^{(n-1)/2} \sum_{j=0}^{(n-1)/2} (-1)^j 2^{2j} \left(\frac{n+1+2j}{2j+1} - 1 \right)$$

$$\frac{\prod_{k=0}^{2j-1}\left(\frac{n-1}{2}+j-k\right)}{(2j)!} \cos^{2j+1}\theta$$

$$= (-1)^{(n-1)/2} n \sum_{j=0}^{(n-1)/2} (-1)^j \frac{\prod_{k=0}^{2j-1}(n-1+2j-2k)}{(2j+1)!} \cos^{2j+1}\theta$$

$$= (-1)^{(n-1)/2} n \sum_{j=0}^{(n-1)/2} (-1)^j \frac{\prod_{k=0}^{j-1}\left(n^2-(2k+1)^2\right)}{(2j+1)!} \cos^{2j+1}\theta.$$

□

2.2 Trigonometrical Products

In this section, we will apply Euler's formula to factor a polynomial. Then, we will apply the factorization to express $\sin\theta$ and $\cos\theta$ as infinite products. As a corollary to the above results, we will also find the value of $\zeta(2)$.

Example 2.10 (Loney, 1928) (*Factorization of* $x^{2n} - 2x^n \cos n\theta + 1$)
Prove that $\forall (x, \theta, n) \in \mathbb{R}^2 \times \mathbb{N}$,

$$x^{2n} - 2x^n \cos n\theta + 1 = \prod_{j=0}^{n-1}\left(x^2 - 2x\cos\left(\theta + \frac{2j\pi}{n}\right) + 1\right)$$

$$= \left(x^2 - 2x\cos\theta + 1\right)\left(x^2 - 2x\cos\left(\theta + \frac{2\pi}{n}\right) + 1\right)$$

$$\left(x^2 - 2x\cos\left(\theta + \frac{4\pi}{n}\right) + 1\right) \ldots \left(x^2 - 2x\cos\left(\theta + \frac{2(n-1)\pi}{n}\right) + 1\right). \tag{2.20}$$

Proof Using Euler's formula, we solve the equation $x^{2n} - 2x^n \cos n\theta + 1 = 0$ as follows:

$$x^{2n} - 2x^n \cos n\theta + 1 = 0$$

$$\iff x^{2n} - x^n \left(e^{in\theta} + e^{-in\theta}\right) + 1 = 0$$

$$\implies x^n = \frac{\left(e^{in\theta} + e^{-in\theta}\right) \pm \sqrt{\left(e^{in\theta} + e^{-in\theta}\right)^2 - 4}}{2} \tag{2.21}$$

$$= \frac{\left(e^{in\theta} + e^{-in\theta}\right) \pm \left(e^{in\theta} - e^{-in\theta}\right)}{2}$$

$$= e^{\pm i(n\theta + 2j\pi)}, \quad j \in [0..n-1]$$

$$\implies x = e^{\pm i(\theta + 2j\pi/n)}, \quad j \in [0..n-1].$$

Now using (2.21), we can factor $x^{2n} - 2x^n \cos n\theta + 1$ as follows:

2.2 Trigonometrical Products

$$x^{2n} - 2x^n \cos n\theta + 1 = \prod_{j=0}^{n-1} \left(x - e^{i(\theta+2j\pi/n)}\right)\left(x - e^{-i(\theta+2j\pi/n)}\right)$$

$$= \prod_{j=0}^{n-1} \left(x^2 - x\left(e^{i(\theta+2j\pi/n)} + e^{-i(\theta+2j\pi/n)}\right) + 1\right)$$

$$= \prod_{j=0}^{n-1} \left(x^2 - 2x \cos\left(\theta + \frac{2j\pi}{n}\right) + 1\right).$$

$\square$

Example 2.11 (Loney, 1928) (*Infinite Product for* $\sin\theta$)
Prove that $\forall \theta \in \mathbb{R}$,

$$\sin\theta = \theta \prod_{j=1}^{\infty} \left(1 - \frac{\theta^2}{j^2\pi^2}\right)$$

$$= \theta \left(1 - \frac{\theta^2}{\pi^2}\right)\left(1 - \frac{\theta^2}{2^2\pi^2}\right)\left(1 - \frac{\theta^2}{3^2\pi^2}\right)\cdots$$
(2.22)

Proof Putting $x = 1$ and replacing θ with $2\theta/n$, where $n \in 2\mathbb{N}$ in (2.20), we get

$$2 - 2\cos 2\theta = \prod_{j=0}^{n-1} \left(2 - 2\cos\frac{2\theta + 2j\pi}{n}\right)$$

$$\iff 4\sin^2\theta = \prod_{j=0}^{n-1} \left(4\sin^2\frac{\theta + j\pi}{n}\right)$$

$$\implies \sin\theta = \pm 2^{n-1} \prod_{j=0}^{n-1} \sin\frac{\theta + j\pi}{n}$$

(2.23)

$$= \pm 2^{n-1} \sin\frac{\theta}{n} \sin\frac{\theta + (n/2)\pi}{n} \prod_{j=1}^{n/2-1} \sin\frac{\theta + j\pi}{n} \sin\frac{\theta + (n-j)\pi}{n}$$

$$= \pm 2^{n-1} \sin\frac{\theta}{n} \cos\frac{\theta}{n} \prod_{j=1}^{n/2-1} \left(\cos^2\frac{(n-2j)\pi}{2n} - \cos^2\frac{2\theta + n\pi}{2n}\right)$$

$$= \pm 2^{n-1} \sin\frac{\theta}{n} \cos\frac{\theta}{n} \prod_{j=1}^{n/2-1} \left(\sin^2\frac{j\pi}{n} - \sin^2\frac{\theta}{n}\right)$$

Dividing (2.23) by $\sin(\theta/n)$ and taking the limit as $\theta \to 0$, we get

$$\lim_{\theta \to 0} \frac{\sin \theta}{\sin \frac{\theta}{n}} = \lim_{\theta \to 0} 2^{n-1} \cos \frac{\theta}{n} \prod_{j=1}^{n/2-1} \left(\sin^2 \frac{j\pi}{n} - \sin^2 \frac{\theta}{n} \right) \quad (2.24)$$

$$\iff n = 2^{n-1} \prod_{j=1}^{n/2-1} \sin^2 \frac{j\pi}{n}.$$

Dividing (2.23) by (2.24), we get

$$\frac{\sin \theta}{n} = \pm \sin \frac{\theta}{n} \cos \frac{\theta}{n} \prod_{j=1}^{n/2-1} \left(1 - \frac{\sin^2 \frac{\theta}{n}}{\sin^2 \frac{j\pi}{n}} \right) \quad (2.25)$$

$$\iff \sin \theta = \pm n \sin \frac{\theta}{n} \cos \frac{\theta}{n} \prod_{j=1}^{n/2-1} \left(1 - \frac{\sin^2 \frac{\theta}{n}}{\sin^2 \frac{j\pi}{n}} \right).$$

Taking the limit as $n \to \infty$ in (2.25), we get

$$\sin \theta = \pm \lim_{n \to \infty} n \sin \frac{\theta}{n} \cos \frac{\theta}{n} \prod_{j=1}^{n/2-1} \left(1 - \frac{\sin^2 \frac{\theta}{n}}{\sin^2 \frac{j\pi}{n}} \right) \quad (2.26)$$

$$= \pm \theta \prod_{j=1}^{\infty} \left(1 - \frac{\theta^2}{j^2 \pi^2} \right).$$

$\forall m \in \mathbb{N} - 1$, we have

$$\sin \theta > 0, \quad 2m\pi < \theta < (2m+1)\pi$$

$$\iff \sin \theta > 0, \quad 1 - \left(\frac{2m}{j} \right)^2 > 1 - \frac{\theta^2}{j^2 \pi^2} > 1 - \left(\frac{2m+1}{j} \right)^2, \quad (2.27)$$

and also

$$\sin \theta < 0, \quad (2m+1)\pi < \theta < (2m+2)\pi$$

$$\iff \sin \theta < 0, \quad 1 - \left(\frac{2m+1}{j} \right)^2 > 1 - \frac{\theta^2}{j^2 \pi^2} > 1 - \left(\frac{2m+2}{j} \right)^2. \quad (2.28)$$

From (2.27), we notice that there are $2m$ negative fractions in (2.26), making its RHS positive. From (2.28), we notice that there are $2m+1$ negative fractions in (2.26), making its RHS negative. Therefore, we drop the negative sign in (2.26) to get (2.22). (2.22) is also true for $\theta \in \mathbb{R}^-$, because we have $\sin(-\theta) = -\sin \theta$, $\forall \theta \in \mathbb{R}^-$. □

2.2 Trigonometrical Products

Example 2.12 (Loney, 1928) (*Infinite Product for* $\cos\theta$)
Prove that $\forall \theta \in \mathbb{R}$,

$$\cos\theta = \prod_{j=0}^{\infty}\left(1 - \frac{4\theta^2}{(2j+1)^2\pi^2}\right)$$
$$= \left(1 - \frac{4\theta^2}{\pi^2}\right)\left(1 - \frac{4\theta^2}{3^2\pi^2}\right)\left(1 - \frac{4\theta^2}{5^2\pi^2}\right)\left(1 - \frac{4\theta^2}{7^2\pi^2}\right)\cdots \quad (2.29)$$

Proof For $\theta \neq \mathbb{Z}\pi$, using (2.22), we have

$$\cos\theta = \frac{\sin 2\theta}{2\sin\theta}$$

$$= \frac{2\theta \prod_{j=1}^{\infty}\left(1 - \frac{4\theta^2}{j^2\pi^2}\right)}{2\theta \prod_{j=1}^{\infty}\left(1 - \frac{\theta^2}{j^2\pi^2}\right)}$$

$$= \prod_{j=0}^{\infty}\left(1 - \frac{4\theta^2}{(2j+1)^2\pi^2}\right),$$

Proving (2.29) to be correct. For $\theta = m\pi$, where $m \in \mathbb{Z}$, we have

$$\prod_{j=0}^{\infty}\left(1 - \frac{4m^2\pi^2}{(2j+1)^2\pi^2}\right) = \prod_{j=0}^{\infty}\frac{(2j+1)^2 - 4m^2}{(2j+1)^2}$$

$$= \prod_{j=0}^{\infty}\frac{(2j+1+2|m|)(2j+1-2|m|)}{(2j+1)^2}$$

$$= \lim_{n\to\infty}\frac{\prod_{j=0}^{n}(2j+1+2|m|)\prod_{j=2|m|}^{n}(2j+1-2|m|)\prod_{j=0}^{2|m|-1}(2j+1-2|m|)}{\prod_{j=0}^{n}(2j+1)^2}$$

$$= \lim_{n\to\infty}\frac{(-1)^{|m|}\prod_{j=0}^{|m|-1}(2j+1)^2 \prod_{j=|m|}^{n-2m}(2j+1)^2 \prod_{j=n-2|m|+1}^{n}(2j+1+2|m|)}{\prod_{j=0}^{n}(2j+1)^2}$$

$$= (-1)^m$$
$$= \cos m\pi,$$

once again proving (2.29) to be correct. □

Example 2.13 (Loney, 1928) (*Value of $\zeta(2)$*)
Prove that
$$\zeta(2) = \sum_{j=1}^{\infty} \frac{1}{j^2}$$
$$= \frac{1}{1^2} + \frac{1}{2^2} + \frac{1}{3^2} + \frac{1}{4^2} + \cdots \qquad (2.30)$$
$$= \frac{\pi^2}{6}.$$

Proof Using Euler's formula and (2.22), we have
$$\sin\theta = \sum_{j=0}^{\infty} (-1)^j \frac{\theta^{2j+1}}{(2j+1)!}$$
$$= \theta \prod_{j=1}^{\infty} \left(1 - \frac{\theta^2}{j^2\pi^2}\right) \qquad (2.31)$$
$$= \theta - \left(\sum_{j=1}^{\infty} \frac{1}{j^2\pi^2}\right) \theta^3 + \cdots$$

Equating the coefficient of θ^3 in (2.31), we get
$$-\frac{1}{6} = -\sum_{j=1}^{\infty} \frac{1}{j^2\pi^2}$$
$$\iff \sum_{j=1}^{\infty} \frac{1}{j^2} = \frac{\pi^2}{6},$$

proving (2.30). □

Problems

[2.1] Prove that $\forall (\theta, n) \in (\mathbb{R} - \mathbb{Z}\pi) \times 2\mathbb{N}$,

$$(-1)^{n/2} \frac{n \cos n\theta \sin \theta - \sin n\theta \cos \theta}{n \sin^3 \theta} = \sum_{j=0}^{n/2-1} (-1)^j \frac{\prod_{k=1}^{j} \left(n^2 - (2k)^2\right)}{(2j)!} \cos^{2j} \theta$$

$$= 1 - \frac{\left(n^2 - 2^2\right)}{2!} \cos^2 \theta + \frac{\left(n^2 - 2^2\right)\left(n^2 - 4^2\right)}{4!} \cos^4 \theta - \cdots$$

$$+ (-1)^{n/2-1} \frac{\left(n^2 - 2^2\right)\left(n^2 - 4^2\right)\left(n^2 - 6^2\right)\cdots\left(n^2 - (n-2)^2\right)}{(n-2)!} \cos^{n-2} \theta.$$

[2.2] Prove that $\forall (\theta, n) \in (\mathbb{R} - \mathbb{Z}\pi) \times (2\mathbb{N} - 1)$,

$$(-1)^{(n-1)/2} \frac{n \cos n\theta \sin \theta - \sin n\theta \cos \theta}{\sin^3 \theta}$$

$$= \sum_{j=1}^{(n-1)/2} (-1)^{j-1} \frac{\prod_{k=0}^{j-1} \left(n^2 - (2k+1)^2\right)}{(2j-1)!} \cos^{2j-1} \theta$$

$$= \frac{\left(n^2 - 1^2\right)}{1!} \cos \theta - \frac{\left(n^2 - 1^2\right)\left(n^2 - 3^2\right)}{3!} \cos^3 \theta$$

$$+ \frac{\left(n^2 - 1^2\right)\left(n^2 - 3^2\right)\left(n^2 - 5^2\right)}{5!} \cos^5 \theta - \cdots$$

$$+ (-1)^{(n-3)/2} \frac{\left(n^2 - 1^2\right)\left(n^2 - 3^2\right)\left(n^2 - 5^2\right)\cdots\left(n^2 - (n-2)^2\right)}{(n-2)!} \cos^{n-2} \theta.$$

[2.3] Prove that $\forall (\theta, n) \in \left(\mathbb{R} - (2\mathbb{Z}+1)\frac{\pi}{2}\right) \times (2\mathbb{N} - 1)$,

$$\frac{n \sin n\theta \cos \theta - \cos n\theta \sin \theta}{\cos^3 \theta} = \sum_{j=1}^{(n-1)/2} (-1)^{j-1} \frac{\prod_{k=0}^{j-1} \left(n^2 - (2k+1)^2\right)}{(2j-1)!} \sin^{2j-1} \theta$$

$$= \frac{\left(n^2 - 1^2\right)}{1!} \sin \theta - \frac{\left(n^2 - 1^2\right)\left(n^2 - 3^2\right)}{3!} \sin^3 \theta$$

$$+ \frac{\left(n^2 - 1^2\right)\left(n^2 - 3^2\right)\left(n^2 - 5^2\right)}{5!} \sin^5 \theta - \cdots$$

$$+ (-1)^{(n-3)/2} \frac{\left(n^2 - 1^2\right)\left(n^2 - 3^2\right)\left(n^2 - 5^2\right)\cdots\left(n^2 - (n-2)^2\right)}{(n-2)!} \sin^{n-2} \theta.$$

[2.4] Prove that $\forall (\theta, n) \in \left(\mathbb{R} - (2\mathbb{Z}+1)\dfrac{\pi}{2}\right) \times 2\mathbb{N}$,

$$\dfrac{n\cos n\theta \cos\theta + \sin n\theta \sin\theta}{n\cos^3\theta} = \sum_{j=0}^{n/2-1} (-1)^j \dfrac{\prod_{k=1}^{j}\left(n^2 - (2k)^2\right)}{(2j)!} \sin^{2j}\theta$$

$$= 1 - \dfrac{(n^2 - 2^2)}{2!}\sin^2\theta + \dfrac{(n^2-2^2)(n^2-4^2)}{4!}\sin^4\theta - \cdots$$

$$+ (-1)^{n/2-1} \dfrac{(n^2-2^2)(n^2-4^2)(n^2-6^2)\cdots(n^2-(n-2)^2)}{(n-2)!}\sin^{n-2}\theta.$$

[2.5] (Loney, 1928) Prove that $\forall x \in \mathbb{R} - \{0\} - \left\{\dfrac{1}{m} \,\middle|\, m \in \mathbb{Z} - \{0\}\right\}$,

$$\dfrac{\pi}{x\sin\dfrac{\pi}{x}} = 1 + 2\sum_{j=1}^{\infty}\dfrac{(-1)^{j-1}}{j^2 x^2 - 1}$$

$$= 1 + 2\left(\dfrac{1}{x^2-1} - \dfrac{1}{2^2 x^2 - 1} + \dfrac{1}{3^2 x^2 - 1} - \cdots\right).$$

[2.6] Prove that $\forall (z, \theta) \in \mathbb{Z} \times \left((4z-1)\dfrac{\pi}{2}, (4z+1)\dfrac{\pi}{2}\right)$,

$$\sec^3\theta = \sum_{j=1}^{\infty} \dfrac{j}{2^{2j-1}} \binom{2j}{j} \sin^{2j-2}\theta$$

$$= 1 + \dfrac{2\binom{4}{2}}{2^3}\sin^2\theta + \dfrac{3\binom{6}{3}}{2^5}\sin^4\theta + \dfrac{4\binom{8}{4}}{2^7}\sin^6\theta + \cdots$$

$$= \sum_{j=0}^{\infty} \dfrac{2j+1}{2^{2j}} \binom{2j}{j} \sin^{2j}\theta$$

$$= 1 + \dfrac{3\binom{2}{1}}{2^2}\sin^2\theta + \dfrac{5\binom{4}{2}}{2^4}\sin^4\theta + \dfrac{7\binom{6}{3}}{2^6}\sin^6\theta + \cdots$$

[2.7] Prove that $\forall (z, \theta) \in \mathbb{Z} \times \left((4z-1)\dfrac{\pi}{2}, (4z+1)\dfrac{\pi}{2} \right)$,

$$\sec \theta = \sum_{j=0}^{\infty} \binom{2j}{j} \frac{\sin^{2j} \theta}{2^{2j}}$$

$$= 1 + \frac{\binom{2}{1}}{2^2} \sin^2 \theta + \frac{\binom{4}{2}}{2^4} \sin^4 \theta + \frac{\binom{6}{3}}{2^6} \sin^6 \theta + \cdots$$

[2.8] (Boros and Moll, 2006) Prove that $\forall \theta \in \left[-\dfrac{\pi}{2}, \dfrac{\pi}{2} \right]$,

$$\theta = \sum_{j=0}^{\infty} \binom{2j}{j} \frac{\sin^{2j+1} \theta}{2^{2j}(2j+1)}$$

$$= \sin \theta + \frac{\binom{2}{1}}{2^2 \cdot 3} \sin^3 \theta + \frac{\binom{4}{2}}{2^4 \cdot 5} \sin^5 \theta + \frac{\binom{6}{3}}{2^6 \cdot 7} \sin^7 \theta + \cdots$$

[2.9] (Lehmer, 1985; Boros and Moll, 2006) Prove that $\forall \theta \in \left(-\dfrac{\pi}{2}, \dfrac{\pi}{2} \right)$,

$$2\theta \tan \theta = \sum_{j=1}^{\infty} \frac{2^{2j}}{j} \binom{2j}{j}^{-1} \sin^{2j} \theta$$

$$= \frac{2^2}{1\binom{2}{1}} \sin^2 \theta + \frac{2^4}{2\binom{4}{2}} \sin^4 \theta + \frac{2^6}{3\binom{6}{3}} \sin^6 \theta + \frac{2^8}{4\binom{8}{4}} \sin^8 \theta + \cdots$$

[2.10] (Lehmer, 1985; Boros and Moll, 2006) Prove that $\forall \theta \in \left[-\dfrac{\pi}{2}, \dfrac{\pi}{2} \right]$,

$$2\theta^2 = \sum_{j=1}^{\infty} \frac{2^{2j}}{j^2} \binom{2j}{j}^{-1} \sin^{2j} \theta$$

$$= \frac{2^2}{1^2 \binom{2}{1}} \sin^2 \theta + \frac{2^4}{2^2 \binom{4}{2}} \sin^4 \theta + \frac{2^6}{3^2 \binom{6}{3}} \sin^6 \theta + \frac{2^8}{4^2 \binom{8}{4}} \sin^8 \theta + \cdots$$

[2.11] Prove that $\forall \theta \in \left[-\dfrac{\pi}{2}, \dfrac{\pi}{2}\right]$,

$$\theta \sin\theta + \cos\theta = 1 + \sum_{j=0}^{\infty} \binom{2j}{j} \frac{\sin^{2j+2}\theta}{2^{2j}(2j+1)(2j+2)}$$

$$= 1 + \frac{\binom{0}{0}}{2^0 \cdot 1 \cdot 2}\sin^2\theta + \frac{\binom{2}{1}}{2^2 \cdot 3 \cdot 4}\sin^4\theta + \frac{\binom{4}{2}}{2^4 \cdot 5 \cdot 6}\sin^6\theta + \cdots$$

[2.12] Prove that $\forall \theta \in \left[-\dfrac{\pi}{2}, \dfrac{\pi}{2}\right]$,

$$\frac{\sin\theta - \theta\cos\theta}{2} = \sum_{j=1}^{\infty} \frac{2^{2j-2}}{j(2j+1)}\binom{2j}{j}^{-1} \sin^{2j+1}\theta$$

$$= \frac{2^0}{1\cdot 3 \cdot \binom{2}{1}}\sin^3\theta + \frac{2^2}{2\cdot 5 \cdot \binom{4}{2}}\sin^5\theta + \frac{2^4}{3\cdot 7 \cdot \binom{6}{3}}\sin^7\theta + \frac{2^6}{4\cdot 9 \cdot \binom{8}{4}}\sin^9\theta + \cdots$$

[2.13] Prove that $\forall \theta \in \left[-\dfrac{\pi}{2}, \dfrac{\pi}{2}\right]$,

$$\frac{3\sin^2\theta - \theta^2 - \theta\sin 2\theta}{8} = \sum_{j=1}^{\infty} \frac{2^{2j-2}}{j(2j+1)(2j+2)}\binom{2j}{j}^{-1} \sin^{2j+2}\theta$$

$$= \frac{2^0}{1\cdot 3 \cdot 4 \cdot \binom{2}{1}}\sin^4\theta + \frac{2^2}{2\cdot 5 \cdot 6 \cdot \binom{4}{2}}\sin^6\theta + \frac{2^4}{3\cdot 7 \cdot 8 \cdot \binom{6}{3}}\sin^8\theta$$

$$+ \frac{2^6}{4\cdot 9 \cdot 10 \cdot \binom{8}{4}}\sin^{10}\theta + \cdots$$

[2.14] Prove that $\forall \theta \in \left(-\dfrac{\pi}{2}, \dfrac{\pi}{2}\right)$,

$$\sec^2\theta(1 + \theta\tan\theta) = \sum_{j=0}^{\infty} 2^{2j} \binom{2j}{j}^{-1} \sin^{2j}\theta$$

$$= 1 + \frac{2^2}{\binom{2}{1}}\sin^2\theta + \frac{2^4}{\binom{4}{2}}\sin^4\theta + \frac{2^6}{\binom{6}{3}}\sin^6\theta + \cdots$$

[2.15] Prove that $\forall \theta \in \mathbb{R} - (2\mathbb{Z}+1)\dfrac{\pi}{2}$,

$$\frac{1+2\sin^2\theta}{\cos^5\theta} = \sum_{j=0}^{\infty}\frac{(2j+1)^2}{2^{2j}}\binom{2j}{j}\sin^{2j}\theta$$

$$= 1 + \frac{3^2\binom{2}{1}}{2^2}\sin^2\theta + \frac{5^2\binom{4}{2}}{2^4}\sin^4\theta + \frac{7^2\binom{6}{3}}{2^6}\sin^6\theta + \cdots$$

[2.16] Prove that $\forall \theta \in \mathbb{R} - (2\mathbb{Z}+1)\dfrac{\pi}{2}$,

$$\frac{1+2\sin^2\theta\left(5+2\sin^2\theta\right)}{\cos^7\theta} = \sum_{j=0}^{\infty}\frac{(2j+1)^3}{2^{2j}}\binom{2j}{j}\sin^{2j}\theta$$

$$= 1 + \frac{3^3\binom{2}{1}}{2^2}\sin^2\theta + \frac{5^3\binom{4}{2}}{2^4}\sin^4\theta + \frac{7^3\binom{6}{3}}{2^6}\sin^6\theta + \cdots$$

[2.17] Prove that $\forall \theta \in \mathbb{R} - (2\mathbb{Z}+1)\dfrac{\pi}{2}$,

$$\frac{1+4\sin^2\theta\left(9+15\sin^2\theta+2\sin^4\theta\right)}{\cos^9\theta} = \sum_{j=0}^{\infty}\frac{(2j+1)^4}{2^{2j}}\binom{2j}{j}\sin^{2j}\theta$$

$$= 1 + \frac{3^4\binom{2}{1}}{2^2}\sin^2\theta + \frac{5^4\binom{4}{2}}{2^4}\sin^4\theta + \frac{7^4\binom{6}{3}}{2^6}\sin^6\theta + \cdots$$

[2.18] Prove that $\forall \theta \in \left[-\dfrac{\pi}{2}, \dfrac{\pi}{2}\right]$,

$$\frac{\theta^2}{2}\sin\theta + \theta\cos\theta - \sin\theta = \sum_{j=1}^{\infty}\frac{2^{2j-2}}{j^2(2j+1)}\binom{2j}{j}^{-1}\sin^{2j+1}\theta$$

$$= \frac{2^0}{1^2\cdot 3 \cdot \binom{2}{1}}\sin^3\theta + \frac{2^2}{2^2\cdot 5 \cdot \binom{4}{2}}\sin^5\theta + \frac{2^4}{3^2\cdot 7 \cdot \binom{6}{3}}\sin^7\theta$$

$$+ \frac{2^6}{4^2\cdot 9 \cdot \binom{8}{4}}\sin^9\theta + \cdots$$

[2.19] Prove that $\forall \theta \in \left[-\dfrac{\pi}{2}, \dfrac{\pi}{2}\right]$,

$$\theta \sin\theta + 2\cos\theta - \log\left(\dfrac{e^2}{2}(1+\cos\theta)\right) = \sum_{j=0}^{\infty} \binom{2j}{j} \dfrac{\sin^{2j+2}\theta}{2^{2j}(2j+1)(2j+2)^2}$$

$$= \dfrac{\binom{0}{0}}{2^0 \cdot 1 \cdot 2^2}\sin^2\theta + \dfrac{\binom{2}{1}}{2^2 \cdot 3 \cdot 4^2}\sin^4\theta + \dfrac{\binom{4}{2}}{2^4 \cdot 5 \cdot 6^2}\sin^6\theta + \dfrac{\binom{6}{3}}{2^6 \cdot 7 \cdot 8^2}\sin^8\theta + \cdots$$

[2.20] (Loney, 1928) Prove that $\forall \theta \in \mathbb{R} - \mathbb{Z}\pi$,

$$\dfrac{1 - \theta\cot\theta}{2\theta^2} = \sum_{j=1}^{\infty} \dfrac{1}{j^2\pi^2 - \theta^2}$$

$$= \dfrac{1}{\pi^2 - \theta^2} + \dfrac{1}{2^2\pi^2 - \theta^2} + \dfrac{1}{3^2\pi^2 - \theta^2} + \dfrac{1}{4^2\pi^2 - \theta^2} + \cdots$$

[2.21] (Loney, 1928) Prove that $\forall \theta \in \mathbb{R} - \{0\} \cup (2\mathbb{Z}+1)\dfrac{\pi}{2}$,

$$\dfrac{\tan\theta}{8\theta} = \sum_{j=0}^{\infty} \dfrac{1}{(2j+1)^2\pi^2 - 4\theta^2}$$

$$= \dfrac{1}{\pi^2 - 4\theta^2} + \dfrac{1}{3^2\pi^2 - 4\theta^2} + \dfrac{1}{5^2\pi^2 - 4\theta^2} + \dfrac{1}{7^2\pi^2 - 4\theta^2} + \cdots$$

[2.22] Prove that $\forall \theta \in \mathbb{R} - \mathbb{Z}\pi$,

$$\dfrac{2\theta^2 + \theta\sin 2\theta - 4\sin^2\theta}{8\theta^4 \sin^2\theta} = \sum_{j=1}^{\infty} \dfrac{1}{\left(j^2\pi^2 - \theta^2\right)^2}$$

$$= \dfrac{1}{\left(\pi^2 - \theta^2\right)^2} + \dfrac{1}{\left(2^2\pi^2 - \theta^2\right)^2} + \dfrac{1}{\left(3^2\pi^2 - \theta^2\right)^2} + \dfrac{1}{\left(4^2\pi^2 - \theta^2\right)^2} + \cdots$$

[2.23] Prove that $\forall \theta \in \mathbb{R} - \{0\} \cup (2\mathbb{Z}+1)\dfrac{\pi}{2}$,

$$\dfrac{2\theta - \sin 2\theta}{128\theta^3 \cos^2\theta} = \sum_{j=0}^{\infty} \dfrac{1}{\left((2j+1)^2\pi^2 - 4\theta^2\right)^2}$$

$$= \dfrac{1}{\left(\pi^2 - 4\theta^2\right)^2} + \dfrac{1}{\left(3^2\pi^2 - 4\theta^2\right)^2} + \dfrac{1}{\left(5^2\pi^2 - 4\theta^2\right)^2} + \dfrac{1}{\left(7^2\pi^2 - 4\theta^2\right)^2} + \cdots$$

[2.24] Prove that $\forall (z, \theta) \in \mathbb{Z} \times \left((4z-1)\frac{\pi}{4}, (4z+1)\frac{\pi}{4} \right)$,

$$\frac{1}{4} \sin 4\theta \cos^2 \theta = \sum_{j=1}^{\infty} (-1)^{j-1} j^2 \tan^{2j-1} \theta$$
$$= \tan \theta - 2^2 \tan^3 \theta + 3^2 \tan^5 \theta - 4^2 \tan^7 \theta + \cdots$$

[2.25] Prove that $\forall (z, \theta) \in \mathbb{Z} \times \left((4z-1)\frac{\pi}{4}, (4z+1)\frac{\pi}{4} \right)$,

$$\frac{1}{8} \cos^2 \theta (\sin 6\theta + \sin 2\theta \cos 4\theta) = \sum_{j=1}^{\infty} (-1)^{j-1} j^3 \tan^{2j-1} \theta$$
$$= \tan \theta - 2^3 \tan^3 \theta + 3^3 \tan^5 \theta - 4^3 \tan^7 \theta + \cdots$$

[2.26] Prove that $\forall (z, \theta) \in \mathbb{Z} \times \left((4z-1)\frac{\pi}{4}, (4z+1)\frac{\pi}{4} \right)$,

$$\frac{1}{16} \cos^2 \theta (3 \sin 8\theta - 2 \sin 4\theta) = \sum_{j=1}^{\infty} (-1)^{j-1} j^4 \tan^{2j-1} \theta$$
$$= \tan \theta - 2^4 \tan^3 \theta + 3^4 \tan^5 \theta - 4^4 \tan^7 \theta + \cdots$$

[2.27] Prove that $\forall (\theta, n) \in (\mathbb{R} - \{0\}) \times 2\mathbb{N}$,

$$(-1)^{n/2} \frac{\sinh n\theta \cosh \theta - n \cosh n\theta \sinh \theta}{n \sinh^3 \theta}$$
$$= \sum_{j=0}^{n/2-1} (-1)^j \frac{\prod_{k=1}^{j} \left(n^2 - (2k)^2 \right)}{(2j)!} \cosh^{2j} \theta$$
$$= 1 - \frac{\left(n^2 - 2^2 \right)}{2!} \cosh^2 \theta + \frac{\left(n^2 - 2^2 \right)\left(n^2 - 4^2 \right)}{4!} \cosh^4 \theta + \cdots$$
$$+ (-1)^{n/2-1} \frac{\left(n^2 - 2^2 \right)\left(n^2 - 4^2 \right)\left(n^2 - 6^2 \right) \cdots \left(n^2 - (n-2)^2 \right)}{(n-2)!} \cosh^{n-2} \theta.$$

[2.28] Prove that $\forall (\theta, n) \in (\mathbb{R} - \{0\}) \times (2\mathbb{N} - 1)$,

$$(-1)^{(n-1)/2} \frac{\sinh n\theta \cosh \theta - n \cosh n\theta \sinh \theta}{\sinh^3 \theta}$$

$$= \sum_{j=1}^{(n-1)/2} (-1)^{j-1} \frac{\prod_{k=0}^{j-1} \left(n^2 - (2k+1)^2\right)}{(2j-1)!} \cosh^{2j-1} \theta$$

$$= \frac{\left(n^2 - 1^2\right)}{1!} \cosh \theta - \frac{\left(n^2 - 1^2\right)\left(n^2 - 3^2\right)}{3!} \cosh^3 \theta$$

$$+ \frac{\left(n^2 - 1^2\right)\left(n^2 - 3^2\right)\left(n^2 - 5^2\right)}{5!} \cosh^5 \theta - \cdots$$

$$+ (-1)^{(n-3)/2} \frac{\left(n^2 - 1^2\right)\left(n^2 - 3^2\right)\left(n^2 - 5^2\right) \ldots \left(n^2 - (n-2)^2\right)}{(n-2)!} \cosh^{n-2} \theta.$$

[2.29] Prove that $\forall (\theta, n) \in \mathbb{R} \times (2\mathbb{N} - 1)$,

$$\frac{n \sinh n\theta \cosh \theta - \cosh n\theta \sinh \theta}{\cosh^3 \theta} = \sum_{j=1}^{(n-1)/2} \frac{\prod_{k=0}^{j-1} \left(n^2 - (2k+1)^2\right)}{(2j-1)!} \sinh^{2j-1} \theta$$

$$= \frac{\left(n^2 - 1^2\right)}{1!} \sinh \theta + \frac{\left(n^2 - 1^2\right)\left(n^2 - 3^2\right)}{3!} \sinh^3 \theta$$

$$+ \frac{\left(n^2 - 1^2\right)\left(n^2 - 3^2\right)\left(n^2 - 5^2\right)}{5!} \sinh^5 \theta + \cdots$$

$$+ \frac{\left(n^2 - 1^2\right)\left(n^2 - 3^2\right)\left(n^2 - 5^2\right) \ldots \left(n^2 - (n-2)^2\right)}{(n-2)!} \sinh^{n-2} \theta.$$

[2.30] Prove that $\forall (\theta, n) \in \mathbb{R} \times 2\mathbb{N}$,

$$\frac{n \cosh n\theta \cosh \theta - \sinh n\theta \sinh \theta}{n \cosh^3 \theta} = \sum_{j=0}^{n/2-1} \frac{\prod_{k=1}^{j} \left(n^2 - (2k)^2\right)}{(2j)!} \sinh^{2j} \theta$$

$$= 1 + \frac{\left(n^2 - 2^2\right)}{2!} \sinh^2 \theta + \frac{\left(n^2 - 2^2\right)\left(n^2 - 4^2\right)}{4!} \sinh^4 \theta + \cdots$$

$$+ \frac{\left(n^2 - 2^2\right)\left(n^2 - 4^2\right)\left(n^2 - 6^2\right) \ldots \left(n^2 - (n-2)^2\right)}{(n-2)!} \sinh^{n-2} \theta.$$

Problems

[2.31] Prove that $\forall x \in \mathbb{R} - \{0\}$,

$$\frac{\pi}{x \sinh \dfrac{\pi}{x}} = 1 - 2 \sum_{j=1}^{\infty} \frac{(-1)^{j-1}}{j^2 x^2 + 1}$$

$$= 1 - 2\left(\frac{1}{x^2+1} - \frac{1}{2^2 x^2 + 1} + \frac{1}{3^2 x^2 + 1} - \cdots\right).$$

[2.32] Prove that $\forall \theta \in \left(-\log\left(\sqrt{2}+1\right), \log\left(\sqrt{2}+1\right)\right)$,

$$\operatorname{sech}^3 \theta = \sum_{j=1}^{\infty} (-1)^{j-1} \frac{j}{2^{2j-1}} \binom{2j}{j} \sinh^{2j-2} \theta$$

$$= 1 - \frac{2\binom{4}{2}}{2^1} \sinh^2 \theta + \frac{3\binom{6}{3}}{2^3} \sinh^4 \theta - \frac{4\binom{8}{4}}{2^5} \sinh^6 \theta + \cdots$$

$$= \sum_{j=0}^{\infty} (-1)^j \frac{(2j+1)}{2^{2j}} \binom{2j}{j} \sinh^{2j} \theta$$

$$= 1 - \frac{3\binom{2}{1}}{2^2} \sinh^2 \theta + \frac{5\binom{4}{2}}{2^4} \sinh^4 \theta - \frac{7\binom{6}{3}}{2^6} \sinh^6 \theta + \cdots$$

[2.33] Prove that $\forall \theta \in \left(-\log\left(\sqrt{2}+1\right), \log\left(\sqrt{2}+1\right)\right)$,

$$\operatorname{sech} \theta = \sum_{j=0}^{\infty} (-1)^j \binom{2j}{j} \frac{\sinh^{2j} \theta}{2^{2j}}$$

$$= 1 - \frac{\binom{2}{1}}{2^2} \sinh^2 \theta + \frac{\binom{4}{2}}{2^4} \sinh^4 \theta - \frac{\binom{6}{3}}{2^6} \sinh^6 \theta + \cdots$$

[2.34] Prove that $\forall \theta \in \left(-\log\left(\sqrt{2}+1\right), \log\left(\sqrt{2}+1\right)\right)$,

$$\theta = \sum_{j=0}^{\infty} (-1)^j \binom{2j}{j} \frac{\sinh^{2j+1} \theta}{2^{2j}(2j+1)}$$

$$= \sinh \theta - \frac{\binom{2}{1}}{2^2 \cdot 3} \sinh^3 \theta + \frac{\binom{4}{2}}{2^4 \cdot 5} \sinh^5 \theta - \frac{\binom{6}{3}}{2^6 \cdot 7} \sinh^7 \theta + \cdots$$

[2.35] Prove that $\forall \theta \in \left(-\log\left(\sqrt{2}+1\right), \log\left(\sqrt{2}+1\right)\right)$,

$$2\theta \tanh \theta = \sum_{j=1}^{\infty}(-1)^{j-1}\frac{2^{2j}}{j}\binom{2j}{j}^{-1}\sinh^{2j}\theta$$

$$= \frac{2^2}{1\binom{2}{1}}\sinh^2\theta - \frac{2^4}{2\binom{4}{2}}\sinh^4\theta + \frac{2^6}{3\binom{6}{3}}\sinh^6\theta - \frac{2^8}{4\binom{8}{4}}\sinh^8\theta + \cdots$$

[2.36] Prove that $\forall \theta \in \left(-\log\left(\sqrt{2}+1\right), \log\left(\sqrt{2}+1\right)\right)$,

$$2\theta^2 = \sum_{j=1}^{\infty}(-1)^{j-1}\frac{2^{2j}}{j^2}\binom{2j}{j}^{-1}\sinh^{2j}\theta$$

$$= \frac{2^2}{1^2\binom{2}{1}}\sinh^2\theta - \frac{2^4}{2^2\binom{4}{2}}\sinh^4\theta + \frac{2^6}{3^2\binom{6}{3}}\sinh^6\theta - \frac{2^8}{4^2\binom{8}{4}}\sinh^8\theta + \cdots$$

[2.37] Prove that $\forall \theta \in \left(-\log\left(\sqrt{2}+1\right), \log\left(\sqrt{2}+1\right)\right)$,

$$\cosh\theta - \theta\sinh\theta = 1 + \sum_{j=0}^{\infty}(-1)^{j+1}\binom{2j}{j}\frac{\sinh^{2j+2}\theta}{2^{2j}(2j+1)(2j+2)}$$

$$= 1 - \frac{\binom{0}{0}}{2^0\cdot 1\cdot 2}\sinh^2\theta + \frac{\binom{2}{1}}{2^2\cdot 3\cdot 4}\sinh^4\theta - \frac{\binom{4}{2}}{2^4\cdot 5\cdot 6}\sinh^6\theta + \cdots$$

[2.38] Prove that $\forall \theta \in \left(-\log\left(\sqrt{2}+1\right), \log\left(\sqrt{2}+1\right)\right)$,

$$\frac{\theta\cosh\theta - \sinh\theta}{2} = \sum_{j=1}^{\infty}(-1)^{j-1}\frac{2^{2j-2}}{j(2j+1)}\binom{2j}{j}^{-1}\sinh^{2j+1}\theta$$

$$= \frac{2^0}{1\cdot 3\cdot\binom{2}{1}}\sinh^3\theta - \frac{2^2}{2\cdot 5\cdot\binom{4}{2}}\sinh^5\theta + \frac{2^4}{3\cdot 7\cdot\binom{6}{3}}\sinh^7\theta$$

$$- \frac{2^6}{4\cdot 9\cdot\binom{8}{4}}\sinh^9\theta + \cdots$$

[2.39] Prove that $\forall \theta \in \left(-\log\left(\sqrt{2}+1\right), \log\left(\sqrt{2}+1\right)\right)$,

$$\frac{\theta \sinh 2\theta + \theta^2 - 3\sinh^2 \theta}{8} = \sum_{j=1}^{\infty} (-1)^{j-1} \frac{2^{2j-2}}{j(2j+1)(2j+2)} \binom{2j}{j}^{-1} \sinh^{2j+2} \theta$$

$$= \frac{2^0}{1 \cdot 3 \cdot 4 \cdot \binom{2}{1}} \sinh^4 \theta - \frac{2^2}{2 \cdot 5 \cdot 6 \cdot \binom{4}{2}} \sinh^6 \theta + \frac{2^4}{3 \cdot 7 \cdot 8 \cdot \binom{6}{3}} \sinh^8 \theta$$

$$- \frac{2^6}{4 \cdot 9 \cdot 10 \cdot \binom{8}{4}} \sinh^{10} \theta + \cdots$$

[2.40] Prove that $\forall \theta \in \left(-\log\left(\sqrt{2}+1\right), \log\left(\sqrt{2}+1\right)\right)$,

$$\operatorname{sech}^2 \theta (1 - \theta \tanh \theta) = \sum_{j=0}^{\infty} (-1)^j 2^{2j} \binom{2j}{j}^{-1} \sinh^{2j} \theta$$

$$= 1 - \frac{2^2}{\binom{2}{1}} \sinh^2 \theta + \frac{2^4}{\binom{4}{2}} \sinh^4 \theta - \frac{2^6}{\binom{6}{3}} \sinh^6 \theta + \cdots$$

[2.41] Prove that $\forall \theta \in \left(-\log\left(\sqrt{2}+1\right), \log\left(\sqrt{2}+1\right)\right)$,

$$\frac{1 - 2\sinh^2 \theta}{\cosh^5 \theta} = \sum_{j=0}^{\infty} (-1)^j \frac{(2j+1)^2}{2^{2j}} \binom{2j}{j} \sinh^{2j} \theta$$

$$= 1 - \frac{3^2 \binom{2}{1}}{2^2} \sinh^2 \theta + \frac{5^2 \binom{4}{2}}{2^4} \sinh^4 \theta - \frac{7^2 \binom{6}{3}}{2^6} \sinh^6 \theta + \cdots$$

[2.42] Prove that $\forall \theta \in \left(-\log\left(\sqrt{2}+1\right), \log\left(\sqrt{2}+1\right)\right)$,

$$\frac{1 - 2\sinh^2 \theta \left(5 - 2\sinh^2 \theta\right)}{\cosh^7 \theta} = \sum_{j=0}^{\infty} (-1)^j \frac{(2j+1)^3}{2^{2j}} \binom{2j}{j} \sinh^{2j} \theta$$

$$= 1 - \frac{3^3 \binom{2}{1}}{2^2} \sinh^2 \theta + \frac{5^3 \binom{4}{2}}{2^4} \sinh^4 \theta - \frac{7^3 \binom{6}{3}}{2^6} \sinh^6 \theta + \cdots$$

[2.43] Prove that $\forall \theta \in \left(-\log\left(\sqrt{2}+1\right), \log\left(\sqrt{2}+1\right)\right)$,

$$\frac{1 - 4\sinh^2\theta\left(9 - 15\sinh^2\theta + 2\sinh^4\theta\right)}{\cosh^9\theta} = \sum_{j=0}^{\infty}(-1)^j \frac{(2j+1)^4}{2^{2j}}\binom{2j}{j}\sinh^{2j}\theta$$

$$= 1 - \frac{3^4\binom{2}{1}}{2^2}\sinh^2\theta + \frac{5^4\binom{4}{2}}{2^4}\sinh^4\theta - \frac{7^4\binom{6}{3}}{2^6}\sin^6\theta + \cdots$$

[2.44] Prove that $\forall \theta \in \left(-\log\left(\sqrt{2}+1\right), \log\left(\sqrt{2}+1\right)\right)$,

$$\frac{\theta^2}{2}\sinh\theta - \theta\cosh\theta + \sinh\theta = \sum_{j=1}^{\infty}(-1)^{j-1}\frac{2^{2j-2}}{j^2(2j+1)}\binom{2j}{j}^{-1}\sinh^{2j+1}\theta$$

$$= \frac{2^0}{1^2 \cdot 3 \cdot \binom{2}{1}}\sinh^3\theta - \frac{2^2}{2^2 \cdot 5 \cdot \binom{4}{2}}\sinh^5\theta + \frac{2^4}{3^2 \cdot 7 \cdot \binom{6}{3}}\sinh^7\theta$$

$$- \frac{2^6}{4^2 \cdot 9 \cdot \binom{8}{4}}\sinh^9\theta + \cdots$$

[2.45] Prove that $\forall \theta \in \left(-\log\left(\sqrt{2}+1\right), \log\left(\sqrt{2}+1\right)\right)$,

$$\theta\sinh\theta - 2\cosh\theta + \log\left(\frac{e^2}{2}(1+\cosh\theta)\right)$$

$$= \sum_{j=0}^{\infty}(-1)^j\binom{2j}{j}\frac{\sinh^{2j+2}\theta}{2^{2j}(2j+1)(2j+2)^2}$$

$$= \frac{\binom{0}{0}}{2^0 \cdot 1 \cdot 2^2}\sinh^2\theta - \frac{\binom{2}{1}}{2^2 \cdot 3 \cdot 4^2}\sinh^4\theta + \frac{\binom{4}{2}}{2^4 \cdot 5 \cdot 6^2}\sinh^6\theta$$

$$- \frac{\binom{6}{3}}{2^6 \cdot 7 \cdot 8^2}\sinh^8\theta + \cdots$$

[2.46] Prove that $\forall \theta \in \mathbb{R} - \{0\}$,

$$\frac{\theta \coth \theta - 1}{2\theta^2} = \sum_{j=1}^{\infty} \frac{1}{j^2 \pi^2 + \theta^2}$$

$$= \frac{1}{\pi^2 + \theta^2} + \frac{1}{2^2 \pi^2 + \theta^2} + \frac{1}{3^2 \pi^2 + \theta^2} + \frac{1}{4^2 \pi^2 + \theta^2} + \cdots$$

[2.47] Prove that $\forall \theta \in \mathbb{R} - \{0\}$,

$$\frac{\tanh \theta}{8\theta} = \sum_{j=0}^{\infty} \frac{1}{(2j+1)^2 \pi^2 + 4\theta^2}$$

$$= \frac{1}{\pi^2 + 4\theta^2} + \frac{1}{3^2 \pi^2 + 4\theta^2} + \frac{1}{5^2 \pi^2 + 4\theta^2} + \frac{1}{7^2 \pi^2 + 4\theta^2} + \cdots$$

[2.48] Prove that $\forall \theta \in \mathbb{R} - \{0\}$,

$$\frac{2\theta^2 + \theta \sinh 2\theta - 4 \sinh^2 \theta}{8\theta^4 \sinh^2 \theta} = \sum_{j=1}^{\infty} \frac{1}{(j^2 \pi^2 + \theta^2)^2}$$

$$= \frac{1}{(\pi^2 + \theta^2)^2} + \frac{1}{(2^2 \pi^2 + \theta^2)^2} + \frac{1}{(3^2 \pi^2 + \theta^2)^2} + \frac{1}{(4^2 \pi^2 + \theta^2)^2} + \cdots$$

[2.49] Prove that $\forall \theta \in \mathbb{R} - \{0\}$,

$$\frac{\sinh 2\theta - 2\theta}{128 \theta^3 \cosh^2 \theta} = \sum_{j=0}^{\infty} \frac{1}{\left((2j+1)^2 \pi^2 + 4\theta^2\right)^2}$$

$$= \frac{1}{(\pi^2 + 4\theta^2)^2} + \frac{1}{(3^2 \pi^2 + 4\theta^2)^2} + \frac{1}{(5^2 \pi^2 + 4\theta^2)^2} + \frac{1}{(7^2 \pi^2 + 4\theta^2)^2} + \cdots$$

[2.50] Prove that $\forall \theta \in \mathbb{R}$,

$$\frac{1}{4} \sinh 4\theta \cosh^2 \theta = \sum_{j=1}^{\infty} j^2 \tanh^{2j-1} \theta$$

$$= \tanh \theta + 2^2 \tanh^3 \theta + 3^2 \tanh^5 \theta + 4^2 \tanh^7 \theta + \cdots$$

[2.51] Prove that $\forall \theta \in \mathbb{R}$,

$$\frac{1}{8}\cosh^2\theta\,(\sinh 6\theta + \sinh 2\theta \cosh 4\theta) = \sum_{j=1}^{\infty} j^3 \tanh^{2j-1}\theta$$
$$= \tanh\theta + 2^3\tanh^3\theta + 3^3\tanh^5\theta + 4^3\tanh^7\theta + \cdots$$

[2.52] Prove that $\forall \theta \in \mathbb{R}$,

$$\frac{1}{16}\cosh^2\theta\,(3\sinh 8\theta - 2\sinh 4\theta) = \sum_{j=1}^{\infty} j^4 \tanh^{2j-1}\theta$$
$$= \tanh\theta + 2^4\tanh^3\theta + 3^4\tanh^5\theta + 4^4\tanh^7\theta + \cdots$$

Chapter 3
Bessel Functions

In this chapter, first, we will learn about Euler's constant γ and the Gamma function $\Gamma(z)$, which is a prerequisite for studying the Bessel functions. We will learn about the Bessel coefficients $J_n(z)$ and $I_n(z)$. Then we will learn about their generalizations: the Bessel functions $J_v(z)$ and $I_v(z)$. The problems given in this chapter study some of the properties of the Bessel functions. Integral representations of the Bessel coefficients will be derived in the later chapters.

3.1 Euler's Constant γ

First, we estimate the sum of a *Harmonic Progression (HP)*.

Example 3.1 Estimating the sum of the HP $\left(\dfrac{1}{j}\right)_{j=1}^{n}$

$\forall n \in \mathbb{N}$, let
$$H(n) = \sum_{j=1}^{n} \frac{1}{j}$$
$$= 1 + \frac{1}{2} + \frac{1}{3} + \cdots + \frac{1}{n}.$$

Prove that $\forall n \in \mathbb{N} + 1$,
$$\frac{1}{n} < H(n) - \log n < 1. \tag{3.1}$$

Proof We have

$$\frac{1}{j+1} < \frac{1}{x} < \frac{1}{j}, \quad \forall x \in (j, j+1)$$

$$\Longrightarrow \int_j^{j+1} \frac{dx}{j+1} < \int_j^{j+1} \frac{dx}{x} < \int_j^{j+1} \frac{dx}{j}$$

$$\Longrightarrow \frac{1}{j+1} < \log(j+1) - \log j < \frac{1}{j} \qquad (3.2)$$

$$\Longrightarrow \sum_{j=1}^{n-1} \frac{1}{j+1} < \sum_{j=1}^{n-1}(\log(j+1) - \log j) < \sum_{j=1}^{n-1} \frac{1}{j}$$

$$\Longrightarrow H(n) - 1 < \log n < H(n) - \frac{1}{n}$$

$$\Longrightarrow -1 < \log n - H(n) < -\frac{1}{n}$$

$$\Longrightarrow \frac{1}{n} < H(n) - \log n < 1.$$

$\square$

Now we show that the limit of (3.1) as $n \to \infty$ exists.

Example 3.2 (*Convergence of the Sequence* $(H(j) - \log j)_{j=1}^{\infty}$)
Prove that the limit $\lim_{n \to \infty}(H(n) - \log n)$ exists.

Proof From (3.2), we have

$$0 < \int_j^{j+1}\left(\frac{1}{j} - \frac{1}{x}\right)dx < \left(\frac{1}{j} - \frac{1}{j+1}\right)dx$$

$$\Longrightarrow 0 < \int_j^{j+1} \frac{(x-j)dx}{jx} < \int_j^{j+1} \frac{dx}{j(j+1)} < \int_j^{j+1} \frac{dx}{j^2}$$

$$\Longrightarrow 0 < \frac{1}{j} - \log(j+1) + \log j < \frac{1}{j^2}.$$

From (2.30), the series $\sum_{j=0}^{\infty} 1/j^2$ converges with sum $\pi^2/6$. Therefore, using the comparison test of convergence (Whittaker and Watson, 2016), the sum $\sum_{j=0}^{\infty}(1/j - \log(j+1) + \log j) = \lim_{n \to \infty}(H(n) - \log n)$ is convergent. $\square$

Definition 3.3 (*Euler's Constant* γ)
Euler's Constant γ is defined as

$$\gamma = \lim_{n \to \infty}(H(n) - \log n)$$

$$= \lim_{n \to \infty}\left(1 + \frac{1}{2} + \frac{1}{3} + \cdots + \frac{1}{n} - \log n\right). \qquad (3.3)$$

3.2 The Gamma Function $\Gamma(z)$

From (3.1), we have
$$0 < \gamma < 1.$$

The decimal representation of γ is
$$\gamma = 0.57721566490153286060651209008240243\ldots$$

3.2 The Gamma Function $\Gamma(z)$

Definition 3.4 (*The Gamma Function $\Gamma(z)$*)
$\forall z \in \mathbb{C} - (\mathbb{Z} - \mathbb{N})$, the Gamma Function, $\Gamma(z)$, is defined as

$$\Gamma(z) = \frac{e^{-\gamma z}}{z} \prod_{j=1}^{\infty} e^{z/j} \left(1 + \frac{z}{j}\right)^{-1}$$
$$= \frac{e^{-\gamma z}}{z} e^{z} (1+z)^{-1} e^{z/2}\left(1+\frac{z}{2}\right)^{-1} e^{z/3}\left(1+\frac{z}{3}\right)^{-1} e^{z/4}\left(1+\frac{z}{4}\right)^{-1}\ldots \quad (3.4)$$

The Gamma function $\Gamma(z)$ is convergent $\forall z \in \mathbb{C} - (\mathbb{Z} - \mathbb{N})$ (Whittaker and Watson, 2016).

Example 3.5 (*Euler's Identity for $\Gamma(z)$*)
Prove that $\forall z \in \mathbb{C} - (\mathbb{Z} - \mathbb{N})$,

$$\Gamma(z) = \frac{1}{z}\prod_{j=1}^{\infty}\left(1+\frac{1}{j}\right)^z\left(1+\frac{z}{j}\right)^{-1}$$
$$= \frac{2^z}{z}(1+z)^{-1}\left(\frac{3}{2}\right)^z\left(1+\frac{z}{2}\right)^{-1}\left(\frac{4}{3}\right)^z\left(1+\frac{z}{3}\right)^{-1}\left(\frac{5}{4}\right)^z\left(1+\frac{z}{4}\right)^{-1}\ldots \quad (3.5)$$

Proof From (3.3) and (3.4), we have

$$\Gamma(z) = \frac{1}{z}\lim_{n\to\infty} e^{-(\sum_{j=1}^{n} 1/j - \log n)z} \prod_{j=1}^{n} e^{z/j}\left(1+\frac{z}{j}\right)^{-1}$$
$$= \frac{1}{z}\lim_{n\to\infty} e^{-z\sum_{j=1}^{n} 1/j + z\log n} e^{z\sum_{j=1}^{n} 1/j} \prod_{j=1}^{n}\left(1+\frac{z}{j}\right)^{-1}$$
$$= \frac{1}{z}\lim_{n\to\infty} n^z \prod_{j=1}^{n}\left(1+\frac{z}{j}\right)^{-1}$$
$$= \frac{1}{z}\lim_{n\to\infty} \left(1+\frac{1}{n}\right)^{-z} \prod_{j=1}^{n}\left(1+\frac{1}{j}\right)^z\left(1+\frac{z}{j}\right)^{-1}$$

$$= \frac{1}{z} \lim_{n\to\infty} \prod_{j=1}^{n} \left(1+\frac{1}{j}\right)^{z} \left(1+\frac{z}{j}\right)^{-1}.$$

□

Example 3.6 (*A Recursive Formula for* $\Gamma(z)$)
Prove that $\forall z \in \mathbb{C} - (\mathbb{Z} - \mathbb{N})$,

$$\Gamma(z+1) = z\Gamma(z). \tag{3.6}$$

Proof From (3.5), we have

$$\Gamma(z+1) = \frac{1}{z+1} \lim_{n\to\infty} \prod_{j=1}^{n} \left(1+\frac{1}{j}\right)^{z+1} \left(1+\frac{z+1}{j}\right)^{-1}$$

$$= \frac{1}{z+1} \lim_{n\to\infty} \prod_{j=1}^{n} \left(1+\frac{1}{j}\right)^{z} \left(1+\frac{z+1}{j}\right)^{-1} \left(1+\frac{1}{j}\right)$$

$$= \frac{1}{z+1} \lim_{n\to\infty} \prod_{j=1}^{n} \left(1+\frac{1}{j}\right)^{z} \left(\frac{j}{z+j+1}\right) \left(\frac{j+1}{j}\right)$$

$$= \frac{1}{z+1} \lim_{n\to\infty} \prod_{j=1}^{n} \left(1+\frac{1}{j}\right)^{z} \left(1+\frac{z}{j+1}\right)^{-1}$$

$$= \lim_{n\to\infty} \left(1+\frac{z}{n+1}\right)^{-1} \prod_{j=1}^{n} \left(1+\frac{1}{j}\right)^{z} \left(1+\frac{z}{j}\right)^{-1}$$

$$= z\Gamma(z).$$

□

3.3 The Bessel Coefficients $J_n(z)$

Definition 3.7 (*The Bessel Coefficients* $J_n(z)$)
$\forall (z, n) \in \mathbb{C} \times \mathbb{Z}$, the Bessel coefficients, $J_n(z)$, are defined by the following generating function:

$$e^{z(t-1/t)/2} = \sum_{n=-\infty}^{\infty} J_n(z) t^n. \tag{3.7}$$

Example 3.8 (*Power Series for* $J_n(z)$)
Prove that $\forall (z, n) \in \mathbb{C} \times (\mathbb{N} - 1)$,

3.3 The Bessel Coefficients $J_n(z)$

$$J_n(z) = \sum_{j=0}^{\infty} (-1)^j \frac{z^{n+2j}}{j!(n+j)!2^{n+2j}}$$

$$= \frac{z^n}{n!2^n} - \frac{z^{n+2}}{1!(n+1)!2^{n+2}} + \frac{z^{n+4}}{2!(n+2)!2^{n+4}} - \frac{z^{n+6}}{3!(n+3)!2^{n+6}} + \cdots$$

(3.8)

Proof From (3.7), we have

$$\sum_{n=-\infty}^{\infty} J_n(z) t^n = e^{zt/2} \cdot e^{-z/(2t)}$$

$$= \sum_{k=0}^{\infty} \frac{z^k}{k! 2^k} t^k \sum_{j=0}^{\infty} (-1)^j \frac{z^j}{j! 2^j} t^{-j}$$

$$= \sum_{k=0}^{\infty} \sum_{j=0}^{\infty} (-1)^j \frac{z^{k+j}}{k! j! 2^{k+j}} t^{k-j}$$

$$= \sum_{n=-\infty}^{\infty} \left(\sum_{j=0}^{\infty} (-1)^j \frac{z^{n+2j}}{j!(n+j)!2^{n+2j}} \right) t^n.$$

(3.9)

We get (3.8) by equating the coefficient of t^n in (3.9). □

Example 3.9 (*A Recursive Formula for $J_n(z)$*)
Prove that $\forall (z, n) \in \mathbb{C} \times \mathbb{Z}$,

$$J_{n-1}(z) + J_{n+1}(z) = \frac{2n}{z} J_n(z).$$

(3.10)

Proof Differentiating (3.7) with respect to t, we get

$$\frac{z}{2}\left(1 + \frac{1}{t^2}\right) e^{z(t-1/t)/2} = \sum_{n=-\infty}^{\infty} n J_n(z) t^{n-1}$$

$$\Longrightarrow \frac{z}{2}\left(1 + \frac{1}{t^2}\right) \sum_{n=-\infty}^{\infty} J_n(z) t^n = \sum_{n=-\infty}^{\infty} n J_n(z) t^{n-1}$$

$$\Longrightarrow \frac{z}{2} \sum_{n=-\infty}^{\infty} (J_{n-1}(z) + J_{n+1}(z)) t^{n-1} = \sum_{n=-\infty}^{\infty} n J_n(z) t^{n-1}$$

$$\Longrightarrow \sum_{n=-\infty}^{\infty} (J_{n-1}(z) + J_{n+1}(z)) t^{n-1} = \sum_{n=-\infty}^{\infty} \frac{2n}{z} J_n(z) t^{n-1}.$$

(3.11)

We get (3.10) by equating the coefficients of t^{n-1} in (3.11). □

Example 3.10 (*A Recursive Formula for $J'_n(z)$*)
Prove that $\forall (z, n) \in \mathbb{C} \times \mathbb{Z}$,

$$J_{n-1}(z) - J_{n+1}(z) = 2J'_n(z). \qquad (3.12)$$

Proof Differentiating (3.7) with respect to z, we get

$$\frac{1}{2}\left(t - \frac{1}{t}\right)e^{z(t-1/t)/2} = \sum_{n=-\infty}^{\infty} J'_n(z)t^n$$

$$\implies \frac{1}{2}\left(t - \frac{1}{t}\right)\sum_{n=-\infty}^{\infty} J_n(z)t^n = \sum_{n=-\infty}^{\infty} J'_n(z)t^n \qquad (3.13)$$

$$\implies \frac{1}{2}\sum_{n=-\infty}^{\infty} (J_{n-1}(z) - J_{n+1}(z))t^n = \sum_{n=-\infty}^{\infty} J'_n(z)t^n$$

$$\implies \sum_{n=-\infty}^{\infty} (J_{n-1}(z) - J_{n+1}(z))t^n = \sum_{n=-\infty}^{\infty} 2J'_n(z)t^n.$$

We get (3.12) by equating the coefficients of t^n in (3.13). □

Example 3.11 (*Bessel's Differential Equation for $J_n(z)$*)
Prove that $\forall (z, n) \in \mathbb{C} \times \mathbb{Z}$,

$$z^2 \frac{d^2 J_n(z)}{dz^2} + z \frac{dJ_n(z)}{dz} + (z^2 - n^2) J_n(z) = 0. \qquad (3.14)$$

Proof Replacing n with $n - 1$ in [3.12], we get

$$\frac{d}{dz}\left(z^{-n+1} J_{n-1}(z)\right) = -z^{-n+1} J_n(z). \qquad (3.15)$$

Multiplying both sides of [3.11] by z^{-2n+1}, we get

$$z^{-2n+1} \frac{d}{dz}\left(z^n J_n(z)\right) = z^{-n+1} J_{n-1}(z). \qquad (3.16)$$

Differentiating (3.16) with respect to z and using (3.15), we get

$$\frac{d}{dz}\left(z^{-2n+1} \frac{d}{dz}\left(z^n J_n(z)\right)\right) = -z^{-n+1} J_n(z)$$

$$\iff \frac{d}{dz}\left(z^{-2n+1}\left(nz^{n-1} J_n(z) + z^n \frac{dJ_n(z)}{dz}\right)\right) = -z^{-n+1} J_n(z)$$

$$\iff \frac{d}{dz}\left(nz^{-n} J_n(z) + z^{-n+1} \frac{dJ_n(z)}{dz}\right) = -z^{-n+1} J_n(z)$$

3.4 The Bessel Coefficients $I_n(z)$

$$\iff n\frac{d}{dz}\left(z^{-n}J_n(z)\right) + \frac{d}{dz}\left(z^{-n+1}\frac{dJ_n(z)}{dz}\right) = -z^{-n+1}J_n(z)$$

$$\iff -n^2 z^{-n-1}J_n(z) + nz^{-n}\frac{dJ_n(z)}{dz} + (-n+1)z^{-n}\frac{dJ_n(z)}{dz}$$

$$+ z^{-n+1}\frac{d^2 J_n(z)}{dz} = -z^{-n+1}J_n(z)$$

$$\iff z^2\frac{d^2 J_n(z)}{dz^2} + z\frac{dJ_n(z)}{dz} + \left(z^2 - n^2\right)J_n(z) = 0.$$

$\square$

3.4 The Bessel Coefficients $I_n(z)$

Definition 3.12 (*The Bessel Coefficients $I_n(z)$*)
$\forall (z, n) \in \mathbb{C} \times \mathbb{Z}$, we define the Bessel coefficients, $I_n(z)$, as

$$I_n(z) = i^{-n} J_n(iz). \qquad (3.17)$$

Example 3.13 (*Generating Function for $I_n(z)$*)
$\forall (z, n) \in \mathbb{C} \times \mathbb{Z}$, prove that the Bessel coefficients $I_n(z)$ are defined by the following generating function

$$e^{z(t+1/t)/2} = \sum_{n=-\infty}^{\infty} I_n(z) t^n. \qquad (3.18)$$

Proof Replacing z with iz and t with $-it$, respectively, in (3.7), and using (3.17), we get

$$e^{iz(-it-1/(-it))/2} = \sum_{n=-\infty}^{\infty} J_n(iz)(-it)^n$$

$$\iff e^{iz(-it-i/t)/2} = \sum_{n=-\infty}^{\infty} i^n I_n(z)(-i)^n t^n$$

$$\iff e^{z(t+1/t)/2} = \sum_{n=-\infty}^{\infty} I_n(z) t^n.$$

$\square$

Example 3.14 (*Power Series for $I_n(z)$*)
Prove that $\forall (z, n) \in \mathbb{C} \times (\mathbb{N} - 1)$,

$$I_n(z) = \sum_{j=0}^{\infty} \frac{z^{n+2j}}{j!(n+j)! 2^{n+2j}}$$

$$= \frac{z^n}{n! 2^n} + \frac{z^{n+2}}{1!(n+1)! 2^{n+2}} + \frac{z^{n+4}}{2!(n+2)! 2^{n+4}} + \frac{z^{n+6}}{3!(n+3)! 2^{n+6}} + \cdots$$

$$(3.19)$$

Proof From (3.18), we have

$$\sum_{n=-\infty}^{\infty} I_n(z) t^n = e^{zt/2} \cdot e^{z/(2t)}$$

$$= \sum_{k=0}^{\infty} \frac{z^k}{k! 2^k} t^k \sum_{j=0}^{\infty} \frac{z^j}{j! 2^j} t^{-j}$$

$$= \sum_{k=0}^{\infty} \sum_{j=0}^{\infty} \frac{z^{k+j}}{k! j! 2^{k+j}} t^{k-j} \qquad (3.20)$$

$$= \sum_{n=-\infty}^{\infty} \left(\sum_{j=0}^{\infty} \frac{z^{n+2j}}{j!(n+j)! 2^{n+2j}} \right) t^n.$$

We get (3.19) by equating the coefficient of t^n in (3.20). □

Example 3.15 (*A Recursive Formula for $I_n(z)$*)
Prove that $\forall (z, n) \in \mathbb{C} \times \mathbb{Z}$,

$$I_{n-1}(z) - I_{n+1}(z) = \frac{2n}{z} I_n(z). \qquad (3.21)$$

Proof Differentiating (3.18) with respect to t, we get

$$\frac{z}{2}\left(1 - \frac{1}{t^2}\right) e^{z(t+1/t)/2} = \sum_{n=-\infty}^{\infty} n I_n(z) t^{n-1}$$

$$\implies \frac{z}{2}\left(1 - \frac{1}{t^2}\right) \sum_{n=-\infty}^{\infty} I_n(z) t^n = \sum_{n=-\infty}^{\infty} n I_n(z) t^{n-1}$$

$$\implies \frac{z}{2} \sum_{n=-\infty}^{\infty} (I_{n-1}(z) - I_{n+1}(z)) t^{n-1} = \sum_{n=-\infty}^{\infty} n I_n(z) t^{n-1} \qquad (3.22)$$

$$\implies \sum_{n=-\infty}^{\infty} (I_{n-1}(z) - I_{n+1}(z)) t^{n-1} = \sum_{n=-\infty}^{\infty} \frac{2n}{z} I_n(z) t^{n-1}.$$

We get (3.21) by equating the coefficients of t^{n-1} in (3.22). □

Example 3.16 (*A Recursive Formula for $I'_n(z)$*)
Prove that $\forall (z, n) \in \mathbb{C} \times \mathbb{Z}$,

$$I_{n-1}(z) + I_{n+1}(z) = 2 I'_n(z). \qquad (3.23)$$

3.4 The Bessel Coefficients $I_n(z)$

Proof Differentiating (3.18) with respect to z, we get

$$\frac{1}{2}\left(t + \frac{1}{t}\right)e^{z(t+1/t)/2} = \sum_{n=-\infty}^{\infty} I'_n(z)t^n$$

$$\implies \frac{1}{2}\left(t + \frac{1}{t}\right)\sum_{n=-\infty}^{\infty} I_n(z)t^n = \sum_{n=-\infty}^{\infty} I'_n(z)t^n$$

$$\implies \frac{1}{2}\sum_{n=-\infty}^{\infty}(I_{n-1}(z) + I_{n+1}(z))t^n = \sum_{n=-\infty}^{\infty} I'_n(z)t^n \qquad (3.24)$$

$$\implies \sum_{n=-\infty}^{\infty}(I_{n-1}(z) + I_{n+1}(z))t^n = \sum_{n=-\infty}^{\infty} 2I'_n(z)t^n.$$

We get (3.23) by equating the coefficients of t^n in (3.24). □

Example 3.17 (*Bessel's Differential Equation for $I_n(z)$*)
Prove that $\forall (z, n) \in \mathbb{C} \times \mathbb{Z}$,

$$z^2 \frac{d^2 I_n(z)}{dz^2} + z \frac{d I_n(z)}{dz} - \left(z^2 + n^2\right) I_n(z) = 0.$$

Proof Replacing n with $n - 1$ in [3.22], we get

$$\frac{d}{dz}\left(z^{-n+1} I_{n-1}(z)\right) = z^{-n+1} I_n(z). \qquad (3.25)$$

Multiplying both sides of [3.21] by z^{-2n+1}, we get

$$z^{-2n+1} \frac{d}{dz}\left(z^n I_n(z)\right) = z^{-n+1} I_{n-1}(z). \qquad (3.26)$$

Differentiating (3.26) with respect to z and using (3.25), we get

$$\frac{d}{dz}\left(z^{-2n+1} \frac{d}{dz}\left(z^n I_n(z)\right)\right) = z^{-n+1} I_n(z)$$

$$\iff \frac{d}{dz}\left(z^{-2n+1}\left(nz^{n-1} I_n(z) + z^n \frac{d I_n(z)}{dz}\right)\right) = z^{-n+1} I_n(z)$$

$$\iff \frac{d}{dz}\left(nz^{-n} I_n(z) + z^{-n+1} \frac{d I_n(z)}{dz}\right) = z^{-n+1} I_n(z)$$

$$\iff n\frac{d}{dz}\left(z^{-n} I_n(z)\right) + \frac{d}{dz}\left(z^{-n+1} \frac{d I_n(z)}{dz}\right) = z^{-n+1} I_n(z)$$

$$\iff -n^2 z^{-n-1} I_n(z) + nz^{-n} \frac{dI_n(z)}{dz} + (-n+1)z^{-n} \frac{dI_n(z)}{dz}$$
$$+ z^{-n+1} \frac{d^2 I_n(z)}{dz} = z^{-n+1} I_n(z)$$
$$\iff z^2 \frac{d^2 I_n(z)}{dz^2} + z \frac{dI_n(z)}{dz} - (z^2 + n^2) I_n(z) = 0.$$

□

3.5 The Bessel Functions $J_v(z)$

From Sect. 3.3, we notice that the Bessel coefficients $J_n(z)$ can be defined in three ways. The original definition of $J_n(z)$ is the generating function (3.7). An alternative way of defining $J_n(z)$ is by the power series (3.8). We can also define $J_n(z)$ as a solution of Bessel's differential equation (3.14).

In this section, we will generalize the power series of $J_n(z)$ (3.8) to define the Bessel functions $J_v(z)$ for all complex values of v. As a special case, when $v = n$ (an integer), the Bessel functions $J_v(z)$ will become the Bessel coefficients $J_n(z)$.

Definition 3.18 (*The Bessel Functions $J_v(z)$*)
$\forall (z, v) \in \mathbb{C} \times (\mathbb{C} - (\mathbb{Z} - \mathbb{N}))$, the Bessel functions, $J_v(z)$, are defined by the following power series:

$$J_v(z) = \sum_{j=0}^{\infty} (-1)^j \frac{z^{v+2j}}{j! \Gamma(v+j+1) 2^{v+2j}}$$
$$= \frac{z^v}{\Gamma(v+1) 2^v} - \frac{z^{v+2}}{1! \Gamma(v+2) 2^{v+2}} + \frac{z^{v+4}}{2! \Gamma(v+3) 2^{v+4}} - \frac{z^{v+6}}{3! \Gamma(v+4) 2^{v+6}} + \cdots$$
(3.27)

$\forall (z, v) \in \mathbb{C} \times (\mathbb{Z} - \mathbb{N})$, we define the Bessel functions, $J_v(z)$, as the corresponding Bessel coefficients $J_v(z)$.

Example 3.19 (*A Recursive Formula for $J_v(z)$*)
Prove that $\forall (z, v) \in \mathbb{C}^2$,

$$J_{v-1}(z) + J_{v+1}(z) = \frac{2v}{z} J_v(z).$$
(3.28)

Proof From [3.29], we have

$$z^v J_{v-1}(z) = \frac{d}{dz} (z^v J_v(z))$$
$$= vz^{v-1} J_v(z) + z^v J'_v(z)$$
$$\iff J_{v-1}(z) = \frac{v}{z} J_v(z) + J'_v(z).$$
(3.29)

3.6 The Bessel Functions $I_v(z)$

From [3.30], we have

$$-z^{-v}J_{v+1}(z) = \frac{d}{dz}\left(z^{-v}J_v(z)\right)$$
$$= -vz^{-v-1}J_v(z) + z^{-v}J_v'(z) \quad (3.30)$$
$$\iff J_{v+1}(z) = \frac{v}{z}J_v(z) - J_v'(z).$$

We get (3.28) by adding (3.29) and (3.30). □

Example 3.20 (*A Recursive Formula for $J_v'(z)$*)
Prove that $\forall (z, v) \in \mathbb{C}^2$,

$$J_{v-1}(z) - J_{v+1}(z) = 2J_v'(z). \quad (3.31)$$

Proof We get (3.31) by subtracting (3.30) from (3.29). □

Example 3.21 (*Bessel's Differential Equation for $J_v(z)$*)
Prove that $\forall (z, v) \in \mathbb{C}^2$,

$$z^2\frac{d^2 J_v(z)}{dz^2} + z\frac{dJ_v(z)}{dz} + \left(z^2 - v^2\right)J_v(z) = 0.$$

Proof Using [3.29] and [3.30], the proof is identical to Example 3.11. □

3.6 The Bessel Functions $I_v(z)$

Definition 3.22 (*The Bessel Functions $I_v(z)$*)
$\forall (z, v) \in \mathbb{C} \times (\mathbb{C} - (\mathbb{Z} - \mathbb{N}))$, the Bessel functions, $I_v(z)$, are defined by the following power series:

$$I_v(z) = \sum_{j=0}^{\infty} \frac{z^{v+2j}}{j!\Gamma(v+j+1)2^{v+2j}}$$
$$= \frac{z^v}{\Gamma(v+1)2^v} + \frac{z^{v+2}}{1!\Gamma(v+2)2^{v+2}} + \frac{z^{v+4}}{2!\Gamma(v+3)2^{v+4}} + \frac{z^{v+6}}{3!\Gamma(v+4)2^{v+6}} + \cdots$$
(3.32)

$\forall (z, v) \in \mathbb{C} \times (\mathbb{Z} - \mathbb{N})$, we define the Bessel functions, $I_v(z)$, as the corresponding Bessel coefficients $I_v(z)$.

Example 3.23 (*A Recursive Formula for $I_v(z)$*)
Prove that $\forall (z, v) \in \mathbb{C}^2$,

$$I_{v-1}(z) - I_{v+1}(z) = \frac{2v}{z}I_v(z). \quad (3.33)$$

Proof From [3.31], we have

$$z^v I_{v-1}(z) = \frac{d}{dz}(z^v I_v(z))$$
$$= vz^{v-1} I_v(z) + z^v I'_v(z) \qquad (3.34)$$
$$\iff I_{v-1}(z) = \frac{v}{z} I_v(z) + I'_v(z).$$

From [3.32], we have

$$z^{-v} I_{v+1}(z) = \frac{d}{dz}\left(z^{-v} I_v(z)\right)$$
$$= -vz^{-v-1} I_v(z) + z^{-v} I'_v(z) \qquad (3.35)$$
$$\iff I_{v+1}(z) = -\frac{v}{z} I_v(z) + I'_v(z).$$

We get (3.33) by subtracting (3.35) from (3.34). □

Example 3.24 (*A Recursive Formula for $I'_v(z)$*)
Prove that $\forall (z, v) \in \mathbb{C}^2$,

$$I_{v-1}(z) + I_{v+1}(z) = 2J'_v(z). \qquad (3.36)$$

Proof We get (3.36) by adding (3.34) and (3.35). □

Example 3.25 (*Bessel's Differential Equation for $I_v(z)$*)
Prove that $\forall (z, v) \in \mathbb{C}^2$,

$$z^2 \frac{d^2 I_v(z)}{dz^2} + z \frac{dI_v(z)}{dz} - (z^2 + v^2) I_v(z) = 0.$$

Proof Using [3.31] and [3.32], the proof is identical to Example 3.17. □

Problems

[3.1] (Whittaker and Watson, 2016) Prove that $\forall n \in \mathbb{N}$,

$$H(n) = \int_0^1 \frac{1 - (1-t)^n}{t} dt.$$

Problems

[3.2] (Whittaker and Watson, 2016) Prove that

$$\gamma = \lim_{n\to\infty} \left(\int_0^1 \frac{1-(1-t/n)^n}{t}\,dt - \int_1^n \frac{(1-t/n)^n}{t}\,dt \right).$$

[3.3] (Whittaker and Watson, 2016) Prove that

$$\gamma = \int_0^1 \left(1-e^{-t}\right) \frac{dt}{t} - \int_1^\infty \frac{e^{-t}}{t}\,dt.$$

[3.4] (Whittaker and Watson, 2016) Prove that

$$\Gamma(1) = 1.$$

[3.5] (Whittaker and Watson, 2016) Prove that

$$\Gamma'(1) = -\gamma.$$

[3.6] (Whittaker and Watson, 2016) Prove that $\forall n \in \mathbb{N}$,

$$\Gamma(n) = (n-1)!.$$

[3.7] (Whittaker and Watson, 2016) Prove that

$$\Gamma\left(\frac{1}{2}\right) = \sqrt{\pi}.$$

[3.8] (Whittaker and Watson, 2016) Prove that $\forall z \in \mathbb{C} - (\mathbb{Z} - \mathbb{N})$,

$$\Gamma(z) = \frac{1}{z} \lim_{n\to\infty} n^z \prod_{j=1}^{n-1} \frac{j}{z+j}.$$

[3.9] (Watson, 2011) Prove that $\forall (z, n) \in \mathbb{C} \times \mathbb{Z}$,
$$J_{-n}(z) = (-1)^n J_n(z).$$

[3.10] (Watson, 2011) Prove that $\forall z \in \mathbb{C}$,
$$J_0'(z) = -J_1(z).$$

[3.11] (Watson, 2011) Prove that $\forall (z, n) \in \mathbb{C} \times \mathbb{Z}$,
$$\frac{d}{dz}\left(z^n J_n(z)\right) = z^n J_{n-1}(z).$$

[3.12] (Watson, 2011) Prove that $\forall (z, n) \in \mathbb{C} \times \mathbb{Z}$,
$$\frac{d}{dz}\left(z^{-n} J_n(z)\right) = -z^{-n} J_{n+1}(z).$$

[3.13] (Watson, 2011) Prove that $\forall (z, \theta) \in \mathbb{C} \times \mathbb{R}$,
$$\begin{aligned}\cos(z \sin \theta) &= J_0(z) + 2 \sum_{j=1}^{\infty} J_{2j}(z) \cos 2j\theta \\ &= J_0(z) + 2\left(J_2(z) \cos 2\theta + J_4(z) \cos 4\theta + J_6(z) \cos 6\theta + J_8(z) \cos 8\theta + \cdots\right).\end{aligned}$$

[3.14] (Watson, 2011) Prove that $\forall (z, \theta) \in \mathbb{C} \times \mathbb{R}$,
$$\begin{aligned}\sin(z \sin \theta) &= 2 \sum_{j=0}^{\infty} J_{2j+1}(z) \sin(2j+1)\theta \\ &= 2\left(J_1(z) \sin \theta + J_3(z) \sin 3\theta + J_5(z) \sin 5\theta + J_7(z) \sin 7\theta + \cdots\right).\end{aligned}$$

[3.15] (Watson, 2011) Prove that $\forall (z, \theta) \in \mathbb{C} \times \mathbb{R}$,

$$\cos(z \cos \theta) = J_0(z) + 2 \sum_{j=1}^{\infty} (-1)^j J_{2j}(z) \cos 2j\theta$$
$$= J_0(z) - 2 \left(J_2(z) \cos 2\theta - J_4(z) \cos 4\theta + J_6(z) \cos 6\theta - J_8(z) \cos 8\theta + \cdots \right).$$

[3.16] (Watson, 2011) Prove that $\forall (z, \theta) \in \mathbb{C} \times \mathbb{R}$,

$$\sin(z \cos \theta) = 2 \sum_{j=0}^{\infty} (-1)^j J_{2j+1}(z) \cos(2j+1)\theta$$
$$= 2 \left(J_1(z) \cos \theta - J_3(z) \cos 3\theta + J_5(z) \cos 5\theta - J_7(z) \cos 7\theta + \cdots \right).$$

[3.17] (Watson, 2011) Prove that $\forall z \in \mathbb{C}$,

$$z \sin z = 2 \sum_{j=1}^{\infty} (-1)^{j-1} (2j)^2 J_{2j}(z)$$
$$= 2 \left(2^2 J_2(z) - 4^2 J_4(z) + 6^2 J_6(z) - 8^2 J_8(z) + \cdots \right).$$

[3.18] (Watson, 2011) Prove that $\forall z \in \mathbb{C}$,

$$z \cos z = 2 \sum_{j=0}^{\infty} (-1)^j (2j+1)^2 J_{2j+1}(z)$$
$$= 2 \left(1^2 J_1(z) - 3^2 J_3(z) + 5^2 J_5(z) - 7^2 J_7(z) + \cdots \right).$$

[3.19] Prove that $\forall (z, n) \in \mathbb{C} \times \mathbb{Z}$,

$$I_{-n}(z) = I_n(z).$$

[3.20] Prove that $\forall z \in \mathbb{C}$,
$$I_0'(z) = I_1(z).$$

[3.21] (Whittaker and Watson, 2016) Prove that $\forall (z, n) \in \mathbb{C} \times \mathbb{Z}$,

$$\frac{d}{dz}\left(z^n I_n(z)\right) = z^n I_{n-1}(z).$$

[3.22] (Whittaker and Watson, 2016) Prove that $\forall (z, n) \in \mathbb{C} \times \mathbb{Z}$,

$$\frac{d}{dz}\left(z^{-n} I_n(z)\right) = z^{-n} I_{n+1}(z).$$

[3.23] Prove that $\forall (z, \theta) \in \mathbb{C} \times \mathbb{R}$,

$$\cosh(z \sin \theta) = I_0(z) + 2 \sum_{j=1}^{\infty} (-1)^j I_{2j}(z) \cos 2j\theta$$
$$= I_0(z) - 2\left(I_2(z) \cos 2\theta - I_4(z) \cos 4\theta + I_6(z) \cos 6\theta - I_8(z) \cos 8\theta + \cdots\right).$$

[3.24] Prove that $\forall (z, \theta) \in \mathbb{C} \times \mathbb{R}$,

$$\sinh(z \sin \theta) = 2 \sum_{j=0}^{\infty} (-1)^j I_{2j+1}(z) \sin(2j+1)\theta$$
$$= 2\left(I_1(z) \sin \theta - I_3(z) \sin 3\theta + I_5(z) \sin 5\theta - I_7(z) \sin 7\theta + \cdots\right).$$

[3.25] Prove that $\forall (z, \theta) \in \mathbb{C} \times \mathbb{R}$,

$$\cosh(z \cos \theta) = I_0(z) + 2 \sum_{j=1}^{\infty} I_{2j}(z) \cos 2j\theta$$
$$= I_0(z) + 2\left(I_2(z) \cos 2\theta + I_4(z) \cos 4\theta + I_6(z) \cos 6\theta + I_8(z) \cos 8\theta + \cdots\right).$$

[3.26] Prove that $\forall (z, \theta) \in \mathbb{C} \times \mathbb{R}$,

$$\sinh(z \cos \theta) = 2 \sum_{j=0}^{\infty} I_{2j+1}(z) \cos(2j+1)\theta$$
$$= 2\left(I_1(z) \cos \theta + I_3(z) \cos 3\theta + I_5(z) \cos 5\theta + I_7(z) \cos 7\theta + \cdots\right).$$

[3.27] Prove that $\forall z \in \mathbb{C}$,

$$z \sinh z = 2 \sum_{j=1}^{\infty} (2j)^2 I_{2j}(z)$$
$$= 2 \left(2^2 I_2(z) + 4^2 I_4(z) + 6^2 I_6(z) + 8^2 I_8(z) + \cdots \right).$$

[3.28] Prove that $\forall z \in \mathbb{C}$,

$$z \cosh z = 2 \sum_{j=0}^{\infty} (2j+1)^2 I_{2j+1}(z)$$
$$= 2 \left(1^2 I_1(z) + 3^2 I_3(z) + 5^2 I_5(z) + 7^2 I_7(z) + \cdots \right).$$

[3.29] (Watson, 2011) Prove that $\forall (z, v) \in \mathbb{C}^2$,

$$\frac{d}{dz} \left(z^v J_v(z) \right) = z^v J_{v-1}(z).$$

[3.30] (Watson, 2011) Prove that $\forall (z, v) \in \mathbb{C}^2$,

$$\frac{d}{dz} \left(z^{-v} J_v(z) \right) = -z^{-v} J_{v+1}(z).$$

[3.31] (Watson, 2011) Prove that $\forall (z, v) \in \mathbb{C}^2$,

$$\frac{d}{dz} \left(z^v I_v(z) \right) = z^v I_{v-1}(z).$$

[3.32] (Watson, 2011) Prove that $\forall (z, v) \in \mathbb{C}^2$,

$$\frac{d}{dz} \left(z^{-v} I_v(z) \right) = z^{-v} I_{v+1}(z).$$

Chapter 4
Special Integrals

In this chapter, we will prove several theorems for finding the integral representation of the coefficients of a series. We will also find integral representations of various sums involving the coefficients of a series. We will also derive an integral related to the Fourier series. The problems list several special integrals and series sums derived using the theorems and corollaries proved in this chapter.

4.1 Special Integrals 1

To begin with, we will use the complex variables, $e^{i\theta}$ and $e^{-i\theta}$, to find an integral representation of the constant term in a series.

Theorem 4.1 (Integral Representation of the Constant of a Series)
Let
$$f(z) = \sum_{j=-\infty}^{\infty} a_j z^j, \tag{4.1}$$

be a function in which $f(z)$ can contain all possible integral powers of z (positive, negative or zero). The series may be finite or infinite. In case it is infinite, $f\left(e^{i\theta}\right)$ and $f\left(e^{-i\theta}\right)$ both should be convergent for $\theta \in (0, \pi)$. Then, the constant a_0 is given by

$$a_0 = \frac{1}{\pi} \int_0^\pi \frac{f\left(e^{i\theta}\right) + f\left(e^{-i\theta}\right)}{2} d\theta.$$

Proof Putting $z = e^{i\theta}$ in (4.1), we get

$$f\left(e^{i\theta}\right) = a_0 + \sum_{j=-\infty, j\neq 0}^{\infty} a_j e^{ij\theta}. \tag{4.2}$$

Putting $z = e^{-i\theta}$ in (4.1), we get

$$f\left(e^{-i\theta}\right) = a_0 + \sum_{j=-\infty, j\neq 0}^{\infty} a_j e^{-ij\theta}. \tag{4.3}$$

Adding (4.2) and (4.3), and dividing the result by 2, we get

$$\frac{f\left(e^{i\theta}\right) + f\left(e^{-i\theta}\right)}{2} = a_0 + \sum_{j=-\infty, j\neq 0}^{\infty} a_j \frac{e^{ij\theta} + e^{-ij\theta}}{2}$$

$$= a_0 + \sum_{j=-\infty, j\neq 0}^{\infty} a_j \cos j\theta. \tag{4.4}$$

Integrating (4.4) with respect to θ from 0 to π, and dividing the result by π, we get

$$\frac{1}{\pi} \int_0^\pi \frac{f\left(e^{i\theta}\right) + f\left(e^{-i\theta}\right)}{2} d\theta = \frac{1}{\pi} \int_0^\pi \left(a_0 + \sum_{j=-\infty, j\neq 0}^{\infty} a_j \cos j\theta \right) d\theta$$

$$= \frac{1}{\pi} \int_0^\pi a_0 \, d\theta + \frac{1}{\pi} \sum_{j=-\infty, j\neq 0}^{\infty} \int_0^\pi a_j \cos j\theta d\theta$$

$$= a_0 + \frac{1}{\pi} \sum_{j=-\infty, j\neq 0}^{\infty} \left.\frac{a_j \sin j\theta}{j}\right|_0^\pi$$

$$= a_0.$$

$\square$

Now, we generalize the previous result to find three equivalent integral representations of the coefficients of a series.

Theorem 4.2 (First Generalization of Theorem 4.1)
Let $f(z)$ be the function as given in (4.1) in which it can contain all possible integral powers of z (positive, negative, or zero). The series may be finite or infinite. In case it is infinite, $f\left(e^{i\theta}\right)$ and $f\left(e^{-i\theta}\right)$ both should be convergent for $\theta \in (0, \pi)$. Then, $\forall n \in \mathbb{Z}$, the coefficient of z^n, a_n is given by

$$a_n = \frac{1}{\pi} \int_0^\pi \frac{e^{-in\theta} f\left(e^{i\theta}\right) + e^{in\theta} f\left(e^{-i\theta}\right)}{2} d\theta. \tag{4.5}$$

Proof Putting $z = e^{i\theta}$ in (4.1), and multiplying by $e^{-in\theta}$, we get

$$e^{-ni\theta} f\left(e^{i\theta}\right) = a_n + \sum_{j=-\infty, j\neq n}^{\infty} a_j e^{i(j-n)\theta}. \tag{4.6}$$

4.1 Special Integrals 1

Putting $z = e^{-i\theta}$ in (4.1), and multiplying by $e^{in\theta}$ we get

$$e^{in\theta} f\left(e^{-i\theta}\right) = a_n + \sum_{j=-\infty, j\neq n}^{\infty} a_j e^{-i(j-n)\theta}. \tag{4.7}$$

Adding (4.6) and (4.7), and dividing the result by 2, we get

$$\frac{e^{-in\theta} f\left(e^{i\theta}\right) + e^{in\theta} f\left(e^{-i\theta}\right)}{2} = a_n + \sum_{j=-\infty, j\neq n}^{\infty} a_j \frac{e^{i(j-n)\theta} + e^{-i(j-n)\theta}}{2}$$

$$= a_n + \sum_{j=-\infty, j\neq n}^{\infty} a_j \cos(j-n)\theta. \tag{4.8}$$

Integrating (4.8) with respect to θ from 0 to π, and dividing the result by π, we get

$$\frac{1}{\pi} \int_0^{\pi} \frac{e^{-in\theta} f\left(e^{i\theta}\right) + e^{in\theta} f\left(e^{-i\theta}\right)}{2} d\theta$$

$$= \frac{1}{\pi} \int_0^{\pi} a_n \, d\theta + \frac{1}{\pi} \sum_{j=-\infty, j\neq n}^{\infty} \int_0^{\pi} a_j \cos(j-n)\theta \, d\theta$$

$$= a_n + \frac{1}{\pi} \sum_{j=-\infty, j\neq n}^{\infty} \frac{a_j \sin(j-n)\theta}{j-n} \Bigg|_0^{\pi}$$

$$= a_n.$$

$\square$

Corollary 4.3 (Corollary to Theorem 4.2)
Let $f(z)$ be the function given in (4.1) in which it can contain all possible non-negative integral powers of z (positive or zero). The series may be finite or infinite. In case it is infinite, $f\left(e^{i\theta}\right)$ and $f\left(e^{-i\theta}\right)$ both should be convergent for $\theta \in (0, \pi)$. Then, $\forall n \in \mathbb{N} - 1$,

$$\frac{f^{(n)}(0)}{n!} = \frac{1}{\pi} \int_0^{\pi} \frac{e^{-in\theta} f(e^{i\theta}) + e^{in\theta} f(e^{-i\theta})}{2} d\theta.$$

Proof In (4.5), we have $a_n = f^{(n)}(0)/n!$ by Maclaurin's series. $\square$

Theorem 4.4 (Second Generalization of Theorem 4.1)
Let $f(z)$ be the function as given in (4.1) in which it can contain all possible integral powers of z (positive, negative, or zero). The series may be finite or infinite. In case it is infinite, $f\left(e^{i\theta}\right)$ and $f\left(e^{-i\theta}\right)$ both should be convergent for $\theta \in (0, \pi)$. Then, $\forall n \in \mathbb{N}$,

$$a_n + a_{-n} = \frac{2}{\pi} \int_0^{\pi} \frac{f\left(e^{i\theta}\right) + f\left(e^{-i\theta}\right)}{2} \cos n\theta \, d\theta.$$

Proof Multiplying (4.4) by $(2/\pi)\cos n\theta$, and integrating the result with respect to θ from 0 to π, we get

$$\frac{2}{\pi}\int_0^\pi \frac{f\left(e^{i\theta}\right)+f\left(e^{-i\theta}\right)}{2}\cos n\theta\, d\theta = \frac{2}{\pi}\int_0^\pi \left(\sum_{j=-\infty}^{\infty} a_j\cos j\theta\cos n\theta\right) d\theta$$

$$= \frac{1}{\pi}\sum_{j=-\infty}^{\infty} a_j \int_0^\pi 2\cos j\theta\cos n\theta\, d\theta$$

$$= \frac{1}{\pi}\sum_{j=-\infty}^{\infty} a_j \int_0^\pi (\cos(j-n)\theta + \cos(j+n)\theta)\, d\theta$$

$$= \frac{a_n}{\pi}\int_0^\pi (1+\cos 2n\theta)\, d\theta + \frac{a_{-n}}{\pi}\int_0^\pi (1+\cos(-2n\theta))\, d\theta$$

$$+ \frac{1}{\pi}\sum_{j=-\infty, j\neq \pm n}^{\infty} a_j \int_0^\pi (\cos(j-n)\theta + \cos(j+n)\theta)\, d\theta$$

$$= \frac{a_n}{\pi}\left(\theta + \frac{\sin 2n\theta}{2n}\right)\Big|_0^\pi + \frac{a_{-n}}{\pi}\left(\theta + \frac{\sin 2n\theta}{2n}\right)\Big|_0^\pi$$

$$+ \sum_{j=-\infty, j\neq \pm n}^{\infty} \frac{a_j}{\pi}\left(\frac{\sin(j-n)\theta}{j-n} + \frac{\sin(j+n)\theta}{j+n}\right)\Big|_0^\pi$$

$$= a_n + a_{-n}.$$

$\square$

After finding integral representations of the coefficients of a series, we will focus our attention on finding sums involving the coefficients of a series. As a first step, we will find the sum of squares of the coefficients. Then, later, we will generalize the result to find a general quadratic sum involving the coefficients.

Theorem 4.5 (Finding Sum of Squares of Terms of a Series)
Let $f(z)$ be the function as given in (4.1) in which it can contain all possible non-negative integral powers of z (positive or zero). The series may be finite or infinite. In case it is infinite, $f\left(e^{i\theta}\right)$ and $f\left(e^{-i\theta}\right)$ both should be convergent for $\theta \in (0,\pi)$. Then

$$\sum_{j=0}^{\infty} a_j^2 = a_0^2 + a_1^2 + a_2^2 + a_3^2 + \cdots$$

$$= \frac{1}{\pi}\int_0^\pi f\left(e^{i\theta}\right) f\left(e^{-i\theta}\right) d\theta.$$

4.1 Special Integrals 1

Proof Multiplying (4.2) and (4.3), we get

$$\begin{aligned}
f\left(e^{i\theta}\right) f\left(e^{-i\theta}\right) &= \sum_{j=-\infty}^{\infty} a_j e^{ij\theta} \sum_{k=-\infty}^{\infty} a_k e^{-ik\theta} \\
&= \sum_{j=0}^{\infty} \sum_{k=0}^{\infty} a_j a_k e^{i(j-k)\theta} \\
&= \sum_{j=0}^{\infty} a_j^2 + \sum_{l=1}^{\infty} \sum_{j=0}^{\infty} \left(a_{l+j} a_j e^{il\theta} + a_j a_{l+j} e^{-il\theta}\right) \\
&= \sum_{j=0}^{\infty} a_j^2 + 2 \sum_{l=1}^{\infty} \sum_{j=0}^{\infty} a_j a_{l+j} \cos l\theta.
\end{aligned} \quad (4.9)$$

Integrating (4.9) with respect to θ from 0 to π, and dividing the result by π, we get

$$\begin{aligned}
&\frac{1}{\pi} \int_0^{\pi} f\left(e^{i\theta}\right) f\left(e^{-i\theta}\right) d\theta \\
&= \frac{1}{\pi} \int_0^{\pi} \left(\sum_{j=0}^{\infty} a_j^2\right) d\theta + \frac{2}{\pi} \int_0^{\pi} \left(\sum_{l=1}^{\infty} \sum_{j=0}^{\infty} a_j a_{l+j} \cos l\theta\right) d\theta \\
&= \sum_{j=0}^{\infty} a_j^2 + \frac{2}{\pi} \sum_{l=1}^{\infty} \sum_{j=0}^{\infty} a_j a_{l+j} \left. \frac{\sin l\theta}{l} \right|_0^{\pi} \\
&= \sum_{j=0}^{\infty} a_j^2.
\end{aligned}$$

$\square$

Theorem 4.6 (Generalization of Theorem 4.5)
Let $f(z)$ be the function as given in (4.1) in which it can contain all possible non-negative integral powers of z (positive or zero). The series may be finite or infinite. In case it is infinite, $f\left(e^{i\theta}\right)$ and $f\left(e^{-i\theta}\right)$ both should be convergent for $\theta \in (0, \pi)$. Then $\forall n \in \mathbb{N} - 1$,

$$\sum_{j=0}^{\infty} a_j a_{n+j} = a_0 a_n + a_1 a_{n+1} + a_2 a_{n+2} + a_3 a_{n+3} + \cdots$$

$$= \frac{1}{\pi} \int_0^{\pi} f\left(e^{i\theta}\right) f\left(e^{-i\theta}\right) \cos n\theta \, d\theta.$$

Proof The $n = 0$ case is Theorem 4.5. We will prove the theorem for $n \in \mathbb{N}$. Assuming $n \in \mathbb{N}$, multiplying (4.9) by $(\cos n\theta)/\pi$, and integrating the result with respect to θ from 0 to π, we get

$$\frac{1}{\pi}\int_0^\pi f\left(e^{i\theta}\right)f\left(e^{-i\theta}\right)\cos n\theta\,d\theta$$

$$=\frac{1}{\pi}\int_0^\pi\left(\sum_{j=0}^\infty a_j^2\cos n\theta+\sum_{l=1}^\infty\sum_{j=0}^\infty 2a_ja_{l+j}\cos l\theta\cos n\theta\right)d\theta$$

$$=\frac{1}{\pi}\int_0^\pi\left(\sum_{j=0}^\infty a_j^2\cos n\theta\right)d\theta+\frac{1}{\pi}\int_0^\pi\left(\sum_{j=0}^\infty 2a_ja_{n+j}\cos^2 n\theta\right)d\theta$$

$$+\frac{1}{\pi}\int_0^\pi\left(\sum_{l=1,l\neq n}^\infty\sum_{j=0}^\infty 2a_ja_{l+j}\cos l\theta\cos n\theta\right)d\theta$$

$$=\frac{1}{\pi}\int_0^\pi\left(\sum_{j=0}^\infty a_j^2\cos n\theta\right)d\theta+\frac{1}{\pi}\int_0^\pi\left(\sum_{j=0}^\infty a_ja_{n+j}(1+\cos 2n\theta)\right)d\theta$$

$$+\frac{1}{\pi}\int_0^\pi\left(\sum_{l=1,l\neq n}^\infty\sum_{j=0}^\infty a_ja_{l+j}(\cos(l-n)\theta+\cos(l+n)\theta)\right)d\theta$$

$$=\frac{1}{\pi}\sum_{j=0}^\infty a_j^2\frac{\sin n\theta}{n}\bigg|_0^\pi+\frac{1}{\pi}\sum_{j=0}^\infty a_ja_{n+j}\left(\theta+\frac{\sin 2n\theta}{2n}\right)\bigg|_0^\pi$$

$$+\frac{1}{\pi}\sum_{l=1,l\neq n}^\infty\sum_{j=0}^\infty a_ja_{l+j}\left(\frac{\sin(l-n)\theta}{l-n}+\frac{\sin(l+n)\theta}{l+n}\right)\bigg|_0^\pi$$

$$=\sum_{j=0}^\infty a_ja_{n+j}.$$

□

Now, we will find series sums involving the central binomial coefficients of the form $\binom{2j}{j}$. Then, later, we will find generalized sums involving the binomial coefficients of the form $\binom{n+2j}{j}$.

Theorem 4.7 (A Series Involving Central Binomial Coefficients)
Let $f(z)$ be the function as given in (4.1) in which it can contain all possible nonnegative integral powers of z (positive or zero). The series may be finite or infinite. In case it is infinite, $f(2x\cos\theta)$ should be convergent for $\theta\in(0,\pi)$. Let

$$A_0(x)=\sum_{j=0}^\infty\binom{2j}{j}a_{2j}x^{2j}=a_0+\binom{2}{1}a_2x^2+\binom{4}{2}a_4x^4+\binom{6}{3}a_6x^6+\cdots,$$

(4.10)

4.1 Special Integrals 1

then, $A_0(x)$ is given by

$$A_0(x) = \frac{1}{\pi} \int_0^\pi f(2x \cos\theta)\, d\theta.$$

Proof Replacing z with $2x\cos\theta = x\left(e^{i\theta} + e^{-i\theta}\right)$ in (4.1), we get

$$f(2x\cos\theta) = f\left(x\left(e^{i\theta} + e^{-i\theta}\right)\right)$$

$$= \sum_{j=0}^\infty a_j \left(x\left(e^{i\theta} + e^{-i\theta}\right)\right)^j$$

$$= \sum_{j=0}^\infty a_{2j}\left(x\left(e^{i\theta}+e^{-i\theta}\right)\right)^{2j} + \sum_{j=0}^\infty a_{2j+1}\left(x\left(e^{i\theta}+e^{-i\theta}\right)\right)^{2j+1}$$

$$= \sum_{j=0}^\infty \binom{2j}{j} a_{2j} x^{2j} + \sum_{j=0}^\infty a_{2j} x^{2j} \sum_{k=0}^{j-1}\left(\binom{2j}{k} e^{ik\theta} e^{-i(2j-k)\theta}\right.$$

$$\left. + \binom{2j}{2j-k} e^{i(2j-k)\theta} e^{-ik\theta}\right) + \sum_{j=0}^\infty a_{2j+1} x^{2j+1} \sum_{k=0}^j \left(\binom{2j+1}{k} e^{ik\theta} e^{-i(2j+1-k)\theta}\right.$$

$$\left. + \binom{2j+1}{2j+1-k} e^{i(2j+1-k)\theta} e^{-ik\theta}\right)$$

$$= \sum_{j=0}^\infty \binom{2j}{j} a_{2j} x^{2j} + \sum_{j=0}^\infty a_{2j} x^{2j} \sum_{k=0}^{j-1} \binom{2j}{k} \left(e^{-i(2j-2k)\theta} + e^{i(2j-2k)\theta}\right)$$

$$+ \sum_{j=0}^\infty a_{2j+1} x^{2j+1} \sum_{k=0}^j \binom{2j+1}{k} \left(e^{-i(2j-2k+1)\theta} + e^{i(2j-2k+1)\theta}\right)$$

$$= \sum_{j=0}^\infty \binom{2j}{j} a_{2j} x^{2j} + 2 \sum_{j=0}^\infty a_{2j} x^{2j} \sum_{k=0}^{j-1} \binom{2j}{k} \cos(2j-2k)\theta$$

$$+ 2 \sum_{j=0}^\infty a_{2j+1} x^{2j+1} \sum_{k=0}^j \binom{2j+1}{k} \cos(2j-2k+1)\theta. \tag{4.11}$$

Dividing (4.11) by π, integrating the result with respect to θ from 0 to π, and using (4.10), we get

$$\frac{1}{\pi}\int_0^\pi f(2x\cos\theta)\,d\theta = \frac{1}{\pi}\int_0^\pi \left(\sum_{j=0}^\infty \binom{2j}{j} a_{2j} x^{2j}\right) d\theta$$

$$+ \frac{1}{\pi}\int_0^\pi \left(2\sum_{j=0}^\infty a_{2j} x^{2j} \sum_{k=0}^{j-1} \binom{2j}{k} \cos(2j-2k)\theta\right) d\theta$$

$$+ \frac{1}{\pi}\int_0^\pi \left(2\sum_{j=0}^\infty a_{2j+1} x^{2j+1} \sum_{k=0}^{j} \binom{2j+1}{k} \cos(2j-2k+1)\theta\right) d\theta$$

$$= \frac{1}{\pi}\sum_{j=0}^\infty \binom{2j}{j} a_{2j} x^{2j} \int_0^\pi d\theta + \frac{2}{\pi}\sum_{j=0}^\infty a_{2j} x^{2j} \sum_{k=0}^{j-1}\binom{2j}{k}\int_0^\pi \cos(2j-2k)\theta\,d\theta$$

$$+ \frac{2}{\pi}\sum_{j=0}^\infty a_{2j+1} x^{2j+1} \sum_{k=0}^{j}\binom{2j+1}{k}\int_0^\pi \cos(2j-2k+1)\theta\,d\theta$$

$$= \sum_{j=0}^\infty \binom{2j}{j} a_{2j} x^{2j} + \frac{2}{\pi}\sum_{j=0}^\infty a_{2j} x^{2j} \sum_{k=0}^{j-1}\binom{2j}{k} \left.\frac{\sin(2j-2k)\theta}{2j-2k}\right|_0^\pi$$

$$+ \frac{2}{\pi}\sum_{j=0}^\infty a_{2j+1} x^{2j+1} \sum_{k=0}^{j}\binom{2j+1}{k} \left.\frac{\sin(2j-2k+1)\theta}{2j-2k+1}\right|_0^\pi$$

$$= \sum_{j=0}^\infty \binom{2j}{j} a_{2j} x^{2j}$$

$$= A_0(x).$$

$\square$

Theorem 4.8 (Generalization of Theorem 4.7)
Let $f(z)$ be the function as given in (4.1) in which it can contain all possible non-negative integral powers of z (positive or zero). The series may be finite or infinite. In case it is infinite, $f(2x\cos\theta)$ should be convergent for $\theta \in (0, \pi)$. $\forall n \in \mathbb{N} - 1$, let

$$A_n(x) = \sum_{j=0}^\infty \binom{n+2j}{j} a_{n+2j} x^{n+2j}$$

$$= \binom{n}{0} a_n x^n + \binom{n+2}{1} a_{n+2} x^{n+2} + \binom{n+4}{2} a_{n+4} x^{n+4} + \binom{n+6}{3} a_{n+6} x^{n+6} + \cdots, \tag{4.12}$$

then, $A_n(x)$ is given by

4.1 Special Integrals 1

$$A_n(x) = \frac{1}{\pi} \int_0^\pi f(2x\cos\theta) \cos n\theta \, d\theta. \tag{4.13}$$

Proof Multiplying (4.11) by $\cos n\theta$, we get

$f(2x\cos\theta) \cos n\theta$

$$= \sum_{j=0}^\infty \binom{2j}{j} a_{2j} x^{2j} \cos n\theta + 2 \sum_{j=0}^\infty a_{2j} x^{2j} \sum_{k=0}^{j-1} \binom{2j}{k} \cos(2j-2k)\theta \cos n\theta$$

$$+ 2 \sum_{j=0}^\infty a_{2j+1} x^{2j+1} \sum_{k=0}^j \binom{2j+1}{k} \cos(2j-2k+1)\theta \cos n\theta$$

$$= \sum_{j=0}^\infty \binom{2j}{j} a_{2j} x^{2j} \cos n\theta + 2 \sum_{k=0}^\infty \binom{n+2k}{k} a_{n+2k} x^{n+2k} \cos^2 n\theta$$

$$+ 2 \sum_{j=0}^\infty a_{2j} x^{2j} \sum_{\substack{k=0, 2j-2k \neq n}}^{j-1} \binom{2j}{k} \cos(2j-2k)\theta \cos n\theta$$

$$+ 2 \sum_{j=0}^\infty a_{2j+1} x^{2j+1} \sum_{\substack{k=0, 2j-2k+1 \neq n}}^j \binom{2j+1}{k} \cos(2j-2k+1)\theta \cos n\theta$$

$$= \sum_{j=0}^\infty \binom{2j}{j} a_{2j} x^{2j} \cos n\theta + \sum_{j=0}^\infty \binom{n+2j}{j} a_{n+2j} x^{n+2j} (1 + \cos 2n\theta)$$

$$+ \sum_{j=0}^\infty a_{2j} x^{2j} \sum_{\substack{k=0, 2j-2k \neq n}}^{j-1} \binom{2j}{k} \big(\cos(2j-2k+n)\theta + \cos(2j-2k-n)\theta\big)$$

$$+ \sum_{j=0}^\infty a_{2j+1} x^{2j+1}$$

$$\sum_{\substack{k=0, 2j-2k+1 \neq n}}^j \binom{2j+1}{k} \big(\cos(2j-2k+1+n)\theta + \cos(2j-2k+1-n)\theta\big).$$

$$\tag{4.14}$$

Dividing (4.14) by π, integrating the result with respect to θ from 0 to π, and using (4.12), we get

$$\frac{1}{\pi}\int_0^\pi f(2x\cos\theta)\cos n\theta\, d\theta$$

$$= \frac{1}{\pi}\int_0^\pi \left(\sum_{j=0}^\infty \binom{2j}{j} a_{2j} x^{2j} \cos n\theta\right) d\theta$$

$$+ \frac{1}{\pi}\int_0^\pi \left(\sum_{j=0}^\infty \binom{n+2j}{j} a_{n+2j} x^{n+2j}(1+\cos 2n\theta)\right) d\theta + \frac{1}{\pi}\int_0^\pi \left(\sum_{j=0}^\infty a_{2j} x^{2j}\right.$$

$$\sum_{k=0, 2j-2k\neq n}^{j-1} \binom{2j}{k}(\cos(2j-2k+n)\theta + \cos(2j-2k-n)\theta)\bigg) d\theta$$

$$+ \frac{1}{\pi}\int_0^\pi \left(\sum_{j=0}^\infty a_{2j+1} x^{2j+1} \sum_{k=0, 2j-2k+1\neq n}^{j} \binom{2j+1}{k}\right.$$

$$(\cos(2j-2k+1+n)\theta + \cos(2j-2k+1-n)\theta))\, d\theta$$

$$= \frac{1}{\pi}\sum_{j=0}^\infty \binom{2j}{j} a_{2j} x^{2j} \int_0^\pi \cos n\theta\, d\theta$$

$$+ \frac{1}{\pi}\sum_{j=0}^\infty \binom{n+2j}{j} a_{n+2j} x^{n+2j} \int_0^\pi (1+\cos 2n\theta)\, d\theta + \frac{1}{\pi}\sum_{j=0}^\infty a_{2j} x^{2j}$$

$$\sum_{k=0, 2j-2k\neq n}^{j-1} \binom{2j}{k} \int_0^\pi (\cos(2j-2k+n)\theta + \cos(2j-2k-n)\theta)\, d\theta$$

$$+ \frac{1}{\pi}\sum_{j=0}^\infty a_{2j+1} x^{2j+1} \sum_{k=0, 2j-2k+1\neq n}^{j} \binom{2j+1}{k}$$

$$\int_0^\pi (\cos(2j-2k+1+n)\theta + \cos(2j-2k+1-n)\theta)\, d\theta$$

$$= \frac{1}{\pi}\sum_{j=0}^\infty \binom{2j}{j} a_{2j} x^{2j} \frac{\sin n\theta}{n}\bigg|_0^\pi + \frac{1}{\pi}\sum_{j=0}^\infty \binom{n+2j}{j} a_{n+2j} x^{n+2j}\left(\theta + \frac{\sin 2n\theta}{2n}\right)\bigg|_0^\pi$$

$$+ \frac{1}{\pi}\sum_{j=0}^\infty a_{2j} x^{2j} \sum_{k=0, 2j-2k\neq n}^{j-1} \binom{2j}{k}\left(\frac{\sin(2j-2k+n)\theta}{2j-2k+n} + \frac{\sin(2j-2k-n)\theta}{2j-2k-n}\right)\bigg|_0^\pi$$

$$+ \frac{1}{\pi}\sum_{j=0}^\infty a_{2j+1} x^{2j+1} \sum_{k=0, 2j-2k+1\neq n}^{j} \binom{2j+1}{k}$$

$$\left(\frac{\sin(2j-2k+1+n)\theta}{2j-2k+1+n} + \frac{\sin(2j-2k+1-n)\theta}{2j-2k+1-n}\right)\bigg|_0^\pi$$

4.1 Special Integrals 1

$$= \sum_{j=0}^{\infty} \binom{n+2j}{j} a_{n+2j} x^{n+2j}$$

$$= A_n(x).$$

□

Corollary 4.9 (Corollary to Theorem 4.8)
Let $f(z)$ and $A_n(x)$ be the functions as defined in Theorem 4.8. Then, $\forall y \in \mathbb{R} - \{0\}$, we have

$$A_n(x) = \frac{y}{\pi} \int_0^{\pi/y} f(2x \cos y\theta) \cos n\theta \, d\theta. \tag{4.15}$$

Proof We use the substitution $\phi = y\theta$ in (4.15). As θ varies from 0 to π/y, ϕ varies from 0 to π. We also have $d\phi = y d\theta \implies d\theta = d\phi/y$. From (4.13), we get

$$\frac{y}{\pi} \int_0^{\pi/y} f(2x \cos y\theta) \cos n\theta \, d\theta = \frac{y}{\pi} \int_0^{\pi} f(2x \cos \phi) \cos n\phi \, \frac{d\phi}{y}$$

$$= \frac{1}{\pi} \int_0^{\pi} f(2x \cos \phi) \cos n\phi \, d\phi$$

$$= A_n(x).$$

□

Theorem 4.10 (Another Series Involving Central Binomial Coefficients)
Let $f(z)$ be the function as given in (4.1) in which it can contain all possible non-negative integral powers of z (positive or zero). The series may be finite or infinite. In case it is infinite, $f(2ix \sin \theta)$ should be convergent for $\theta \in (0, \pi)$. Let

$$B_0(x) = \sum_{j=0}^{\infty} (-1)^j \binom{2j}{j} a_{2j} x^{2j}$$

$$= a_0 - \binom{2}{1} a_2 x^2 + \binom{4}{2} a_4 x^4 - \binom{6}{3} a_6 x^6 + \cdots, \tag{4.16}$$

then, if f is an even function, then $B_0(x)$ is given by

$$B_0(x) = \frac{1}{\pi} \int_0^{\pi} f(2ix \sin \theta) \, d\theta.$$

Proof By replacing z with $2ix \sin \theta = x \left(e^{i\theta} - e^{-i\theta}\right)$ in (4.1), we get

$$f(2ix\sin\theta) = f\left(x\left(e^{i\theta} - e^{-i\theta}\right)\right) = \sum_{j=0}^{\infty} a_j \left(x\left(e^{i\theta} - e^{-i\theta}\right)\right)^j$$

$$= \sum_{j=0}^{\infty} a_{2j} \left(x\left(e^{i\theta} - e^{-i\theta}\right)\right)^{2j} + \sum_{j=0}^{\infty} a_{2j+1} \left(x\left(e^{i\theta} - e^{-i\theta}\right)\right)^{2j+1}$$

$$= \sum_{j=0}^{\infty} (-1)^j \binom{2j}{j} a_{2j} x^{2j}$$

$$+ \sum_{j=0}^{\infty} a_{2j} x^{2j} \sum_{k=0}^{j-1} \left((-1)^{2j-k} \binom{2j}{k} e^{ik\theta} e^{-i(2j-k)\theta} + (-1)^k \binom{2j}{2j-k} e^{i(2j-k)\theta} e^{-ik\theta}\right)$$

$$+ \sum_{j=0}^{\infty} a_{2j+1} x^{2j+1} \sum_{k=0}^{j} \left((-1)^{2j+1-k} \binom{2j+1}{k} e^{ik\theta} e^{-i(2j+1-k)\theta}\right.$$

$$\left. + (-1)^k \binom{2j+1}{2j+1-k} e^{i(2j+1-k)\theta} e^{-ik\theta}\right)$$

$$= \sum_{j=0}^{\infty} (-1)^j \binom{2j}{j} a_{2j} x^{2j} + \sum_{j=0}^{\infty} a_{2j} x^{2j} \sum_{k=0}^{j-1} (-1)^k \binom{2j}{k} \left(e^{-i(2j-2k)\theta} + e^{i(2j-2k)\theta}\right)$$

$$+ \sum_{j=0}^{\infty} a_{2j+1} x^{2j+1} \sum_{k=0}^{j} (-1)^k \binom{2j+1}{k} \left(-e^{-i(2j-2k+1)\theta} + e^{i(2j-2k+1)\theta}\right)$$

$$= \sum_{j=0}^{\infty} (-1)^j \binom{2j}{j} a_{2j} x^{2j} + 2 \sum_{j=0}^{\infty} a_{2j} x^{2j} \sum_{k=0}^{j-1} (-1)^k \binom{2j}{k} \cos(2j-2k)\theta$$

$$+ 2i \sum_{j=0}^{\infty} a_{2j+1} x^{2j+1} \sum_{k=0}^{j} (-1)^k \binom{2j+1}{k} \sin(2j-2k+1)\theta. \tag{4.17}$$

Since f is an even function, we have $a_{2j+1} = 0$, $\forall j \in \mathbb{N} - 1$. Dividing (4.17) by π, integrating the result with respect to θ from 0 to π, and using (4.16), we get

4.1 Special Integrals 1

$$\frac{1}{\pi}\int_0^\pi f(2ix\sin\theta)\,d\theta = \frac{1}{\pi}\int_0^\pi \left(\sum_{j=0}^\infty (-1)^j \binom{2j}{j} a_{2j} x^{2j}\right) d\theta$$

$$+ \frac{1}{\pi}\int_0^\pi \left(2\sum_{j=0}^\infty a_{2j}x^{2j} \sum_{k=0}^{j-1}(-1)^k \binom{2j}{k}\cos(2j-2k)\theta\right) d\theta$$

$$= \frac{1}{\pi}\sum_{j=0}^\infty (-1)^j \binom{2j}{j} a_{2j} x^{2j} \int_0^\pi d\theta$$

$$+ \frac{2}{\pi}\sum_{j=0}^\infty a_{2j}x^{2j} \sum_{k=0}^{j-1}(-1)^k \binom{2j}{k} \int_0^\pi \cos(2j-2k)\theta\,d\theta$$

$$= \sum_{j=0}^\infty (-1)^j \binom{2j}{j} a_{2j} x^{2j} + \frac{2}{\pi}\sum_{j=0}^\infty a_{2j} x^{2j} \sum_{k=0}^{j-1}(-1)^k \binom{2j}{k} \frac{\sin(2j-2k)\theta}{2j-2k}\bigg|_0^\pi$$

$$= \sum_{j=0}^\infty (-1)^j \binom{2j}{j} a_{2j} x^{2j}$$

$$= B_0(x).$$

□

Theorem 4.11 (First Generalization of Theorem 4.10)
Let $f(z)$ be the function as given in (4.1) in which it can contain all possible non-negative integral powers of z (positive or zero). The series may be finite or infinite. In case it is infinite, $f(2ix\sin\theta)$ should be convergent for $\theta \in (0,\pi)$. $\forall n \in \mathbb{N}-1$, let

$$B_n(x) = \sum_{j=0}^\infty (-1)^j \binom{n+2j}{j} a_{n+2j} x^{n+2j}$$

$$= \binom{n}{0}a_n x^n - \binom{n+2}{1}a_{n+2}x^{n+2} + \binom{n+4}{2}a_{n+4}x^{n+4} - \binom{n+6}{3}a_{n+6}x^{n+6} + \cdots,$$
(4.18)

then, if f is an even function, then $B_n(x)$ is given by

$$B_n(x) = \frac{1}{\pi}\int_0^\pi f(2ix\sin\theta)\cos n\theta\,d\theta.$$

Proof Since f is an even function, we have $a_{2j+1} = 0$, $\forall j \in \mathbb{N}-1$. Multiplying (4.17) by $\cos n\theta$, we get

$$f(2ix\sin\theta)\cos n\theta = \sum_{j=0}^{\infty}(-1)^j\binom{2j}{j}a_{2j}x^{2j}\cos n\theta$$

$$+2\sum_{j=0}^{\infty}a_{2j}x^{2j}\sum_{k=0}^{j-1}(-1)^k\binom{2j}{k}\cos(2j-2k)\theta\cos n\theta$$

$$+2i\sum_{j=0}^{\infty}a_{2j+1}x^{2j+1}\sum_{k=0}^{j}(-1)^k\binom{2j+1}{k}\sin(2j-2k+1)\theta\cos n\theta$$

$$=\sum_{j=0}^{\infty}(-1)^j\binom{2j}{j}a_{2j}x^{2j}\cos n\theta + 2\sum_{k=0}^{\infty}(-1)^k\binom{n+2k}{k}a_{n+2k}x^{n+2k}\cos^2 n\theta$$

$$+2\sum_{j=0}^{\infty}a_{2j}x^{2j}\sum_{\substack{k=0,2j-2k\neq n}}^{j-1}(-1)^k\binom{2j}{k}\cos(2j-2k)\theta\cos n\theta$$

$$=\sum_{j=0}^{\infty}(-1)^j\binom{2j}{j}a_{2j}x^{2j}\cos n\theta + \sum_{j=0}^{\infty}(-1)^j\binom{n+2j}{j}a_{n+2j}x^{n+2j}(1+\cos 2n\theta)$$

$$+\sum_{j=0}^{\infty}a_{2j}x^{2j}\sum_{\substack{k=0,2j-2k\neq n}}^{j-1}(-1)^k\binom{2j}{k}(\cos(2j-2k+n)\theta+\cos(2j-2k-n)\theta).$$

(4.19)

Dividing (4.19) by π, integrating the result with respect to θ from 0 to π, and using (4.18), we get

$$\frac{1}{\pi}\int_0^{\pi} f(2ix\sin\theta)\cos n\theta\, d\theta$$

$$=\frac{1}{\pi}\int_0^{\pi}\left(\sum_{j=0}^{\infty}(-1)^j\binom{2j}{j}a_{2j}x^{2j}\cos n\theta\right)d\theta$$

$$+\frac{1}{\pi}\int_0^{\pi}\left(\sum_{j=0}^{\infty}(-1)^j\binom{n+2j}{j}a_{n+2j}x^{n+2j}(1+\cos 2n\theta)\right)d\theta$$

$$+\frac{1}{\pi}\int_0^{\pi}\left(\sum_{j=0}^{\infty}a_{2j}x^{2j}\right.$$

$$\left.\sum_{\substack{k=0,2j-2k\neq n}}^{j-1}(-1)^k\binom{2j}{k}(\cos(2j-2k+n)\theta+\cos(2j-2k-n)\theta)\right)d\theta$$

4.1 Special Integrals 1

$$= \frac{1}{\pi}\sum_{j=0}^{\infty}(-1)^j\binom{2j}{j}a_{2j}x^{2j}\int_0^{\pi}\cos n\theta\, d\theta$$

$$+ \frac{1}{\pi}\sum_{j=0}^{\infty}(-1)^j\binom{n+2j}{j}a_{n+2j}x^{n+2j}\int_0^{\pi}(1+\cos 2n\theta)\, d\theta$$

$$+ \frac{1}{\pi}\sum_{j=0}^{\infty}a_{2j}x^{2j}$$

$$\sum_{k=0,2j-2k\neq n}^{j-1}(-1)^k\binom{2j}{k}\int_0^{\pi}(\cos(2j-2k+n)\theta + \cos(2j-2k-n)\theta)\, d\theta$$

$$= \frac{1}{\pi}\sum_{j=0}^{\infty}(-1)^j\binom{2j}{j}a_{2j}x^{2j}\left.\frac{\sin n\theta}{n}\right|_0^{\pi}$$

$$+ \frac{1}{\pi}\sum_{j=0}^{\infty}(-1)^j\binom{n+2j}{j}a_{n+2j}x^{n+2j}\left.\left(\theta+\frac{\sin 2n\theta}{2n}\right)\right|_0^{\pi}$$

$$+ \frac{1}{\pi}\sum_{j=0}^{\infty}a_{2j}x^{2j}$$

$$\sum_{k=0,2j-2k\neq n}^{j-1}(-1)^k\binom{2j}{k}\left.\left(\frac{\sin(2j-2k+n)\theta}{2j-2k+n}+\frac{\sin(2j-2k-n)\theta}{2j-2k-n}\right)\right|_0^{\pi}$$

$$= \sum_{j=0}^{\infty}(-1)^j\binom{n+2j}{j}a_{n+2j}x^{n+2j}$$

$$= B_n(x).$$

$\square$

Corollary 4.12 (Corollary to Theorem 4.11)
Let $f(z)$ be the function as given in (4.1) in which it can contain all possible non-negative integral powers of z (positive or zero). The series may be finite or infinite. In case it is infinite, $f(2ix\sin\theta)$ should be convergent for $\theta \in (0,\pi)$. If f is an even function, then $\forall (x,n) \in \mathbb{R} \times (2\mathbb{N}-1)$,

$$\int_0^{\pi} f(2ix\sin\theta)\cos n\theta = 0.$$

Proof Since f is an even function and $n \in 2\mathbb{N}-1$, we have $a_{n+2j}=0, \forall j \in \mathbb{N}-1$. By Theorem 4.11, we get

$$\int_0^\pi f(2ix\sin\theta)\cos n\theta = \pi B_n(x)$$

$$= \pi \sum_{j=0}^{\infty} (-1)^j \binom{n+2j}{j} a_{n+2j} x^{n+2j}$$

$$= 0.$$

$\square$

Theorem 4.13 (Second Generalization of Theorem 4.10)
Let $f(z)$ be the function as given in (4.1) in which it can contain all possible non-negative integral powers of z (positive or zero). The series may be finite or infinite. In case it is infinite, $f(2ix\sin\theta)$ should be convergent for $\theta \in (0, \pi)$. If f is an odd function, then $\forall n \in \mathbb{N} - 1$, $B_n(x)$ is given by

$$B_n(x) = \frac{1}{\pi} \int_0^\pi -if(2ix\sin\theta)\sin n\theta\, d\theta.$$

Proof Since f is an odd function, we have $a_{2j} = 0$, $\forall j \in \mathbb{N} - 1$. Multiplying (4.17) by $-i\sin n\theta$, we get

$$-if(2ix\sin\theta)\sin n\theta = -i\sum_{j=0}^{\infty}(-1)^j\binom{2j}{j}a_{2j}x^{2j}\sin n\theta$$

$$-2i\sum_{j=0}^{\infty}a_{2j}x^{2j}\sum_{k=0}^{j-1}(-1)^k\binom{2j}{k}\cos(2j-2k)\theta\sin n\theta$$

$$+2\sum_{j=0}^{\infty}a_{2j+1}x^{2j+1}\sum_{k=0}^{j}(-1)^k\binom{2j+1}{k}\sin(2j-2k+1)\theta\sin n\theta$$

$$= 2\sum_{k=0}^{\infty}(-1)^k\binom{n+2k}{k}a_{n+2k}x^{n+2k}\sin^2 n\theta$$

$$+2\sum_{j=0}^{\infty}a_{2j+1}x^{2j+1}\sum_{k=0,2j-2k+1\neq n}^{j}(-1)^k\binom{2j+1}{k}\sin(2j-2k+1)\theta\sin n\theta$$

$$= \sum_{j=0}^{\infty}(-1)^j\binom{n+2j}{j}a_{n+2j}x^{n+2j}(1-\cos 2n\theta)$$

$$+\sum_{j=0}^{\infty}a_{2j+1}x^{2j+1}\sum_{k=0,2j-2k+1\neq n}^{j}(-1)^k\binom{2j+1}{k}$$
$$(\cos(2j-2k+1-n)\theta - \cos(2j-2k+1+n)\theta).$$

(4.20)

4.1 Special Integrals 1

Dividing (4.20) by π, integrating the result with respect to θ from 0 to π, and using (4.18), we get

$$\frac{1}{\pi}\int_0^\pi -if(2ix\sin\theta)\sin n\theta\, d\theta$$

$$= \frac{1}{\pi}\int_0^\pi \left(\sum_{j=0}^\infty (-1)^j \binom{n+2j}{j} a_{n+2j} x^{n+2j}(1-\cos 2n\theta)\right) d\theta$$

$$+ \frac{1}{\pi}\int_0^\pi \left(\sum_{j=0}^\infty a_{2j+1} x^{2j+1} \sum_{k=0, 2j-2k+1\neq n}^{j} (-1)^k \binom{2j+1}{k}\right)$$
$$(\cos(2j-2k+1-n)\theta - \cos(2j-2k+1+n)\theta))\, d\theta$$

$$= \frac{1}{\pi}\sum_{j=0}^\infty (-1)^j \binom{n+2j}{j} a_{n+2j} x^{n+2j} \int_0^\pi (1-\cos 2n\theta)\, d\theta$$

$$+ \frac{1}{\pi}\sum_{j=0}^\infty a_{2j+1} x^{2j+1} \sum_{k=0, 2j-2k+1\neq n}^{j} (-1)^k \binom{2j+1}{k}$$
$$\int_0^\pi (\cos(2j-2k+1-n)\theta - \cos(2j-2k+1+n)\theta)\, d\theta$$

$$= \frac{1}{\pi}\sum_{j=0}^\infty (-1)^j \binom{n+2j}{j} a_{n+2j} x^{n+2j} \left(\theta - \frac{\sin 2n\theta}{2n}\right)\Big|_0^\pi$$

$$+ \frac{1}{\pi}\sum_{j=0}^\infty a_{2j+1} x^{2j+1} \sum_{k=0, 2j-2k+1\neq n}^{j} (-1)^k \binom{2j+1}{k}$$
$$\left(\frac{\sin(2j-2k+1-n)\theta}{2j-2k+1-n} - \frac{\sin(2j-2k+1+n)\theta}{2j-2k+1+n}\right)\Big|_0^\pi$$

$$= \sum_{j=0}^\infty (-1)^j \binom{n+2j}{j} a_{n+2j} x^{n+2j}$$

$$= B_n(x).$$

$\square$

Corollary 4.14 (Corollary to Theorem 4.13)
Let $f(z)$ be the function as given in (4.1) in which it can contain all possible non-negative integral powers of z (positive or zero). The series may be finite or infinite. In case it is infinite, $f(2ix\sin\theta)$ should be convergent for $\theta \in (0, \pi)$. If f is an odd function, then $\forall(x, n) \in \mathbb{R} \times (2\mathbb{N} - 2)$,

$$\int_0^\pi f(2ix\sin\theta)\sin n\theta = 0.$$

Proof Since f is an odd function and $n \in 2\mathbb{N} - 2$, we have $a_{n+2j} = 0$, $\forall j \in \mathbb{N} - 1$. By Theorem 4.13, we get

$$\int_0^\pi f(2ix\sin\theta)\sin n\theta = i\pi B_n(x)$$

$$= i\pi \sum_{j=0}^\infty (-1)^j \binom{n+2j}{j} a_{n+2j} x^{n+2j}$$

$$= 0.$$

$\square$

Now, we will derive a double integral representation of the integral of a product of two functions with Fourier series.

Theorem 4.15 (A Double Integral Representation)
Let f and g be any two functions that can be expanded by Fourier's series in the interval $[0, 2\pi]$, then

$$2\pi \int_0^{2\pi} f(\theta)g(\theta)\,d\theta = \int_0^{2\pi}\int_0^{2\pi} f(\phi)g(\psi)\left(1 + 2\sum_{j=1}^\infty \cos j(\phi-\psi)\right) d\phi\,d\psi.$$

Proof We write the Fourier series expansion of $f(\theta)$ as

$$f(\theta)$$
$$= \frac{1}{2\pi}\int_0^{2\pi} f(\phi)\,d\phi$$
$$+ \frac{1}{\pi}\sum_{j=1}^\infty \left(\left(\int_0^{2\pi} f(\phi)\cos j\phi\,d\phi\right)\cos j\theta + \left(\int_0^{2\pi} f(\phi)\sin j\phi\,d\phi\right)\sin j\theta\right),$$
(4.21)

and the Fourier series expansion of $g(\theta)$ as

$$g(\theta)$$
$$= \frac{1}{2\pi}\int_0^{2\pi} g(\psi)\,d\psi$$
$$+ \frac{1}{\pi}\sum_{k=1}^\infty \left(\left(\int_0^{2\pi} g(\psi)\cos k\psi\,d\psi\right)\cos k\theta + \left(\int_0^{2\pi} g(\psi)\sin k\psi\,d\psi\right)\sin k\theta\right).$$
(4.22)

Multiplying (4.21) and (4.22), we get

4.1 Special Integrals 1

$f(\theta)g(\theta)$

$$= \frac{1}{4\pi^2} \int_0^{2\pi} f(\phi)\,d\phi \int_0^{2\pi} g(\psi)\,d\psi$$

$$+ \frac{1}{\pi^2} \sum_{j=1}^{\infty} \sum_{k=1}^{\infty} \left(\int_0^{2\pi} f(\phi)\cos j\phi\,d\phi \right) \left(\int_0^{2\pi} g(\psi)\cos k\psi\,d\psi \right) \cos j\theta \cos k\theta$$

$$+ \frac{1}{\pi^2} \sum_{j=1}^{\infty} \sum_{k=1}^{\infty} \left(\int_0^{2\pi} f(\phi)\cos j\phi\,d\phi \right) \left(\int_0^{2\pi} g(\psi)\sin k\psi\,d\psi \right) \cos j\theta \sin k\theta$$

$$+ \frac{1}{\pi^2} \sum_{j=1}^{\infty} \sum_{k=1}^{\infty} \left(\int_0^{2\pi} f(\phi)\sin j\phi\,d\phi \right) \left(\int_0^{2\pi} g(\psi)\cos k\psi\,d\psi \right) \sin j\theta \cos k\theta$$

$$+ \frac{1}{\pi^2} \sum_{j=1}^{\infty} \sum_{k=1}^{\infty} \left(\int_0^{2\pi} f(\phi)\sin j\phi\,d\phi \right) \left(\int_0^{2\pi} g(\psi)\sin k\psi\,d\psi \right) \sin j\theta \sin k\theta$$

$$= \frac{1}{4\pi^2} \int_0^{2\pi} \int_0^{2\pi} f(\phi)g(\psi)\,d\phi\,d\psi$$

$$+ \frac{1}{\pi^2} \sum_{j=1}^{\infty} \sum_{k=1}^{\infty} \left(\int_0^{2\pi} \int_0^{2\pi} f(\phi)\cos j\phi\, g(\psi)\cos k\psi\,d\phi\,d\psi \right) \cos j\theta \cos k\theta$$

$$+ \frac{1}{\pi^2} \sum_{j=1}^{\infty} \sum_{k=1}^{\infty} \left(\int_0^{2\pi} \int_0^{2\pi} f(\phi)\cos j\phi\, g(\psi)\sin k\psi\,d\phi\,d\psi \right) \cos j\theta \sin k\theta$$

$$+ \frac{1}{\pi^2} \sum_{j=1}^{\infty} \sum_{k=1}^{\infty} \left(\int_0^{2\pi} \int_0^{2\pi} f(\phi)\sin j\phi\, g(\psi)\cos k\psi\,d\phi\,d\psi \right) \sin j\theta \cos k\theta$$

$$+ \frac{1}{\pi^2} \sum_{j=1}^{\infty} \sum_{k=1}^{\infty} \left(\int_0^{2\pi} \int_0^{2\pi} f(\phi)\sin j\phi\, g(\psi)\sin k\psi\,d\phi\,d\psi \right) \sin j\theta \sin k\theta$$

$$= \frac{1}{4\pi^2} \int_0^{2\pi} \int_0^{2\pi} f(\phi)g(\psi)\,d\phi\,d\psi$$

$$+ \frac{1}{\pi^2} \sum_{j=1}^{\infty} \left(\int_0^{2\pi} \int_0^{2\pi} f(\phi)\cos j\phi\, g(\psi)\cos j\psi\,d\phi\,d\psi \right) \cos^2 j\theta$$

$$+ \frac{1}{\pi^2} \sum_{j=1}^{\infty} \left(\int_0^{2\pi} \int_0^{2\pi} f(\phi)\cos j\phi\, g(\psi)\sin j\psi\,d\phi\,d\psi \right) \cos j\theta \sin j\theta$$

$$+ \frac{1}{\pi^2} \sum_{j=1}^{\infty} \left(\int_0^{2\pi} \int_0^{2\pi} f(\phi)\sin j\phi\, g(\psi)\cos j\psi\,d\phi\,d\psi \right) \sin j\theta \cos j\theta$$

$$+ \frac{1}{\pi^2} \sum_{j=1}^{\infty} \left(\int_0^{2\pi} \int_0^{2\pi} f(\phi)\sin j\phi\, g(\psi)\sin j\psi\,d\phi\,d\psi \right) \sin^2 j\theta$$

$$+ \frac{1}{\pi^2} \sum_{j=1}^{\infty} \sum_{k=1, k \neq j}^{\infty} \left(\int_0^{2\pi} \int_0^{2\pi} f(\phi) \cos j\phi \, g(\psi) \cos k\psi \, d\phi \, d\psi \right) \cos j\theta \cos k\theta$$

$$+ \frac{1}{\pi^2} \sum_{j=1}^{\infty} \sum_{k=1, k \neq j}^{\infty} \left(\int_0^{2\pi} \int_0^{2\pi} f(\phi) \cos j\phi \, g(\psi) \sin k\psi \, d\phi \, d\psi \right) \cos j\theta \sin k\theta$$

$$+ \frac{1}{\pi^2} \sum_{j=1}^{\infty} \sum_{k=1, k \neq j}^{\infty} \left(\int_0^{2\pi} \int_0^{2\pi} f(\phi) \sin j\phi \, g(\psi) \cos k\psi \, d\phi \, d\psi \right) \sin j\theta \cos k\theta$$

$$+ \frac{1}{\pi^2} \sum_{j=1}^{\infty} \sum_{k=1, k \neq j}^{\infty} \left(\int_0^{2\pi} \int_0^{2\pi} f(\phi) \sin j\phi \, g(\psi) \sin k\psi \, d\phi \, d\psi \right) \sin j\theta \sin k\theta$$

$$= \frac{1}{4\pi^2} \int_0^{2\pi} \int_0^{2\pi} f(\phi) g(\psi) \, d\phi \, d\psi$$

$$+ \frac{1}{2\pi^2} \sum_{j=1}^{\infty} \left(\int_0^{2\pi} \int_0^{2\pi} f(\phi) \cos j\phi \, g(\psi) \cos j\psi \, d\phi \, d\psi \right) (1 + \cos 2j\theta)$$

$$+ \frac{1}{2\pi^2} \sum_{j=1}^{\infty} \left(\int_0^{2\pi} \int_0^{2\pi} f(\phi) \cos j\phi \, g(\psi) \sin j\psi \, d\phi \, d\psi \right) \sin 2j\theta$$

$$+ \frac{1}{2\pi^2} \sum_{j=1}^{\infty} \left(\int_0^{2\pi} \int_0^{2\pi} f(\phi) \sin j\phi \, g(\psi) \cos j\psi \, d\phi \, d\psi \right) \sin 2j\theta$$

$$+ \frac{1}{2\pi^2} \sum_{j=1}^{\infty} \left(\int_0^{2\pi} \int_0^{2\pi} f(\phi) \sin j\phi \, g(\psi) \sin j\psi \, d\phi \, d\psi \right) (1 - \cos 2j\theta)$$

$$+ \frac{1}{2\pi^2} \sum_{j=1}^{\infty} \sum_{k=1, k \neq j}^{\infty} \left(\int_0^{2\pi} \int_0^{2\pi} f(\phi) \cos j\phi \, g(\psi) \cos k\psi \, d\phi \, d\psi \right)$$
$$(\cos(j+k)\theta + \cos(j-k)\theta)$$

$$+ \frac{1}{2\pi^2} \sum_{j=1}^{\infty} \sum_{k=1, k \neq j}^{\infty} \left(\int_0^{2\pi} \int_0^{2\pi} f(\phi) \cos j\phi \, g(\psi) \sin k\psi \, d\phi \, d\psi \right)$$
$$(\sin(j+k)\theta - \sin(j-k)\theta)$$

$$+ \frac{1}{2\pi^2} \sum_{j=1}^{\infty} \sum_{k=1, k \neq j}^{\infty} \left(\int_0^{2\pi} \int_0^{2\pi} f(\phi) \sin j\phi \, g(\psi) \cos k\psi \, d\phi \, d\psi \right)$$
$$(\sin(j+k)\theta + \sin(j-k)\theta)$$

$$+ \frac{1}{2\pi^2} \sum_{j=1}^{\infty} \sum_{k=1, k \neq j}^{\infty} \left(\int_0^{2\pi} \int_0^{2\pi} f(\phi) \sin j\phi \, g(\psi) \sin k\psi \, d\phi \, d\psi \right)$$
$$(\cos(j-k)\theta - \cos(j+k)\theta).$$

(4.23)

4.1 Special Integrals 1

Multiplying (4.23) by 2π, and integrating the result with respect to θ from 0 to 2π, we get

$$2\pi \int_0^{2\pi} f(\theta)g(\theta)\,d\theta$$

$$= \frac{1}{2\pi} \int_0^{2\pi}\int_0^{2\pi} f(\phi)g(\psi)\,d\phi\,d\psi \int_0^{2\pi} d\theta$$

$$+ \frac{1}{\pi}\sum_{j=1}^{\infty} \left(\int_0^{2\pi}\int_0^{2\pi} f(\phi)\cos j\phi\, g(\psi)\cos j\psi\,d\phi\,d\psi\right) \int_0^{2\pi} (1+\cos 2j\theta)\,d\theta$$

$$+ \frac{1}{\pi}\sum_{j=1}^{\infty} \left(\int_0^{2\pi}\int_0^{2\pi} f(\phi)\cos j\phi\, g(\psi)\sin j\psi\,d\phi\,d\psi\right) \int_0^{2\pi} \sin 2j\theta\,d\theta$$

$$+ \frac{1}{\pi}\sum_{j=1}^{\infty} \left(\int_0^{2\pi}\int_0^{2\pi} f(\phi)\sin j\phi\, g(\psi)\cos j\psi\,d\phi\,d\psi\right) \int_0^{2\pi} \sin 2j\theta\,d\theta$$

$$+ \frac{1}{\pi}\sum_{j=1}^{\infty} \left(\int_0^{2\pi}\int_0^{2\pi} f(\phi)\sin j\phi\, g(\psi)\sin j\psi\,d\phi\,d\psi\right) \int_0^{2\pi} (1-\cos 2j\theta)\,d\theta$$

$$+ \frac{1}{\pi}\sum_{j=1}^{\infty}\sum_{k=1, k\neq j}^{\infty} \left(\int_0^{2\pi}\int_0^{2\pi} f(\phi)\cos j\phi\, g(\psi)\cos k\psi\,d\phi\,d\psi\right)$$
$$\int_0^{2\pi} (\cos(j+k)\theta + \cos(j-k)\theta)\,d\theta$$

$$+ \frac{1}{\pi}\sum_{j=1}^{\infty}\sum_{k=1, k\neq j}^{\infty} \left(\int_0^{2\pi}\int_0^{2\pi} f(\phi)\cos j\phi\, g(\psi)\sin k\psi\,d\phi\,d\psi\right)$$
$$\int_0^{2\pi} (\sin(j+k)\theta - \sin(j-k)\theta)\,d\theta$$

$$+ \frac{1}{\pi}\sum_{j=1}^{\infty}\sum_{k=1, k\neq j}^{\infty} \left(\int_0^{2\pi}\int_0^{2\pi} f(\phi)\sin j\phi\, g(\psi)\cos k\psi\,d\phi\,d\psi\right)$$
$$\int_0^{2\pi} (\sin(j+k)\theta + \sin(j-k)\theta)\,d\theta$$

$$+ \frac{1}{\pi}\sum_{j=1}^{\infty}\sum_{k=1, k\neq j}^{\infty} \left(\int_0^{2\pi}\int_0^{2\pi} f(\phi)\sin j\phi\, g(\psi)\sin k\psi\,d\phi\,d\psi\right)$$
$$\int_0^{2\pi} (\cos(j-k)\theta - \cos(j+k)\theta)\,d\theta$$

$$= \int_0^{2\pi}\int_0^{2\pi} f(\phi)g(\psi)\,d\phi\,d\psi$$

$$+ \frac{1}{\pi} \sum_{j=1}^{\infty} \left(\int_0^{2\pi} \int_0^{2\pi} f(\phi) \cos j\phi \, g(\psi) \cos j\psi \, d\phi \, d\psi \right) \left(\theta + \frac{\sin 2j\theta}{2j} \right) \Bigg|_0^{2\pi}$$

$$+ \frac{1}{\pi} \sum_{j=1}^{\infty} \left(\int_0^{2\pi} \int_0^{2\pi} f(\phi) \cos j\phi \, g(\psi) \sin j\psi \, d\phi \, d\psi \right) \left(-\frac{\cos 2j\theta}{2j} \right) \Bigg|_0^{2\pi}$$

$$+ \frac{1}{\pi} \sum_{j=1}^{\infty} \left(\int_0^{2\pi} \int_0^{2\pi} f(\phi) \sin j\phi \, g(\psi) \cos j\psi \, d\phi \, d\psi \right) \left(-\frac{\cos 2j\theta}{2j} \right) \Bigg|_0^{2\pi}$$

$$+ \frac{1}{\pi} \sum_{j=1}^{\infty} \left(\int_0^{2\pi} \int_0^{2\pi} f(\phi) \sin j\phi \, g(\psi) \sin j\psi \, d\phi \, d\psi \right) \left(\theta - \frac{\sin 2j\theta}{2j} \right) \Bigg|_0^{2\pi}$$

$$+ \frac{1}{\pi} \sum_{j=1}^{\infty} \sum_{k=1, k \neq j}^{\infty} \left(\int_0^{2\pi} \int_0^{2\pi} f(\phi) \cos j\phi \, g(\psi) \cos k\psi \, d\phi \, d\psi \right)$$
$$\left(\frac{\sin(j+k)\theta}{j+k} + \frac{\sin(j-k)\theta}{j-k} \right) \Bigg|_0^{2\pi}$$

$$+ \frac{1}{\pi} \sum_{j=1}^{\infty} \sum_{k=1, k \neq j}^{\infty} \left(\int_0^{2\pi} \int_0^{2\pi} f(\phi) \cos j\phi \, g(\psi) \sin k\psi \, d\phi \, d\psi \right)$$
$$\left(-\frac{\cos(j+k)\theta}{j+k} + \frac{\cos(j-k)\theta}{j-k} \right) \Bigg|_0^{2\pi}$$

$$+ \frac{1}{\pi} \sum_{j=1}^{\infty} \sum_{k=1, k \neq j}^{\infty} \left(\int_0^{2\pi} \int_0^{2\pi} f(\phi) \sin j\phi \, g(\psi) \cos k\psi \, d\phi \, d\psi \right)$$
$$\left(-\frac{\cos(j+k)\theta}{j+k} - \frac{\cos(j-k)\theta}{j-k} \right) \Bigg|_0^{2\pi}$$

$$+ \frac{1}{\pi} \sum_{j=1}^{\infty} \sum_{k=1, k \neq j}^{\infty} \left(\int_0^{2\pi} \int_0^{2\pi} f(\phi) \sin j\phi \, g(\psi) \sin k\psi \, d\phi \, d\psi \right)$$
$$\left(\frac{\sin(j-k)\theta}{j-k} - \frac{\sin(j+k)\theta}{j+k} \right) \Bigg|_0^{2\pi}$$

$$= \int_0^{2\pi} \int_0^{2\pi} f(\phi) g(\psi) \, d\phi \, d\psi$$
$$+ \left(2 \sum_{j=1}^{\infty} \left(\int_0^{2\pi} \int_0^{2\pi} f(\phi) \cos j\phi \, g(\psi) \cos j\psi \, d\phi \, d\psi \right) \right)$$
$$\left(2 \sum_{j=1}^{\infty} \left(\int_0^{2\pi} \int_0^{2\pi} f(\phi) \sin j\phi \, g(\psi) \sin j\psi \, d\phi \, d\psi \right) \right)$$

4.1 Special Integrals 1

$$= \int_0^{2\pi} \int_0^{2\pi} f(\phi)g(\psi) \left(1 + 2\sum_{j=1}^{\infty} \cos j(\phi - \psi)\right) d\phi\, d\psi.$$

$\square$

Finally, we conclude the section by deriving another double integral representation.

Theorem 4.16 (Another Double Integral Representation)
Let f and g be any two functions which can be expanded by Maclaurin's series, such that $f(e^{i\theta})$, $f(e^{-i\theta})$, $g(e^{i\theta})$, and $g(e^{-i\theta})$ all are convergent for $\theta \in (0, \pi)$. Then

$$2\pi \int_0^{\pi} \left(f(e^{i\theta})g(e^{i\theta})e^{-in\theta} + f(e^{-i\theta})g(e^{-i\theta})e^{in\theta}\right) d\theta$$

$$= \sum_{j=0}^{n} \int_0^{\pi} \int_0^{\pi} \left(f(e^{i\phi})e^{-ij\phi} + f(e^{-i\phi})e^{ij\phi}\right)$$

$$\left(g(e^{i\psi})e^{-i(n-j)\psi} + g(e^{-i\psi})e^{i(n-j)\phi}\right) d\phi\, d\psi.$$

Proof Let

$$f(z) = \sum_{j=0}^{\infty} a_j z^j, \qquad (4.24)$$

and

$$g(z) = \sum_{k=0}^{\infty} b_k z^k. \qquad (4.25)$$

Putting $z = e^{i\theta}$ in (4.24), we get

$$f(e^{i\theta}) = \sum_{j=0}^{\infty} a_j e^{ij\theta}. \qquad (4.26)$$

Putting $z = e^{i\theta}$ in (4.25), we get

$$g(e^{i\theta}) = \sum_{k=0}^{\infty} b_k e^{ik\theta}. \qquad (4.27)$$

Putting $z = e^{-i\theta}$ in (4.24), we get

$$f(e^{-i\theta}) = \sum_{j=0}^{\infty} a_j e^{-ij\theta}. \qquad (4.28)$$

Putting $z = e^{-i\theta}$ in (4.25), we get

$$g\left(e^{-i\theta}\right) = \sum_{k=0}^{\infty} b_k e^{-ik\theta}. \tag{4.29}$$

Multiplying (4.26) and (4.27), we get

$$f\left(e^{i\theta}\right) g\left(e^{i\theta}\right) = \sum_{j=0}^{\infty} \sum_{k=0}^{\infty} a_j b_k e^{i(j+k)\theta}. \tag{4.30}$$

Multiplying (4.28) and (4.29), we get

$$f\left(e^{-i\theta}\right) g\left(e^{-i\theta}\right) = \sum_{j=0}^{\infty} \sum_{k=0}^{\infty} a_j b_k e^{-i(j+k)\theta}. \tag{4.31}$$

Multiplying (4.30) by $e^{-in\theta}$, multiplying (4.31) by $e^{in\theta}$, and adding the results, we get

$$\begin{aligned}
f\left(e^{i\theta}\right) g\left(e^{i\theta}\right) e^{-in\theta} &+ f\left(e^{-i\theta}\right) g\left(e^{-i\theta}\right) e^{in\theta} \\
&= \sum_{j=0}^{\infty} \sum_{k=0}^{\infty} a_j b_k \left(e^{i(j+k-n)\theta} + e^{-i(j+k-n)\theta}\right) \\
&= 2 \sum_{j=0}^{\infty} \sum_{k=0}^{\infty} a_j b_k \cos(j+k-n)\theta.
\end{aligned} \tag{4.32}$$

Multiplying (4.32) by 2π, integrating the result with respect to θ from 0 to π, and using Theorem 4.2, we get

$$2\pi \int_0^{\pi} \left(f\left(e^{-i\theta}\right) g\left(e^{-i\theta}\right) e^{-in\theta} + f\left(e^{-i\theta}\right) g\left(e^{-i\theta}\right) e^{in\theta} \right) d\theta$$

$$= 4\pi \sum_{j=0}^{\infty} \sum_{k=0}^{\infty} a_j b_k \int_0^{\pi} \cos(j+k-n)\theta \, d\theta$$

$$= 4\pi \sum_{j+k=n, j\in[0..n], k\in[0..n]} a_j b_k \int_0^{\pi} d\theta$$

$$+ 4\pi \sum_{j=0}^{\infty} \sum_{k=0, j+k\neq n}^{\infty} a_j b_k \int_0^{\pi} \cos(j+k-n)\theta \, d\theta$$

$$= 4\pi^2 \sum_{j+k=n, j\in[0..n], k\in[0..n]} a_j b_k + 4\pi \sum_{j=0}^{\infty} \sum_{k=0, j+k\neq n}^{\infty} a_j b_k \left.\frac{\sin(j+k-n)\theta}{j+k-n}\right|_0^{\pi}$$

$$= 4\pi^2 \sum_{j+k=n, j\in[0..n], k\in[0..n]} a_j b_k$$

$$= 4\pi^2 \sum_{j+k=n, j\in[0..n], k\in[0..n]} \left(\frac{1}{\pi}\int_0^\pi \frac{e^{-ij\phi}f\left(e^{i\phi}\right) + e^{ij\phi}f\left(e^{-i\phi}\right)}{2} d\phi\right)$$

$$\left(\frac{1}{\pi}\int_0^\pi \frac{e^{-ik\psi}g\left(e^{i\psi}\right) + e^{ik\psi}g\left(e^{-i\psi}\right)}{2} d\psi\right)$$

$$= \sum_{j=0}^n \int_0^\pi \int_0^\pi \left(f\left(e^{i\phi}\right)e^{-ij\phi} + f\left(e^{-i\phi}\right)e^{ij\phi}\right)$$

$$\left(g\left(e^{i\psi}\right)e^{-i(n-j)\psi} + g\left(e^{-i\psi}\right)e^{i(n-j)\phi}\right) d\phi\, d\psi.$$

□

4.2 Special Integrals 2

In this section, we will continue our development of the previous section. We will prove some theorems about series sums involving the binomial coefficients and several of their corollaries. We will see that Theorems 4.1, 4.4, and 4.8 can be derived as a corollary to these theorems.

We begin with an integral representation of sums involving the coefficients of a series and the binomial coefficients.

Theorem 4.17 (A Binomial Series)
Let

$$f(z) = \sum_{j=0}^\infty a_j z^j = a_0 + a_1 z + a_2 z^2 + a_3 z^3 + \cdots, \qquad (4.33)$$

be a function which can be expanded by Maclaurin's series such that $f\left(xe^{i\theta} + ye^{-i\theta}\right)$ and $f\left(xe^{-i\theta} + ye^{i\theta}\right)$ both are convergent for $\theta \in (0, \pi)$, and $\forall (x, y, n) \in \mathbb{C}^2 \times \times (\mathbb{N} - 1)$, let the function $T_n(x, y)$ be defined as

$$T_n(x, y) = (x^n + y^n) \sum_{j=0}^\infty \binom{n+2j}{j} a_{n+2j}(xy)^j$$

$$= (x^n + y^n) \left(\binom{n}{0} a_n + \binom{n+2}{1} a_{n+2}(xy) + \binom{n+4}{2} a_{n+4}(xy)^2 \right.$$

$$\left. + \binom{n+6}{3} a_{n+6}(xy)^3 + \cdots \right),$$

(4.34)

then

$$T_n(x, y) = \frac{2}{\pi} \int_0^\pi \frac{f\left(xe^{i\theta} + ye^{-i\theta}\right) + f\left(xe^{-i\theta} + ye^{i\theta}\right)}{2} \cos n\theta \, d\theta. \quad (4.35)$$

Proof Putting $z = xe^{i\theta} + ye^{-i\theta}$ in (4.33), we get

$$f\left(xe^{i\theta} + ye^{-i\theta}\right) = \sum_{j=0}^\infty a_j \left(xe^{i\theta} + ye^{-i\theta}\right)^j$$

$$= \sum_{j=0}^\infty a_j \sum_{k=0}^j \binom{j}{k} \left(xe^{i\theta}\right)^k \left(ye^{-i\theta}\right)^{j-k} \quad (4.36)$$

$$= \sum_{j=0}^\infty a_j \sum_{k=0}^j \binom{j}{k} x^k y^{j-k} e^{i(2k-j)\theta}.$$

Putting $z = xe^{-i\theta} + ye^{i\theta}$ in (4.33), we get

$$f\left(xe^{-i\theta} + ye^{i\theta}\right) = \sum_{j=0}^\infty a_j \left(xe^{-i\theta} + ye^{i\theta}\right)^j$$

$$= \sum_{j=0}^\infty a_j \sum_{k=0}^j \binom{j}{k} \left(xe^{-i\theta}\right)^k \left(ye^{i\theta}\right)^{j-k} \quad (4.37)$$

$$= \sum_{j=0}^\infty a_j \sum_{k=0}^j \binom{j}{k} x^k y^{j-k} e^{-i(2k-j)\theta}.$$

Adding (4.36) and (4.37), dividing the result by 2, and using (4.34) we get

4.2 Special Integrals 2

$$\frac{f\left(xe^{i\theta} + ye^{-i\theta}\right) + f\left(xe^{-i\theta} + ye^{i\theta}\right)}{2}$$

$$= \sum_{j=0}^{\infty} a_j \sum_{k=0}^{j} \binom{j}{k} x^k y^{j-k} \frac{e^{i(2k-j)\theta} + e^{-i(2k-j)\theta}}{2}$$

$$= \sum_{j=0}^{\infty} a_j \sum_{k=0}^{j} \binom{j}{k} x^k y^{j-k} \cos(2k-j)\theta$$

$$= \sum_{j=0}^{\infty} a_{n+2j} \binom{n+2j}{n+j} x^{n+j} y^j \cos n\theta + \sum_{j=0}^{\infty} a_{n+2j} \binom{n+2j}{j} x^j y^{n+j} \cos(-n\theta)$$

$$+ \sum_{j=0}^{\infty} a_j \sum_{k=0, 2k-j \neq \pm n}^{j} \binom{j}{k} x^k y^{j-k} \cos(2k-j)\theta$$

$$= (x^n + y^n) \sum_{j=0}^{\infty} a_{n+2j} \binom{n+2j}{j} (xy)^j \cos n\theta$$

$$+ \sum_{j=0}^{\infty} a_j \sum_{k=0, 2k-j \neq \pm n}^{j} \binom{j}{k} x^k y^{j-k} \cos(2k-j)\theta$$

$$= T_n(x, y) \cos n\theta + \sum_{j=0}^{\infty} a_j \sum_{k=0, 2k-j \neq \pm n}^{j} \binom{j}{k} x^k y^{j-k} \cos(2k-j)\theta.$$

(4.38)

Multiplying (4.38) by $(2/\pi) \cos n\theta$, and integrating the result with respect to θ from 0 to π, we get

$$\frac{2}{\pi} \int_0^\pi \frac{f\left(xe^{i\theta} + ye^{-i\theta}\right) + f\left(xe^{-i\theta} + ye^{i\theta}\right)}{2} \cos n\theta \, d\theta$$

$$= \frac{2}{\pi} \int_0^\pi T_n(x, y) \cos^2 n\theta \, d\theta$$

$$+ \frac{2}{\pi} \int_0^\pi \left(\sum_{j=0}^{\infty} a_j \sum_{k=0, 2k-j \neq \pm n}^{j} \binom{j}{k} x^k y^{j-k} \cos(2k-j)\theta \cos n\theta \right) d\theta$$

$$= \frac{1}{\pi} \int_0^\pi T_n(x, y)(1 + \cos 2n\theta) \, d\theta + \frac{1}{\pi} \sum_{j=0}^{\infty} a_j$$

$$\sum_{k=0, 2k-j \neq \pm n}^{j} \binom{j}{k} x^k y^{j-k} \int_0^\pi (\cos(2k-j+n)\theta + \cos(2k-j-n)\theta) \, d\theta$$

$$= \frac{1}{\pi} T_n(x, y) \left(\theta + \frac{\sin 2n\theta}{2n} \right) \Big|_0^\pi$$

$$+ \frac{1}{\pi} \sum_{j=0}^{\infty} a_j \sum_{k=0, 2k-j \neq \pm n}^{j} \binom{j}{k} x^k y^{j-k} \left(\frac{\sin(2k-j+n)\theta}{2k-j+n} + \frac{\sin(2k-j-n)\theta}{2k-j-n} \right) \Big|_0^{\pi}$$

$$= T_n(x, y).$$

$\square$

Corollary 4.18 (Corollary to Theorem 4.17)
Let $f(z)$ be the function as defined in Theorem 4.17. Then, $\forall (x, n) \in (\mathbb{R} - \{0\}) \times (\mathbb{N} - 1)$,

$$T_n\left(x, \frac{1}{x}\right) = \left(x^n + \frac{1}{x^n}\right) \sum_{j=0}^{\infty} \binom{n+2j}{j} a_{n+2j}$$

$$= \left(x^n + \frac{1}{x^n}\right) \left(\binom{n}{0} a_n + \binom{n+2}{1} a_{n+2} + \binom{n+4}{2} a_{n+4} + \binom{n+6}{3} a_{n+6} + \cdots \right)$$

$$= \frac{2}{\pi} \int_0^{\pi} \frac{f\left(xe^{i\theta} + x^{-1}e^{-i\theta}\right) + f\left(xe^{-i\theta} + x^{-1}e^{i\theta}\right)}{2} \cos n\theta \, d\theta.$$

Proof Put $y = x^{-1}$ in (4.35). $\square$

Corollary 4.19 (Corollary to Theorem 4.17)
Let $f(z)$ be the function as defined in Theorem 4.1. Then, $\forall (x, n) \in (\mathbb{R} - \{0\}) \times (\mathbb{N} - 1)$,

$$T_n\left(x, -\frac{1}{x}\right) = \left(x^n + \frac{(-1)^n}{x^n}\right) \sum_{j=0}^{\infty} (-1)^j \binom{n+2j}{j} a_{n+2j}$$

$$= \left(x^n + \frac{(-1)^n}{x^n}\right) \left(\binom{n}{0} a_n - \binom{n+2}{1} a_{n+2} + \binom{n+4}{2} a_{n+4} - \binom{n+6}{3} a_{n+6} \right.$$
$$\left. + \cdots \right)$$

$$= \frac{2}{\pi} \int_0^{\pi} \frac{f\left(xe^{i\theta} - x^{-1}e^{-i\theta}\right) + f\left(xe^{-i\theta} - x^{-1}e^{i\theta}\right)}{2} \cos n\theta \, d\theta.$$

Proof Put $y = -x^{-1}$ in (4.35). $\square$

Corollary 4.20 (Corollary to Theorem 4.17. This is also Theorem 4.8)
Let $f(z)$ be the function as defined in Theorem 4.17. Then, $\forall (x, n) \in \mathbb{R} \times (\mathbb{N} - 1)$,

4.2 Special Integrals 2

$$T_n(x,x) = 2\sum_{j=0}^{\infty} \binom{n+2j}{j} a_{n+2j} x^{n+2j}$$

$$= 2\left(\binom{n}{0} a_n x^n + \binom{n+2}{1} a_{n+2} x^{n+2} + \binom{n+4}{2} a_{n+4} x^{n+4}\right.$$

$$\left. + \binom{n+6}{3} a_{n+6} x^{n+6} + \cdots \right)$$

$$= \frac{2}{\pi} \int_0^\pi f(2x\cos\theta) \cos n\theta \, d\theta.$$

Proof Put $y = x$ in (4.35). $\square$

Corollary 4.21 (Corollary to Theorem 4.17)
Let $f(z)$ be the function as defined in Theorem 4.17. Then, $\forall (x, y, n) \in \mathbb{R}^2 \times (\mathbb{N} - 1)$,

$$T_n(ix, iy) = i^n (x^n + y^n) \sum_{j=0}^{\infty} (-1)^j \binom{n+2j}{j} a_{n+2j}(xy)^j$$

$$= i^n (x^n + y^n) \left(\binom{n}{0} a_n - \binom{n+2}{1} a_{n+2}(xy) + \binom{n+4}{2} a_{n+4}(xy)^2 \right.$$

$$\left. - \binom{n+6}{3} a_{n+6}(xy)^3 + \cdots \right)$$

$$= \frac{2}{\pi} \int_0^\pi \frac{f\left(i\left(xe^{i\theta} + ye^{-i\theta}\right)\right) + f\left(i\left(xe^{-i\theta} + ye^{i\theta}\right)\right)}{2} \cos n\theta \, d\theta.$$

Proof Replace x by ix and y by iy in (4.35). $\square$

Corollary 4.22 (Corollary to Theorem 4.17)
Let $f(z)$ be the function as defined in Theorem 4.17. Then, $\forall (x, y, n) \in \mathbb{R}^2 \times (\mathbb{N} - 1)$,

$$T_n(-ix, -iy) = (-i)^n (x^n + y^n) \sum_{j=0}^{\infty} (-1)^j \binom{n+2j}{j} a_{n+2j}(xy)^j$$

$$= (-i)^n (x^n + y^n) \left(\binom{n}{0} a_n - \binom{n+2}{1} a_{n+2}(xy) + \binom{n+4}{2} a_{n+4}(xy)^2 \right.$$

$$\left. - \binom{n+6}{3} a_{n+6}(xy)^3 + \cdots \right)$$

$$= \frac{2}{\pi} \int_0^\pi \frac{f\left(-i\left(xe^{i\theta} + ye^{-i\theta}\right)\right) + f\left(-i\left(xe^{-i\theta} + ye^{i\theta}\right)\right)}{2} \cos n\theta \, d\theta.$$

Proof Replace x by $-ix$ and y by $-iy$ in (4.35). $\square$

Corollary 4.23 (Corollary to Theorem 4.17)
Let $f(z)$ be the function as defined in Theorem 4.17. Then, $\forall (x, y, n) \in \mathbb{R}^2 \times (\mathbb{N} - 1)$,

$$T_n(ix, -iy) = i^n \left(x^n + (-y)^n\right) \sum_{j=0}^{\infty} \binom{n+2j}{j} a_{n+2j}(xy)^j$$

$$= i^n \left(x^n + (-y)^n\right) \left(\binom{n}{0} a_n + \binom{n+2}{1} a_{n+2}(xy) + \binom{n+4}{2} a_{n+4}(xy)^2 \right.$$

$$\left. + \binom{n+6}{3} a_{n+6}(xy)^3 + \cdots \right)$$

$$= \frac{2}{\pi} \int_0^\pi \frac{f\left(i\left(xe^{i\theta} - ye^{-i\theta}\right)\right) + f\left(i\left(xe^{-i\theta} - ye^{i\theta}\right)\right)}{2} \cos n\theta \, d\theta.$$

Proof Replace x by ix and y by $-iy$ in (4.35). □

Corollary 4.24 (Corollary to Theorem 4.17)
Let $f(z)$ be the function as defined in Theorem 4.17. Then, $\forall (x, y, n) \in \mathbb{R}^2 \times (\mathbb{N} - 1)$,

$$T_n(-x, -y) = (-1)^n \left(x^n + y^n\right) \sum_{j=0}^{\infty} \binom{n+2j}{j} a_{n+2j}(xy)^j$$

$$= (-1)^n \left(x^n + y^n\right) \left(\binom{n}{0} a_n + \binom{n+2}{1} a_{n+2}(xy) + \binom{n+4}{2} a_{n+4}(xy)^2 \right.$$

$$\left. + \binom{n+6}{3} a_{n+6}(xy)^3 + \cdots \right)$$

$$= \frac{2}{\pi} \int_0^\pi \frac{f\left(-\left(xe^{i\theta} + ye^{-i\theta}\right)\right) + f\left(-\left(xe^{-i\theta} + ye^{i\theta}\right)\right)}{2} \cos n\theta \, d\theta.$$

Proof Replace x by $-x$ and y by $-y$ in (4.35). □

Corollary 4.25 (Corollary to Theorem 4.17)
Let $f(z)$ be the function as defined in Theorem 4.17. Then, $\forall (x, n) \in \mathbb{R} \times (\mathbb{N} - 1)$,

$$T_n(ix, ix) = 2i^n \sum_{j=0}^{\infty} (-1)^j \binom{n+2j}{j} a_{n+2j} x^{n+2j}$$

$$= 2i^n \left(\binom{n}{0} a_n x^n - \binom{n+2}{1} a_{n+2} x^{n+2} + \binom{n+4}{2} a_{n+4} x^{n+4} \right.$$

$$\left. - \binom{n+6}{3} a_{n+6} x^{n+6} + \cdots \right)$$

$$= \frac{2}{\pi} \int_0^\pi f(2ix \cos \theta) \cos n\theta \, d\theta.$$

Proof Replace x by ix and y by ix in (4.35). □

Corollary 4.26 (Corollary to Theorem 4.17)
Let $f(z)$ be the function as defined in Theorem 4.17. Then, $\forall (x, n) \in \mathbb{R} \times (\mathbb{N} - 1)$,

4.2 Special Integrals 2

$$T_n(ix, -ix) = i^n \left(1 + (-1)^n\right) \sum_{j=0}^{\infty} \binom{n+2j}{j} a_{n+2j} x^{n+2j}$$

$$= i^n \left(1 + (-1)^n\right) \left(\binom{n}{0} a_n x^n + \binom{n+2}{1} a_{n+2} x^{n+2} + \binom{n+4}{2} a_{n+4} x^{n+4}\right.$$

$$\left. + \binom{n+6}{3} a_{n+6} x^{n+6} + \cdots \right)$$

$$= \frac{2}{\pi} \int_0^\pi \frac{f(2x \sin \theta) + f(-2x \sin \theta)}{2} \cos n\theta \, d\theta.$$

Proof Replace x by ix and y by $-ix$ in (4.35). □

Corollary 4.27 (Corollary to Theorem 4.17)
Let $f(z)$ be the function as defined in Theorem 4.17. Then, $\forall (x, n) \in \mathbb{R} \times (\mathbb{N} - 1)$,

$$T_n(x, -x) = \left(1 + (-1)^n\right) \sum_{j=0}^{\infty} (-1)^j \binom{n+2j}{j} a_{n+2j} x^{n+2j}$$

$$= \left(1 + (-1)^n\right) \left(\binom{n}{0} a_n x^n - \binom{n+2}{1} a_{n+2} x^{n+2} + \binom{n+4}{2} a_{n+4} x^{n+4}\right.$$

$$\left. - \binom{n+6}{3} a_{n+6} x^{n+6} + \cdots \right)$$

$$= \frac{2}{\pi} \int_0^\pi \frac{f(2ix \sin \theta) + f(-2ix \sin \theta)}{2} \cos n\theta \, d\theta.$$

Proof Replace y by $-x$ in (4.35). □

Now, we will derive an integral representation of any function in terms of its own integral involving some complex variables.

Theorem 4.28 (Another Binomial Series)
Let

$$f(z) = \sum_{j=0}^{\infty} a_j z^j = a_0 + a_1 z + a_2 z^2 + a_3 z^3 + \cdots, \qquad (4.39)$$

be a function that can be expanded by Maclaurin's series such that $f\left(x + ye^{i\theta}\right)$ and $f\left(x + ye^{-i\theta}\right)$ both are convergent for $\theta \in (0, \pi)$. Then, $\forall (x, y) \in \mathbb{C}^2$,

$$f(x) = \frac{1}{\pi} \int_0^\pi \frac{f\left(x + ye^{i\theta}\right) + f\left(x + ye^{-i\theta}\right)}{2} d\theta. \qquad (4.40)$$

Proof Putting $z = x + ye^{i\theta}$ in (4.39), we get

$$f\left(x+ye^{i\theta}\right) = \sum_{j=0}^{\infty} a_j \left(x+ye^{i\theta}\right)^j$$

$$= \sum_{j=0}^{\infty} a_j x^j + \sum_{j=1}^{\infty} a_j \sum_{k=1}^{j} \binom{j}{k} x^{j-k} y^k e^{ik\theta} \qquad (4.41)$$

$$= f(x) + \sum_{j=1}^{\infty} a_j \sum_{k=1}^{j} \binom{j}{k} x^{j-k} y^k e^{ik\theta}.$$

Putting $z = x + ye^{-i\theta}$ in (4.39), we get

$$f\left(x+ye^{-i\theta}\right) = \sum_{j=0}^{\infty} a_j \left(x+ye^{-i\theta}\right)^j$$

$$= \sum_{j=0}^{\infty} a_j x^j + \sum_{j=1}^{\infty} a_j \sum_{k=1}^{j} \binom{j}{k} x^{j-k} y^k e^{-ik\theta} \qquad (4.42)$$

$$= f(x) + \sum_{j=1}^{\infty} a_j \sum_{k=1}^{j} \binom{j}{k} x^{j-k} y^k e^{-ik\theta}.$$

Adding (4.41) and (4.42), and dividing the result by 2, we get

$$\frac{f\left(x+ye^{i\theta}\right)+f\left(x+ye^{-i\theta}\right)}{2} = f(x) + \sum_{j=1}^{\infty} a_j \sum_{k=1}^{j} \binom{j}{k} x^{j-k} y^k \frac{e^{ik\theta}+e^{-ik\theta}}{2}$$

$$= f(x) + \sum_{j=1}^{\infty} a_j \sum_{k=1}^{j} \binom{j}{k} x^{j-k} y^k \cos k\theta.$$

$$(4.43)$$

Dividing (4.43) by π, and integrating the result with respect to θ from 0 to π, we get

$$\frac{1}{\pi}\int_0^{\pi} \frac{f\left(x+ye^{i\theta}\right)+f\left(x+ye^{-i\theta}\right)}{2} d\theta$$

$$= \frac{1}{\pi}\int_0^{\pi} f(x)\, d\theta + \frac{1}{\pi}\int_0^{\pi} \left(\sum_{j=1}^{\infty} a_j \sum_{k=1}^{j} \binom{j}{k} x^{j-k} y^k \cos k\theta\right) d\theta$$

$$= \frac{f(x)}{\pi}\int_0^{\pi} d\theta + \frac{1}{\pi} \sum_{j=1}^{\infty} a_j \sum_{k=1}^{j} \binom{j}{k} x^{j-k} y^k \int_0^{\pi} \cos k\theta\, d\theta$$

$$= f(x) + \frac{1}{\pi} \sum_{j=1}^{\infty} a_j \sum_{k=1}^{j} \binom{j}{k} x^{j-k} y^k \left.\frac{\sin k\theta}{k}\right|_0^{\pi}$$

$$= f(x).$$

$\square$

4.2 Special Integrals 2

Corollary 4.29 (Corollary to Theorem 4.28)
Let $f(z)$ be the function as defined in Theorem 4.28. Then, $\forall x \in \mathbb{R}$,

$$f(x) = \frac{1}{\pi} \int_0^\pi \frac{f\left(2x \cos \frac{\theta}{2} e^{i\theta/2}\right) + f\left(2x \cos \frac{\theta}{2} e^{-i\theta/2}\right)}{2} \, d\theta.$$

Proof Put $y = x$ in (4.40). □

Corollary 4.30 (Corollary to Theorem 4.28)
Let $f(z)$ be the function as defined in Theorem 4.28. Then, $\forall x \in \mathbb{R}$,

$$f(x) = \frac{1}{\pi} \int_0^\pi \frac{f\left(2x \sin \frac{\theta}{2} e^{i(\pi-\theta)/2}\right) + f\left(2x \sin \frac{\theta}{2} e^{-i(\pi-\theta)/2}\right)}{2} \, d\theta.$$

Proof Put $y = -x$ in (4.40). □

Corollary 4.31 (Corollary to Theorem 4.28)
Let $f(z)$ be the function as defined in Theorem 4.28. Then

$$f(1) = \frac{1}{\pi} \int_0^\pi \frac{f\left(2 \cos \frac{\theta}{2} e^{i\theta/2}\right) + f\left(2 \cos \frac{\theta}{2} e^{-i\theta/2}\right)}{2} \, d\theta.$$

Proof Put $x = y = 1$ in (4.40). □

Corollary 4.32 (Corollary to Theorem 4.28)
Let $f(z)$ be the function as defined in Theorem 4.28. Then

$$f(1) = \frac{1}{\pi} \int_0^\pi \frac{f\left(2 \sin \frac{\theta}{2} e^{i(\pi-\theta)/2}\right) + f\left(2 \sin \frac{\theta}{2} e^{-i(\pi-\theta)/2}\right)}{2} \, d\theta.$$

Proof Put $x = 1$ and $y = -1$ in (4.40). □

Corollary 4.33 (Corollary to Theorem 4.28)
Let $f(z)$ be the function as defined in Theorem 4.28. Then

$$f(-1) = \frac{1}{\pi} \int_0^\pi \frac{f\left(-2 \cos \frac{\theta}{2} e^{i\theta/2}\right) + f\left(-2 \cos \frac{\theta}{2} e^{-i\theta/2}\right)}{2} \, d\theta.$$

Proof Put $x = y = -1$ in (4.40). □

Corollary 4.34 (Corollary to Theorem 4.28)
Let $f(z)$ be the function as defined in Theorem 4.28. Then

$$f(-1) = \frac{1}{\pi} \int_0^\pi \frac{f\left(-2\sin\frac{\theta}{2} e^{i(\pi-\theta)/2}\right) + f\left(-2\sin\frac{\theta}{2} e^{-i(\pi-\theta)/2}\right)}{2} d\theta.$$

Proof Put $x = -1$ and $y = 1$ in (4.40). □

Corollary 4.35 (Corollary to Theorem 4.28)
Let $f(z)$ be the function as defined in Theorem 4.28. Then, $\forall y \in \mathbb{R}$,

$$f(1) = \frac{1}{\pi} \int_0^\pi \frac{f\left(1 + ye^{i\theta}\right) + f\left(1 + ye^{-i\theta}\right)}{2} d\theta.$$

Proof Put $x = 1$ in (4.40). □

Corollary 4.36 (Corollary to Theorem 4.28)
Let $f(z)$ be the function as defined in Theorem 4.28. Then, $\forall y \in \mathbb{R}$,

$$f(-1) = \frac{1}{\pi} \int_0^\pi \frac{f\left(-1 + ye^{i\theta}\right) + f\left(-1 + ye^{-i\theta}\right)}{2} d\theta.$$

Proof Put $x = -1$ in (4.40). □

Corollary 4.37 (Corollary to Theorem 4.28)
Let $f(z)$ be the function as defined in Theorem 4.28. Then, $\forall x \in \mathbb{R}$,

$$f(x) = \frac{1}{\pi} \int_0^\pi \frac{f\left(x + e^{i\theta}\right) + f\left(x + e^{-i\theta}\right)}{2} d\theta.$$

Proof Put $y = 1$ in (4.40). □

Corollary 4.38 (Corollary to Theorem 4.28)
Let $f(z)$ be the function as defined in Theorem 4.28. Then, $\forall x \in \mathbb{R}$,

$$f(x) = \frac{1}{\pi} \int_0^\pi \frac{f\left(x - e^{i\theta}\right) + f\left(x - e^{-i\theta}\right)}{2} d\theta.$$

Proof Put $y = -1$ in (4.40). □

Corollary 4.39 (Corollary to Theorem 4.28)
Let $f(z)$ be the function as defined in Theorem 4.28. Then, $\forall y \in \mathbb{R}$,

$$f(0) = \frac{1}{\pi} \int_0^\pi \frac{f\left(ye^{i\theta}\right) + f\left(ye^{-i\theta}\right)}{2} d\theta.$$

Proof Put $x = 0$ in (4.40). □

4.2 Special Integrals 2

Corollary 4.40 (Corollary to Theorem 4.28. This is also Theorem 4.1)
Let $f(z)$ be the function as defined in Theorem 4.28. Then

$$f(0) = \frac{1}{\pi} \int_0^\pi \frac{f\left(e^{i\theta}\right) + f\left(e^{-i\theta}\right)}{2} d\theta.$$

Proof Put $x = 0$ and $y = 1$ in (4.40). □

The following theorem is a generalization of the previous theorem.

Theorem 4.41 (Generalization of Theorem 4.28)
Let

$$f(z) = \sum_{j=0}^{\infty} a_j z^j = a_0 + a_1 z + a_2 z^2 + a_3 z^3 + \cdots$$

be a function that can be expanded by Maclaurin's series such that $f\left(x + ye^{-i\theta}\right)$ and $f\left(x + ye^{i\theta}\right)$ both are convergent for $\theta \in (0, \pi)$. $\forall (x, y, n) \in \mathbb{C}^2 \times \mathbb{N}$, let the function $S_n(x, y)$ be defined as

$$S_n(x, y) = y^n \sum_{j=0}^{\infty} \binom{n+j}{j} a_{n+j} x^j$$

$$= y^n \left(\binom{n}{0} a_n + \binom{n+1}{1} a_{n+1} x + \binom{n+2}{2} a_{n+2} x^2 + \binom{n+3}{3} a_{n+3} x^3 + \cdots \right). \tag{4.44}$$

Then, the function $S_n(x, y)$ is given by

$$S_n(x, y) = \frac{2}{\pi} \int_0^\pi \frac{f\left(x + ye^{i\theta}\right) + f\left(x + ye^{-i\theta}\right)}{2} \cos n\theta \, d\theta. \tag{4.45}$$

Proof Multiplying (4.43) by $(2/\pi) \cos n\theta$, integrating the result with respect to θ from 0 to π, and using (4.44), we get

$$\frac{2}{\pi} \int_0^\pi \frac{f\left(x + ye^{i\theta}\right) + f\left(x + ye^{-i\theta}\right)}{2} \cos n\theta \, d\theta$$

$$= \frac{2}{\pi} \int_0^\pi f(x) \cos n\theta \, d\theta + \frac{2}{\pi} \int_0^\pi \left(\sum_{j=1}^{\infty} a_j \sum_{k=1}^{j} \binom{j}{k} x^{j-k} y^k \cos k\theta \cos n\theta \right) d\theta$$

$$= \frac{2}{\pi} f(x) \int_0^\pi \cos n\theta \, d\theta + \frac{2}{\pi} \sum_{j=n}^\infty a_j \binom{j}{n} x^{j-n} y^n \int_0^\pi \cos^2 n\theta \, d\theta$$

$$+ \frac{2}{\pi} \sum_{j=1}^\infty a_j \sum_{k=1, k \neq n}^j \binom{j}{k} x^{j-k} y^k \int_0^\pi \cos k\theta \cos n\theta \, d\theta$$

$$= \frac{2}{\pi} f(x) \int_0^\pi \cos n\theta \, d\theta + \frac{1}{\pi} \sum_{j=n}^\infty a_j \binom{j}{n} x^{j-n} y^n \int_0^\pi (1 + \cos 2n\theta) \, d\theta$$

$$+ \frac{1}{\pi} \sum_{j=1}^\infty a_j \sum_{k=1, k \neq n}^j \binom{j}{k} x^{j-k} y^k \int_0^\pi (\cos(k+n)\theta + \cos(k-n)\theta) \, d\theta$$

$$= \frac{2}{\pi} f(x) \frac{\sin n\theta}{n} \bigg|_0^\pi + \frac{1}{\pi} \sum_{j=n}^\infty a_j \binom{j}{n} x^{j-n} y^n \left(\theta + \frac{\sin 2n\theta}{2n} \right) \bigg|_0^\pi$$

$$+ \frac{1}{\pi} \sum_{j=1}^\infty a_j \sum_{k=1, k \neq n}^j \binom{j}{k} x^{j-k} y^k \left(\frac{\sin(k+n)\theta}{k+n} + \frac{\sin(k-n)\theta}{k-n} \right) \bigg|_0^\pi$$

$$= y^n \sum_{j=0}^\infty \binom{n+j}{j} a_{n+j} x^j$$

$$= S_n(x, y).$$

$\square$

Corollary 4.42 (Corollary to Theorem 4.41)
Let $f(z)$ be the function as defined in Theorem 4.41. Then, $\forall (x, n) \in \mathbb{R} \times \mathbb{N}$,

$$S_n(x, x) = \sum_{j=0}^\infty \binom{n+j}{j} a_{n+j} x^{n+j}$$

$$= \binom{n}{0} a_n x^n + \binom{n+1}{1} a_{n+1} x^{n+1} + \binom{n+2}{2} a_{n+2} x^{n+2} + \binom{n+3}{3} a_{n+3} x^{n+3} + \cdots$$

$$= \frac{2}{\pi} \int_0^\pi \frac{f\left(2x \cos \frac{\theta}{2} e^{i\theta/2}\right) + f\left(2x \cos \frac{\theta}{2} e^{-i\theta/2}\right)}{2} \cos n\theta \, d\theta.$$

Proof Put $y = x$ in (4.45). $\square$

Corollary 4.43 (Corollary to Theorem 4.41)
Let $f(z)$ be the function as defined in Theorem 4.41. Then, $\forall (x, n) \in \mathbb{R} \times \mathbb{N}$,

4.2 Special Integrals 2

$$S_n(x, -x) = (-1)^n \sum_{j=0}^{\infty} \binom{n+j}{j} a_{n+j} x^{n+j}$$

$$= (-1)^n \left(\binom{n}{0} a_n x^n + \binom{n+1}{1} a_{n+1} x^{n+1} + \binom{n+2}{2} a_{n+2} x^{n+2} \right.$$

$$\left. + \binom{n+3}{3} a_{n+3} x^{n+3} + \cdots \right)$$

$$= \frac{2}{\pi} \int_0^\pi \frac{f\left(2x \sin \frac{\theta}{2} e^{i(\pi-\theta)/2}\right) + f\left(2x \sin \frac{\theta}{2} e^{-i(\pi-\theta)/2}\right)}{2} \cos n\theta \, d\theta.$$

Proof Put $y = -x$ in (4.45). □

Corollary 4.44 (Corollary to Theorem 4.41)
Let $f(z)$ be the function as defined in Theorem 4.41. Then, $\forall n \in \mathbb{N}$,

$$S_n(1, 1) = \sum_{j=0}^{\infty} \binom{n+j}{j} a_{n+j}$$

$$= \binom{n}{0} a_n + \binom{n+1}{1} a_{n+1} + \binom{n+2}{2} a_{n+2} + \binom{n+3}{3} a_{n+3} + \cdots$$

$$= \frac{2}{\pi} \int_0^\pi \frac{f\left(2\cos \frac{\theta}{2} e^{i\theta/2}\right) + f\left(2\cos \frac{\theta}{2} e^{-i\theta/2}\right)}{2} \cos n\theta \, d\theta.$$

Proof Put $x = y = 1$ in (4.45). □

Corollary 4.45 (Corollary to Theorem 4.41)
Let $f(z)$ be the function as defined in Theorem 4.41. Then, $\forall n \in \mathbb{N}$,

$$S_n(1, -1) = (-1)^n \sum_{j=0}^{\infty} \binom{n+j}{j} a_{n+j}$$

$$= (-1)^n \left(\binom{n}{0} a_n + \binom{n+1}{1} a_{n+1} + \binom{n+2}{2} a_{n+2} + \binom{n+3}{3} a_{n+3} + \cdots \right)$$

$$= \frac{2}{\pi} \int_0^\pi \frac{f\left(2\sin \frac{\theta}{2} e^{i(\pi-\theta)/2}\right) + f\left(2\sin \frac{\theta}{2} e^{-i(\pi-\theta)/2}\right)}{2} \cos n\theta \, d\theta.$$

Proof Put $x = 1$ and $y = -1$ in (4.45). □

Corollary 4.46 (Corollary to Theorem 4.41)
Let $f(z)$ be the function as defined in Theorem 4.41. Then, $\forall n \in \mathbb{N}$,

$$S_n(-1,-1) = (-1)^n \sum_{j=0}^{\infty} (-1)^j \binom{n+j}{j} a_{n+j}$$

$$= (-1)^n \left(\binom{n}{0} a_n - \binom{n+1}{1} a_{n+1} + \binom{n+2}{2} a_{n+2} - \binom{n+3}{3} a_{n+3} + \cdots \right)$$

$$= \frac{2}{\pi} \int_0^{\pi} \frac{f\left(-2\cos\frac{\theta}{2} e^{i\theta/2}\right) + f\left(-2\cos\frac{\theta}{2} e^{-i\theta/2}\right)}{2} \cos n\theta \, d\theta.$$

Proof Put $x = y = -1$ in (4.45). □

Corollary 4.47 (Corollary to Theorem 4.41)
Let $f(z)$ be the function as defined in Theorem 4.41. Then, $\forall n \in \mathbb{N}$,

$$S_n(-1, 1) = \sum_{j=0}^{\infty} (-1)^j \binom{n+j}{j} a_{n+j}$$

$$= \binom{n}{0} a_n - \binom{n+1}{1} a_{n+1} + \binom{n+2}{2} a_{n+2} - \binom{n+3}{3} a_{n+3} + \cdots$$

$$= \frac{2}{\pi} \int_0^{\pi} \frac{f\left(-2\sin\frac{\theta}{2} e^{i(\pi-\theta)/2}\right) + f\left(-2\sin\frac{\theta}{2} e^{-i(\pi-\theta)/2}\right)}{2} \cos n\theta \, d\theta.$$

Proof Put $x = -1$ and $y = 1$ in (4.45). □

Corollary 4.48 (Corollary to Theorem 4.41)
Let $f(z)$ be the function as defined in Theorem 4.41. Then, $\forall (y, n) \in \mathbb{R} \times \mathbb{N}$,

$$S_n(1, y) = y^n \sum_{j=0}^{\infty} \binom{n+j}{j} a_{n+j}$$

$$= y^n \left(\binom{n}{0} a_n + \binom{n+1}{1} a_{n+1} + \binom{n+2}{2} a_{n+2} + \binom{n+3}{3} a_{n+3} + \cdots \right)$$

$$= \frac{2}{\pi} \int_0^{\pi} \frac{f\left(1 + ye^{i\theta}\right) + f\left(1 + ye^{-i\theta}\right)}{2} \cos n\theta \, d\theta.$$

Proof Put $x = 1$ in (4.45). □

Corollary 4.49 (Corollary to Theorem 4.41)
Let $f(z)$ be the function as defined in Theorem 4.41. Then, $\forall (y, n) \in \mathbb{R} \times \mathbb{N}$,

4.2 Special Integrals 2

$$S_n(-1, y) = y^n \sum_{j=0}^{\infty}(-1)^j \binom{n+j}{j} a_{n+j}$$

$$= y^n \left(\binom{n}{0}a_n - \binom{n+1}{1}a_{n+1} + \binom{n+2}{2}a_{n+2} - \binom{n+3}{3}a_{n+3} + \cdots \right)$$

$$= \frac{2}{\pi} \int_0^\pi \frac{f\left(-1 + ye^{i\theta}\right) + f\left(-1 + ye^{-i\theta}\right)}{2} \cos n\theta \, d\theta.$$

Proof Put $x = -1$ in (4.45). □

Corollary 4.50 (Corollary to Theorem 4.41)
Let $f(z)$ be the function as defined in Theorem 4.41. Then, $\forall (x, n) \in \mathbb{R} \times \mathbb{N}$,

$$S_n(x, 1) = \sum_{j=0}^{\infty} \binom{n+j}{j} a_{n+j} x^j$$

$$= \binom{n}{0}a_n + \binom{n+1}{1}a_{n+1}x + \binom{n+2}{2}a_{n+2}x^2 + \binom{n+3}{3}a_{n+3}x^3 + \cdots$$

$$= \frac{2}{\pi} \int_0^\pi \frac{f\left(x + e^{i\theta}\right) + f\left(x + e^{-i\theta}\right)}{2} \cos n\theta \, d\theta.$$

Proof Put $y = 1$ in (4.45). □

Corollary 4.51 (Corollary to Theorem 4.41)
Let $f(z)$ be the function as defined in Theorem 4.41. Then, $\forall (x, n) \in \mathbb{R} \times \mathbb{N}$,

$$S_n(x, -1) = (-1)^n \sum_{j=0}^{\infty} \binom{n+j}{j} a_{n+j} x^j$$

$$= (-1)^n \left(\binom{n}{0}a_n + \binom{n+1}{1}a_{n+1}x + \binom{n+2}{2}a_{n+2}x^2 + \binom{n+3}{3}a_{n+3}x^3 + \cdots \right)$$

$$= \frac{2}{\pi} \int_0^\pi \frac{f\left(x - e^{i\theta}\right) + f\left(x - e^{-i\theta}\right)}{2} \cos n\theta \, d\theta.$$

Proof Put $y = -1$ in (4.45). □

Corollary 4.52 (Corollary to Theorem 4.41)
Let $f(z)$ be the function as defined in Theorem 4.41. Then, $\forall (y, n) \in \mathbb{R} \times \mathbb{N}$,

$$S_n(0, y) = y^n a_n = \frac{2}{\pi} \int_0^\pi \frac{f\left(ye^{i\theta}\right) + f\left(ye^{-i\theta}\right)}{2} \cos n\theta \, d\theta.$$

Proof Put $x = 0$ in (4.45). □

Corollary 4.53 (Corollary to Theorem 4.41. This is also Theorem 4.4)
Let $f(z)$ be the function as defined in Theorem 4.41. Then, $\forall n \in \mathbb{N}$,

$$S_n(0, 1) = a_n = \frac{2}{\pi} \int_0^\pi \frac{f\left(e^{i\theta}\right) + f\left(e^{-i\theta}\right)}{2} \cos n\theta \, d\theta.$$

Proof Put $x = 0$ and $y = 1$ in (4.45). □

Problems

[4.1] Prove that
$$\int_0^\pi \cosh(\cos\theta) \cos(\sin\theta) \, d\theta = \pi.$$

[4.2] (Whittaker and Watson, 2016) Prove that $\forall n \in \mathbb{N} - 1$,
$$\int_0^\pi e^{\cos\theta} \cos(n\theta - \sin\theta) \, d\theta = \frac{\pi}{n!}.$$

[4.3] Prove that $\forall n \in -\mathbb{N}$,
$$\int_0^\pi e^{\cos\theta} \cos(n\theta - \sin\theta) \, d\theta = 0.$$

[4.4] Prove that $\forall n \in \mathbb{N} - 1$,
$$\int_0^\pi e^{-\cos\theta} \cos(n\theta + \sin\theta) \, d\theta = (-1)^n \frac{\pi}{n!}.$$

[4.5] Prove that $\forall n \in -\mathbb{N}$,
$$\int_0^\pi e^{-\cos\theta} \cos(n\theta + \sin\theta) \, d\theta = 0.$$

[4.6] Prove that $\forall (x, n) \in \mathbb{R} \times (\mathbb{N} - 1)$,
$$\int_0^\pi e^{x\cos\theta} \cos(n\theta - x\sin\theta) \, d\theta = \frac{\pi}{n!} x^n.$$

[4.7] Prove that $\forall (x, n) \in \mathbb{R} \times (-\mathbb{N})$,

$$\int_0^\pi e^{x \cos \theta} \cos(n\theta - x \sin \theta)\, d\theta = 0.$$

[4.8] Prove that $\forall (x, n) \in \mathbb{R} \times (2\mathbb{N} - 1)$,

$$\int_0^\pi (\cos(n\theta) \sin(x \cos \theta) \cosh(x \sin \theta) + \sin(n\theta) \cos(x \cos \theta) \sinh(x \sin \theta))\, d\theta$$
$$= (-1)^{(n-1)/2} \frac{\pi}{n!} x^n.$$

[4.9] Prove that $\forall (x, n) \in \mathbb{R} \times (-\mathbb{N} \cup (2\mathbb{N} - 2))$,

$$\int_0^\pi (\cos(n\theta) \sin(x \cos \theta) \cosh(x \sin \theta) + \sin(n\theta) \cos(x \cos \theta) \sinh(x \sin \theta))\, d\theta = 0.$$

[4.10] Prove that $\forall (x, n) \in \mathbb{R} \times (2\mathbb{N} - 2)$,

$$\int_0^\pi (\cos(n\theta) \cos(x \cos \theta) \cosh(x \sin \theta) - \sin(n\theta) \sin(x \cos \theta) \sinh(x \sin \theta))\, d\theta$$
$$= (-1)^{n/2} \frac{\pi}{n!} x^n.$$

[4.11] Prove that $\forall (x, n) \in \mathbb{R} \times (-\mathbb{N} \cup (2\mathbb{N} - 1))$,

$$\int_0^\pi (\cos(n\theta) \cos(x \cos \theta) \cosh(x \sin \theta) - \sin(n\theta) \sin(x \cos \theta) \sinh(x \sin \theta))\, d\theta = 0.$$

[4.12] Prove that $\forall (x, n) \in \mathbb{R} \times (2\mathbb{N} - 1)$,

$$\int_0^\pi (\cos(n\theta) \sinh(x \cos \theta) \cos(x \sin \theta) + \sin(n\theta) \cosh(x \cos \theta) \sin(x \sin \theta))\, d\theta = \frac{\pi}{n!} x^n.$$

[4.13] Prove that $\forall (x,n) \in \mathbb{R} \times (-\mathbb{N} \cup (2\mathbb{N} - 2))$,

$$\int_0^\pi (\cos(n\theta) \sinh(x \cos\theta) \cos(x \sin\theta) + \sin(n\theta) \cosh(x \cos\theta) \sin(x \sin\theta)) \, d\theta = 0.$$

[4.14] Prove that $\forall (x,n) \in \mathbb{R} \times (2\mathbb{N} - 2)$,

$$\int_0^\pi (\cos(n\theta) \cosh(x \cos\theta) \cos(x \sin\theta) + \sin(n\theta) \sinh(x \cos\theta) \sin(x \sin\theta)) \, d\theta = \frac{\pi}{n!} x^n.$$

[4.15] Prove that $\forall (x,n) \in \mathbb{R} \times (-\mathbb{N} \cup (2\mathbb{N} - 1))$,

$$\int_0^\pi (\cos(n\theta) \cosh(x \cos\theta) \cos(x \sin\theta) + \sin(n\theta) \sinh(x \cos\theta) \sin(x \sin\theta)) \, d\theta = 0.$$

[4.16] Prove that $\forall (m,n) \in (\mathbb{N} - 1) \times [0..m]$,

$$\int_0^\pi \cos \frac{(2n-m)\theta}{2} \cos^m \frac{\theta}{2} \, d\theta = \frac{\pi}{2^m} \binom{m}{n}.$$

[4.17] Prove that $\forall (m,n) \in (\mathbb{N} - 1) \times (\mathbb{Z} - [0..m])$,

$$\int_0^\pi \cos \frac{(2n-m)\theta}{2} \cos^m \frac{\theta}{2} \, d\theta = 0.$$

[4.18] Prove that $\forall (m,n) \in (\mathbb{N} - 1) \times [0..m]$,

$$\int_0^\pi \cos \frac{(2n-m)\theta + m\pi}{2} \sin^m \frac{\theta}{2} \, d\theta = (-1)^n \frac{\pi}{2^m} \binom{m}{n}.$$

[4.19] Prove that $\forall (m,n) \in (\mathbb{N}-1) \times (\mathbb{Z}-[0..m])$,

$$\int_0^\pi \cos\frac{(2n-m)\theta + m\pi}{2} \sin^m \frac{\theta}{2}\, d\theta = 0.$$

[4.20] Prove that $\forall (x,m,n) \in [-1,1] \times (\mathbb{N}-1) \times [0..m]$,

$$\int_0^\pi (1+x^2+2x\cos\theta)^{m/2} \cos\left(n\theta - m\tan^{-1}\left(\frac{x\sin\theta}{1+x\cos\theta}\right)\right) d\theta = \pi \binom{m}{n} x^n.$$

[4.21] Prove that $\forall (x,m,n) \in [-1,1] \times (\mathbb{N}-1) \times (\mathbb{Z}-[0..m])$,

$$\int_0^\pi (1+x^2+2x\cos\theta)^{m/2} \cos\left(n\theta - m\tan^{-1}\left(\frac{x\sin\theta}{1+x\cos\theta}\right)\right) d\theta = 0.$$

[4.22] (Watson, 2011) Prove that

$$\int_0^\pi e^{2\cos\theta}\, d\theta = \pi I_0(2)$$
$$= \pi \sum_{j=0}^\infty \frac{1}{(j!)^2}$$
$$= \pi \left(\frac{1}{(0!)^2} + \frac{1}{(1!)^2} + \frac{1}{(2!)^2} + \frac{1}{(3!)^2} + \cdots\right).$$

[4.23] Prove that $\forall n \in \mathbb{N}-1$,

$$\int_0^\pi e^{2\cos\theta} \cos n\theta\, d\theta = \pi I_n(2)$$
$$= \pi \sum_{j=0}^\infty \frac{1}{j!(n+j)!}$$
$$= \pi \left(\frac{1}{0!n!} + \frac{1}{1!(n+1)!} + \frac{1}{2!(n+2)!} + \frac{1}{3!(n+3)!} + \cdots\right).$$

[4.24] (Watson, 2011) Prove that $\forall x \in \mathbb{R}$,

$$\int_0^\pi e^{2x\cos\theta}\, d\theta = \pi I_0(2x)$$

$$= \pi \sum_{j=0}^\infty \frac{x^{2j}}{(j!)^2}$$

$$= \pi \left(\frac{1}{(0!)^2} + \frac{x^2}{(1!)^2} + \frac{x^4}{(2!)^2} + \frac{x^6}{(3!)^2} + \cdots \right).$$

[4.25] Prove that $\forall (x, n) \in \mathbb{R} \times (\mathbb{N} - 1)$,

$$\int_0^\pi e^{2x\cos\theta} \cos n\theta\, d\theta = \pi I_n(2x)$$

$$= \pi x^n \sum_{j=0}^\infty \frac{x^{2j}}{j!(n+j)!}$$

$$= \pi x^n \left(\frac{1}{0!n!} + \frac{x^2}{1!(n+1)!} + \frac{x^4}{2!(n+2)!} + \frac{x^6}{3!(n+3)!} + \cdots \right).$$

[4.26] $\forall(\theta, n, i, a_i) \in \mathbb{R} \times (\mathbb{N} - 1) \times [0..n] \times \mathbb{R}$, let

$$C(\theta) = \sum_{j=0}^n a_j \cos j\theta \tag{4.46}$$
$$= a_0 + a_1 \cos\theta + a_2 \cos 2\theta + \cdots + a_n \cos n\theta,$$

and

$$S(\theta) = \sum_{j=1}^n a_j \sin j\theta \tag{4.47}$$
$$= a_1 \sin\theta + a_2 \sin 2\theta + \cdots + a_n \sin n\theta.$$

Prove that

$$\int_0^\pi \left(C^2(\theta) + S^2(\theta) \right) d\theta = \pi \sum_{j=0}^n a_j^2$$
$$= \pi \left(a_0^2 + a_1^2 + a_2^2 + \cdots + a_n^2 \right).$$

[4.27] Let $C(\theta)$ and $S(\theta)$ be as defined in (4.46) and (4.47) respectively. Prove that $\forall m \in \mathbb{N} - 1$,

$$\int_0^\pi \left(C^2(\theta) + S^2(\theta)\right) \cos m\theta \, d\theta = \pi \sum_{j=0}^{n-m} a_j a_{j+m}$$

$$= \pi \left(a_0 a_m + a_1 a_{1+m} + a_2 a_{2+m} + \cdots + a_{n-m} a_n\right).$$

[4.28] Prove that $\forall n \in \mathbb{N} - 1$,

$$\int_0^\pi \sin^{2n} \frac{\theta}{2} \, d\theta = \frac{\pi}{2^{2n}} \sum_{j=0}^n \binom{n}{j}^2$$

$$= \frac{\pi}{2^{2n}} \left(\binom{n}{0}^2 + \binom{n}{1}^2 + \binom{n}{2}^2 + \cdots + \binom{n}{n}^2\right).$$

[4.29] Prove that $\forall (n, m) \in \mathbb{N} \times [0..n]$,

$$\int_0^\pi \sin^{2n} \frac{\theta}{2} \cos m\theta \, d\theta = (-1)^m \frac{\pi}{2^{2n}} \sum_{j=0}^{n-m} \binom{n}{j}\binom{n}{m+j}$$

$$= (-1)^m \frac{\pi}{2^{2n}} \left(\binom{n}{0}\binom{n}{m} + \binom{n}{1}\binom{n}{m+1} + \binom{n}{2}\binom{n}{m+2}\right.$$

$$\left. + \cdots + \binom{n}{n-m}\binom{n}{n}\right).$$

[4.30] Prove that $\forall (n, m) \in \mathbb{N} \times ((\mathbb{N} - 1) - [0..n])$,

$$\int_0^\pi \sin^{2n} \frac{\theta}{2} \cos m\theta \, d\theta = 0.$$

[4.31] Prove that $\forall x \in \mathbb{R}$,

$$\int_0^\pi \left(\sin^2(x\cos\theta)\cosh^2(x\sin\theta) + \cos^2(x\cos\theta)\sinh^2(x\sin\theta)\right) d\theta$$

$$= \pi \sum_{j=0}^\infty \frac{x^{4j+2}}{((2j+1)!)^2}$$

$$= \pi \left(\frac{x^2}{(1!)^2} + \frac{x^6}{(3!)^2} + \frac{x^{10}}{(5!)^2} + \frac{x^{14}}{(7!)^2} + \cdots\right).$$

[4.32] Prove that $\forall x \in \mathbb{R}$,

$$\int_0^\pi \left(\sinh^2(x\cos\theta)\cos^2(x\sin\theta) + \cosh^2(x\cos\theta)\sin^2(x\sin\theta)\right) d\theta$$

$$= \pi \sum_{j=0}^\infty \frac{x^{4j+2}}{((2j+1)!)^2}$$

$$= \pi \left(\frac{x^2}{(1!)^2} + \frac{x^6}{(3!)^2} + \frac{x^{10}}{(5!)^2} + \frac{x^{14}}{(7!)^2} + \cdots\right).$$

[4.33] Prove that $\forall (x, n) \in \mathbb{R} \times (2\mathbb{N} - 2)$,

$$\int_0^\pi \left(\sin^2(x\cos\theta)\cosh^2(x\sin\theta) + \cos^2(x\cos\theta)\sinh^2(x\sin\theta)\right)\cos n\theta \, d\theta$$

$$= (-1)^{n/2}\pi x^n \sum_{j=0}^\infty \frac{x^{4j+2}}{(2j+1)!(n+2j+1)!}$$

$$= (-1)^{n/2}\pi x^n \left(\frac{x^2}{1!(n+1)!} + \frac{x^6}{3!(n+3)!} + \frac{x^{10}}{5!(n+5)!} + \frac{x^{14}}{7!(n+7)!} + \cdots\right).$$

[4.34] Prove that $\forall (x, n) \in \mathbb{R} \times (2\mathbb{Z} + 1)$,

$$\int_0^\pi \left(\sin^2(x\cos\theta)\cosh^2(x\sin\theta) + \cos^2(x\cos\theta)\sinh^2(x\sin\theta)\right)\cos n\theta \, d\theta$$

$$= 0.$$

[4.35] Prove that $\forall (x, n) \in \mathbb{R} \times (2\mathbb{N} - 2)$,

$$\int_0^\pi \left(\sinh^2(x\cos\theta)\cos^2(x\sin\theta) + \cosh^2(x\cos\theta)\sin^2(x\sin\theta) \right) \cos n\theta \, d\theta$$

$$= \pi x^n \sum_{j=0}^\infty \frac{x^{4j+2}}{(2j+1)!(n+2j+1)!}$$

$$= \pi x^n \left(\frac{x^2}{1!(n+1)!} + \frac{x^6}{3!(n+3)!} + \frac{x^{10}}{5!(n+5)!} + \frac{x^{14}}{7!(n+7)!} + \cdots \right).$$

[4.36] Prove that $\forall (x, n) \in \mathbb{R} \times (2\mathbb{Z} + 1)$,

$$\int_0^\pi \left(\sinh^2(x\cos\theta)\cos^2(x\sin\theta) + \cosh^2(x\cos\theta)\sin^2(x\sin\theta) \right) \cos n\theta \, d\theta = 0.$$

[4.37] Prove that $\forall x \in \mathbb{R}$,

$$\int_0^\pi \left(\cos^2(x\cos\theta)\cosh^2(x\sin\theta) + \sin^2(x\cos\theta)\sinh^2(x\sin\theta) \right) d\theta$$

$$= \pi \sum_{j=0}^\infty \frac{x^{4j}}{((2j)!)^2} = \pi \left(\frac{1}{(0!)^2} + \frac{x^4}{(2!)^2} + \frac{x^8}{(4!)^2} + \frac{x^{12}}{(6!)^2} + \cdots \right).$$

[4.38] Prove that $\forall x \in \mathbb{R}$,

$$\int_0^\pi \left(\cosh^2(x\cos\theta)\cos^2(x\sin\theta) + \sinh^2(x\cos\theta)\sin^2(x\sin\theta) \right) d\theta$$

$$= \pi \sum_{j=0}^\infty \frac{x^{4j}}{((2j)!)^2} = \pi \left(\frac{1}{(0!)^2} + \frac{x^4}{(2!)^2} + \frac{x^8}{(4!)^2} + \frac{x^{12}}{(6!)^2} + \cdots \right).$$

[4.39] Prove that $\forall (x, n) \in \mathbb{R} \times (2\mathbb{N} - 2)$,

$$\int_0^\pi \left(\cos^2(x \cos\theta) \cosh^2(x \sin\theta) + \sin^2(x \cos\theta) \sinh^2(x \sin\theta) \right) \cos n\theta \, d\theta$$

$$= (-1)^{n/2} \pi x^n \sum_{j=0}^\infty \frac{x^{4j}}{(2j)!(n+2j)!}$$

$$= (-1)^{n/2} \pi x^n \left(\frac{1}{0!n!} + \frac{x^4}{2!(n+2)!} + \frac{x^8}{4!(n+4)!} + \frac{x^{12}}{6!(n+6)!} + \cdots \right).$$

[4.40] Prove that $\forall (x, n) \in \mathbb{R} \times (2\mathbb{Z} + 1)$,

$$\int_0^\pi \left(\cos^2(x \cos\theta) \cosh^2(x \sin\theta) + \sin^2(x \cos\theta) \sinh^2(x \sin\theta) \right) \cos n\theta \, d\theta$$

$$= 0.$$

[4.41] Prove that $\forall (x, n) \in \mathbb{R} \times (2\mathbb{N} - 2)$,

$$\int_0^\pi \left(\cosh^2(x \cos\theta) \cos^2(x \sin\theta) + \sinh^2(x \cos\theta) \sin^2(x \sin\theta) \right) \cos n\theta \, d\theta$$

$$= \pi x^n \sum_{j=0}^\infty \frac{x^{4j}}{(2j)!(n+2j)!}$$

$$= \pi x^n \left(\frac{1}{0!n!} + \frac{x^4}{2!(n+2)!} + \frac{x^8}{4!(n+4)!} + \frac{x^{12}}{6!(n+6)!} + \cdots \right).$$

[4.42] Prove that $\forall (x, n) \in \mathbb{R} \times (2\mathbb{Z} + 1)$,

$$\int_0^\pi \left(\cosh^2(x \cos\theta) \cos^2(x \sin\theta) + \sinh^2(x \cos\theta) \sin^2(x \sin\theta) \right) \cos n\theta \, d\theta$$

$$= 0.$$

[4.43] Prove that $\forall n \in \mathbb{N}$,

$$\int_0^\pi \left(\cos n\theta \log\left(2 \cos \frac{\theta}{2}\right) + \frac{\theta}{2} \sin n\theta \right) d\theta = (-1)^{n+1} \frac{\pi}{n}.$$

[4.44] Prove that $\forall n \in -\mathbb{N} \cup \{0\}$,

$$\int_0^\pi \left(\cos n\theta \log\left(2\cos\frac{\theta}{2}\right) + \frac{\theta}{2}\sin n\theta\right) d\theta = 0.$$

[4.45] Prove that $\forall (x, n) \in (\mathbb{R} - (-1, 1)) \times \mathbb{N}$,

$$\int_0^\pi \left(\cos n\theta \log\sqrt{1 + \frac{1}{x^2} + \frac{2}{x}\cos\theta} + \sin n\theta \tan^{-1}\left(\frac{\sin\theta}{x + \cos\theta}\right)\right) d\theta$$
$$= (-1)^{n+1}\frac{\pi}{nx^n}.$$

[4.46] Prove that $\forall (x, n) \in (\mathbb{R} - (-1, 1)) \times (-\mathbb{N} \cup \{0\})$,

$$\int_0^\pi \left(\cos n\theta \log\sqrt{1 + \frac{1}{x^2} + \frac{2}{x}\cos\theta} + \sin n\theta \tan^{-1}\left(\frac{\sin\theta}{x + \cos\theta}\right)\right) d\theta = 0.$$

[4.47] Prove that $\forall (y, n) \in \mathbb{R} \times (\mathbb{N} - 1)$,

$$\int_0^\pi \cos\frac{(2n-y)\theta}{2} \cos^y \frac{\theta}{2} d\theta = \frac{\pi}{2^y n!}\prod_{j=0}^{n-1}(y-j)$$
$$= \frac{\pi y(y-1)(y-2)\ldots(y-n+1)}{2^y n!}.$$

[4.48] Prove that $\forall (y, n) \in \mathbb{R} \times (-\mathbb{N})$,

$$\int_0^\pi \cos\frac{(2n-y)\theta}{2} \cos^y \frac{\theta}{2} d\theta = 0.$$

[4.49] Prove that $\forall (x, y, n) \in (\mathbb{R} - [-1, 1]) \times \mathbb{R} \times (\mathbb{N} - 1)$,

$$\int_0^\pi \left(1 + \frac{1}{x^2} + \frac{2}{x}\cos\theta\right)^{y/2} \cos\left(n\theta - y\tan^{-1}\left(\frac{\sin\theta}{x + \cos\theta}\right)\right) d\theta$$
$$= \frac{\pi}{n!x^n}\prod_{j=0}^{n-1}(y-j) = \frac{\pi y(y-1)(y-2)\ldots(y-n+1)}{n!x^n}.$$

[4.50] Prove that $\forall (x, y, n) \in (\mathbb{R} - [-1, 1]) \times \mathbb{R} \times (-\mathbb{N})$,

$$\int_0^\pi \left(1 + \frac{1}{x^2} + \frac{2}{x}\cos\theta\right)^{y/2} \cos\left(n\theta - y\tan^{-1}\left(\frac{\sin\theta}{x + \cos\theta}\right)\right) d\theta = 0.$$

[4.51] Prove that

$$\int_0^\pi \log^2\left(2\cos\frac{\theta}{2}\right) d\theta = \int_0^\pi \log^2\left(2\sin\frac{\theta}{2}\right) d\theta = \frac{\pi^3}{12}.$$

[4.52] Prove that $\forall n \in 2\mathbb{N} - 2$,

$$\int_0^\pi \left(\log^2\left(2\cos\frac{\theta}{2}\right) + \frac{\theta^2}{4}\right) \cos n\theta \, d\theta = \pi \sum_{j=1}^\infty \frac{1}{j(n+j)}$$
$$= \pi \left(\frac{1}{1(n+1)} + \frac{1}{2(n+2)} + \frac{1}{3(n+3)} + \frac{1}{4(n+4)} + \cdots\right).$$

[4.53] Prove that $\forall n \in 2\mathbb{N} - 1$,

$$\int_0^\pi \left(\log^2\left(2\cos\frac{\theta}{2}\right) + \frac{\theta^2}{4}\right) \cos n\theta \, d\theta = -\pi \sum_{j=1}^\infty \frac{1}{j(n+j)}$$
$$= -\pi \left(\frac{1}{1(n+1)} + \frac{1}{2(n+2)} + \frac{1}{3(n+3)} + \frac{1}{4(n+4)} + \cdots\right).$$

[4.54] Prove that $\forall x \in [1, \infty)$,

$$\int_0^\pi \left(\log^2\sqrt{1 + \frac{1}{x^2} + \frac{2}{x}\cos\theta} + \left(\tan^{-1}\left(\frac{\sin\theta}{x + \cos\theta}\right)\right)^2\right) d\theta = \pi \sum_{j=1}^\infty \frac{1}{j^2 x^{2j}}$$
$$= \pi \left(\frac{1}{x^2} + \frac{1}{2^2 x^4} + \frac{1}{3^2 x^6} + \frac{1}{4^2 x^8} + \cdots\right).$$

[4.55] Prove that $\forall (x, n) \in [1, \infty) \times (2\mathbb{N} - 2)$,

$$\int_0^\pi \left(\log^2 \sqrt{1 + \frac{1}{x^2} + \frac{2}{x} \cos\theta} + \left(\tan^{-1}\left(\frac{\sin\theta}{x + \cos\theta} \right) \right)^2 \right) \cos n\theta \, d\theta$$

$$= \frac{\pi}{x^n} \sum_{j=1}^\infty \frac{1}{j(n+j)x^{2j}}$$

$$= \frac{\pi}{x^n} \left(\frac{1}{1(n+1)x^2} + \frac{1}{2(n+2)x^4} + \frac{1}{3(n+3)x^6} + \frac{1}{4(n+4)x^8} + \cdots \right).$$

[4.56] Prove that $\forall (x, n) \in [1, \infty) \times (2\mathbb{N} - 1)$,

$$\int_0^\pi \left(\log^2 \sqrt{1 + \frac{1}{x^2} + \frac{2}{x} \cos\theta} + \left(\tan^{-1}\left(\frac{\sin\theta}{x + \cos\theta} \right) \right)^2 \right) \cos n\theta \, d\theta$$

$$= -\frac{\pi}{x^n} \sum_{j=1}^\infty \frac{1}{j(n+j)x^{2j}}$$

$$= -\frac{\pi}{x^n} \left(\frac{1}{1(n+1)x^2} + \frac{1}{2(n+2)x^4} + \frac{1}{3(n+3)x^6} + \frac{1}{4(n+4)x^8} + \cdots \right).$$

[4.57] Prove that $\forall (x, y) \in ((1, \infty) \times \mathbb{R}) \cup (\mathbb{R} \times \mathbb{N})$,

$$\int_0^\pi \left(1 + \frac{1}{x^2} + \frac{2}{x} \cos\theta \right)^y d\theta = \pi \sum_{j=0}^\infty \left(\frac{\prod_{k=0}^{j-1}(y-k)}{j! x^j} \right)^2$$

$$= \pi \left((1)^2 + \left(\frac{y}{x} \right)^2 + \left(\frac{y(y-1)}{2! x^2} \right)^2 + \left(\frac{y(y-1)(y-2)}{3! x^3} \right)^2 + \cdots \right).$$

[4.58] Prove that $\forall (x, y, n) \in (1, \infty) \times \mathbb{R} \times (\mathbb{N} - 1)$,

$$\int_0^\pi \left(1 + \frac{1}{x^2} + \frac{2}{x}\cos\theta\right)^y \cos n\theta\, d\theta = \pi \sum_{j=0}^\infty \frac{\prod_{k=0}^{j-1}(y-k)}{j!\, x^j} \cdot \frac{\prod_{k=0}^{n+j-1}(y-k)}{(n+j)!\, x^{n+j}}$$

$$= \pi \left(1\frac{y(y-1)(y-2)\ldots(y-n+1)}{n!\, x^n} + \frac{y}{x} \cdot \frac{y(y-1)(y-2)\ldots(y-n)}{(n+1)!\, x^{n+1}}\right.$$

$$+ \frac{y(y-1)}{2!\, x^2} \cdot \frac{y(y-1)(y-2)\ldots(y-n-1)}{(n+2)!\, x^{n+2}}$$

$$\left. + \frac{y(y-1)(y-2)}{3!\, x^3} \cdot \frac{y(y-1)(y-2)\ldots(y-n-2)}{(n+3)!\, x^{n+3}} + \cdots \right).$$

[4.59] Prove that $\forall (x, m, n) \in (\mathbb{R} - \{0\}) \times (2\mathbb{N} - 2) \times ((2\mathbb{N} - 2) \cap [0\mathinner{.\,.} m]) \cup (\mathbb{R} - \{0\}) \times (2\mathbb{N} - 1) \times ((2\mathbb{N} - 1) \cap [0\mathinner{.\,.} m])$,

$$\int_0^{\pi/x} \cos(nx\theta)\cos^m(x\theta)\, d\theta = \frac{\pi}{2^m x}\binom{m}{(m-n)/2}.$$

[4.60] Prove that $\forall (x, m, n) \in (\mathbb{R} - \{0\}) \times (2\mathbb{N} - 2) \times (2\mathbb{N} - 1) \cup (\mathbb{R} - \{0\}) \times (2\mathbb{N} - 1) \times (2\mathbb{N} - 2)$,

$$\int_0^{\pi/x} \cos(nx\theta)\cos^m(x\theta)\, d\theta = 0.$$

[4.61] Prove that $\forall (x, n) \in \left[-\frac{1}{2}, \frac{1}{2}\right] \times (2\mathbb{N} - 1)$,

$$\int_0^\pi \sin^{-1}(2x\cos\theta)\cos n\theta\, d\theta$$

$$= \pi \sum_{j=0}^\infty \binom{n+2j}{j}\binom{n+2j-1}{(n+2j-1)/2}\frac{x^{n+2j}}{(n+2j)2^{n+2j-1}}$$

$$= \pi \left(\frac{\binom{n}{0}\binom{n-1}{(n-1)/2}}{n \cdot 2^{n-1}} x^n + \frac{\binom{n+2}{1}\binom{n+1}{(n+1)/2}}{(n+2)2^{n+1}} x^{n+2} \right.$$

$$\left. + \frac{\binom{n+4}{2}\binom{n+3}{(n+3)/2}}{(n+4)2^{n+3}} x^{n+4} + \frac{\binom{n+6}{3}\binom{n+5}{(n+5)/2}}{(n+6)2^{n+5}} x^{n+6} + \cdots \right).$$

[4.62] Prove that $\forall (x, n) \in \left[-\frac{1}{2}, \frac{1}{2}\right] \times 2\mathbb{Z}$,

$$\int_0^\pi \sin^{-1}(2x \cos \theta) \cos n\theta \, d\theta = 0.$$

[4.63] Prove that $\forall (x, n) \in \left[-\frac{1}{2}, \frac{1}{2}\right] \times (2\mathbb{N} - 1)$,

$$\int_0^\pi \sinh^{-1}(2x \sin \theta) \sin n\theta \, d\theta$$

$$= \pi \sum_{j=0}^\infty (-1)^j \binom{n+2j}{j} \binom{n+2j-1}{(n+2j-1)/2} \frac{x^{n+2j}}{(n+2j)2^{n+2j-1}}$$

$$= \pi \left(\frac{\binom{n}{0}\binom{n-1}{(n-1)/2}}{n \cdot 2^{n-1}} x^n - \frac{\binom{n+2}{1}\binom{n+1}{(n+1)/2}}{(n+2)2^{n+1}} x^{n+2} \right.$$

$$\left. + \frac{\binom{n+4}{2}\binom{n+3}{(n+3)/2}}{(n+4)2^{n+3}} x^{n+4} - \frac{\binom{n+6}{3}\binom{n+5}{(n+5)/2}}{(n+6)2^{n+5}} x^{n+6} + \cdots \right).$$

[4.64] Prove that $\forall (x, n) \in \mathbb{R} \times 2\mathbb{Z}$,

$$\int_0^\pi \sinh^{-1}(2x \sin \theta) \sin n\theta \, d\theta = 0.$$

[4.65] Prove that $\forall (x,n) \in \left[-\frac{1}{2}, \frac{1}{2}\right] \times (2\mathbb{N} - 1)$,

$$\int_0^\pi \sinh^{-1}(2x\cos\theta)\cos n\theta\, d\theta$$

$$= (-1)^{(n-1)/2}\pi \sum_{j=0}^\infty (-1)^j \binom{n+2j}{j}\binom{n+2j-1}{(n+2j-1)/2}\frac{x^{n+2j}}{(n+2j)2^{n+2j-1}}$$

$$= (-1)^{(n-1)/2}\pi \left(\frac{\binom{n}{0}\binom{n-1}{(n-1)/2}}{n\cdot 2^{n-1}} x^n - \frac{\binom{n+2}{1}\binom{n+1}{(n+1)/2}}{(n+2)2^{n+1}} x^{n+2} \right.$$

$$\left. + \frac{\binom{n+4}{2}\binom{n+3}{(n+3)/2}}{(n+4)2^{n+3}} x^{n+4} - \frac{\binom{n+6}{3}\binom{n+5}{(n+5)/2}}{(n+6)2^{n+5}} x^{n+6} + \cdots \right).$$

[4.66] Prove that $\forall (x, n) \in \mathbb{R} \times 2\mathbb{Z}$,

$$\int_0^\pi \sinh^{-1}(2x\cos\theta)\cos n\theta\, d\theta = 0.$$

[4.67] Prove that $\forall (x,n) \in \left[-\frac{1}{2}, \frac{1}{2}\right] \times (2\mathbb{N} - 1)$,

$$\int_0^\pi \sin^{-1}(2x\sin\theta)\sin n\theta\, d\theta$$

$$= (-1)^{(n-1)/2}\pi \sum_{j=0}^\infty \binom{n+2j}{j}\binom{n+2j-1}{(n+2j-1)/2}\frac{x^{n+2j}}{(n+2j)2^{n+2j-1}}$$

$$= (-1)^{(n-1)/2}\pi \left(\frac{\binom{n}{0}\binom{n-1}{(n-1)/2}}{n\cdot 2^{n-1}} x^n + \frac{\binom{n+2}{1}\binom{n+1}{(n+1)/2}}{(n+2)2^{n+1}} x^{n+2} \right.$$

$$\left. + \frac{\binom{n+4}{2}\binom{n+3}{(n+3)/2}}{(n+4)2^{n+3}} x^{n+4} + \frac{\binom{n+6}{3}\binom{n+5}{(n+5)/2}}{(n+6)2^{n+5}} x^{n+6} + \cdots \right).$$

Problems

[4.68] Prove that $\forall (x,n) \in \left[-\frac{1}{2}, \frac{1}{2}\right] \times 2\mathbb{Z}$,

$$\int_0^\pi \sin^{-1}(2x\sin\theta) \sin n\theta \, d\theta = 0.$$

[4.69] Prove that $\forall (x,n) \in \left[-\frac{1}{2}, \frac{1}{2}\right] \times (2\mathbb{N} - 1)$,

$$\int_0^\pi \tan^{-1}(2x\cos\theta) \cos n\theta \, d\theta = (-1)^{(n-1)/2} \pi \sum_{j=0}^\infty (-1)^j \binom{n+2j}{j} \frac{x^{n+2j}}{n+2j}$$

$$= (-1)^{(n-1)/2} \pi \left(\frac{\binom{n}{0}}{n} x^n - \frac{\binom{n+2}{1}}{n+2} x^{n+2} \right.$$

$$\left. + \frac{\binom{n+4}{2}}{n+4} x^{n+4} - \frac{\binom{n+6}{3}}{n+6} x^{n+6} + \cdots \right).$$

[4.70] Prove that $\forall (x,n) \in \mathbb{R} \times 2\mathbb{Z}$,

$$\int_0^\pi \tan^{-1}(2x\cos\theta) \cos n\theta \, d\theta = 0.$$

[4.71] Prove that $\forall (x,n) \in \left[-\frac{1}{2}, \frac{1}{2}\right] \times (2\mathbb{N} - 1)$,

$$\int_0^\pi \tanh^{-1}(2x\sin\theta) \sin n\theta \, d\theta = (-1)^{(n-1)/2} \pi \sum_{j=0}^\infty \binom{n+2j}{j} \frac{x^{n+2j}}{n+2j}$$

$$= (-1)^{(n-1)/2} \pi \left(\frac{\binom{n}{0}}{n} x^n + \frac{\binom{n+2}{1}}{n+2} x^{n+2} \right.$$

$$\left. + \frac{\binom{n+4}{2}}{n+4} x^{n+4} + \frac{\binom{n+6}{3}}{n+6} x^{n+6} + \cdots \right).$$

[4.72] Prove that $\forall (x, n) \in \mathbb{R} \times 2\mathbb{Z}$,

$$\int_0^\pi \tanh^{-1}(2x \sin \theta) \sin n\theta \, d\theta = 0.$$

[4.73] Prove that $\forall (x, n) \in \left[-\dfrac{1}{2}, \dfrac{1}{2}\right] \times (2\mathbb{N} - 1)$,

$$\int_0^\pi \tanh^{-1}(2x \cos \theta) \cos n\theta \, d\theta = \pi \sum_{j=0}^\infty \binom{n+2j}{j} \frac{x^{n+2j}}{n+2j}$$

$$= \pi \left(\frac{\binom{n}{0}}{n} x^n + \frac{\binom{n+2}{1}}{n+2} x^{n+2} + \frac{\binom{n+4}{2}}{n+4} x^{n+4} + \frac{\binom{n+6}{3}}{n+6} x^{n+6} + \cdots \right).$$

[4.74] Prove that $\forall (x, n) \in \mathbb{R} \times 2\mathbb{Z}$,

$$\int_0^\pi \tanh^{-1}(2x \cos \theta) \cos n\theta \, d\theta = 0.$$

[4.75] Prove that $\forall (x, n) \in \left[-\dfrac{1}{2}, \dfrac{1}{2}\right] \times (2\mathbb{N} - 1)$,

$$\int_0^\pi \tan^{-1}(2x \sin \theta) \sin n\theta \, d\theta = \pi \sum_{j=0}^\infty (-1)^j \binom{n+2j}{j} \frac{x^{n+2j}}{n+2j}$$

$$= \pi \left(\frac{\binom{n}{0}}{n} x^n - \frac{\binom{n+2}{1}}{n+2} x^{n+2} + \frac{\binom{n+4}{2}}{n+4} x^{n+4} - \frac{\binom{n+6}{3}}{n+6} x^{n+6} + \cdots \right).$$

[4.76] Prove that $\forall (x, n) \in \mathbb{R} \times 2\mathbb{Z}$,

$$\int_0^\pi \tan^{-1}(2x \sin \theta) \sin n\theta \, d\theta = 0.$$

[4.77] (Watson, 2011) Prove that $\forall (x, n) \in \mathbb{R} \times (2\mathbb{N} - 1)$,

$$\int_0^\pi \sin(2x\cos\theta)\cos n\theta\, d\theta = (-1)^{(n-1)/2}\pi J_n(2x)$$

$$= (-1)^{(n-1)/2}\pi x^n \sum_{j=0}^\infty (-1)^j \frac{x^{2j}}{j!(n+j)!}$$

$$= (-1)^{(n-1)/2}\pi x^n \left(\frac{1}{0!n!} - \frac{x^2}{1!(n+1)!} + \frac{x^4}{2!(n+2)!} - \frac{x^6}{3!(n+3)!} + \cdots\right).$$

[4.78] Prove that $\forall (x, n) \in \mathbb{R} \times 2\mathbb{Z}$,

$$\int_0^\pi \sin(2x\cos\theta)\cos n\theta\, d\theta = 0.$$

[4.79] Prove that $\forall (x, n) \in \mathbb{R} \times (2\mathbb{N} - 1)$,

$$\int_0^\pi \sinh(2x\sin\theta)\sin n\theta\, d\theta = (-1)^{(n-1)/2}\pi I_n(2x)$$

$$= (-1)^{(n-1)/2}\pi x^n \sum_{j=0}^\infty \frac{x^{2j}}{j!(n+j)!}$$

$$= (-1)^{(n-1)/2}\pi x^n \left(\frac{1}{0!n!} + \frac{x^2}{1!(n+1)!} + \frac{x^4}{2!(n+2)!} + \frac{x^6}{3!(n+3)!} + \cdots\right).$$

[4.80] Prove that $\forall (x, n) \in \mathbb{R} \times 2\mathbb{Z}$,

$$\int_0^\pi \sinh(2x\sin\theta)\sin n\theta\, d\theta = 0.$$

[4.81] Prove that $\forall (x, n) \in \mathbb{R} \times (2\mathbb{N} - 1)$,

$$\int_0^\pi \sinh(2x\cos\theta)\cos n\theta\, d\theta = \pi I_n(2x)$$

$$= \pi x^n \sum_{j=0}^\infty \frac{x^{2j}}{j!(n+j)!}$$

$$= \pi x^n \left(\frac{1}{0!n!} + \frac{x^2}{1!(n+1)!} + \frac{x^4}{2!(n+2)!} + \frac{x^6}{3!(n+3)!} + \cdots\right).$$

[4.82] Prove that $\forall (x, n) \in \mathbb{R} \times 2\mathbb{Z}$,

$$\int_0^\pi \sinh(2x \cos\theta) \cos n\theta \, d\theta = 0.$$

[4.83] (Watson, 2011) Prove that $\forall (x, n) \in \mathbb{R} \times (2\mathbb{N} - 1)$,

$$\int_0^\pi \sin(2x \sin\theta) \sin n\theta \, d\theta = \pi J_n(2x)$$

$$= \pi x^n \sum_{j=0}^\infty (-1)^j \frac{x^{2j}}{j!(n+j)!}$$

$$= \pi x^n \left(\frac{1}{0!n!} - \frac{x^2}{1!(n+1)!} + \frac{x^4}{2!(n+2)!} - \frac{x^6}{3!(n+3)!} + \cdots \right).$$

[4.84] Prove that $\forall (x, n) \in \mathbb{R} \times 2\mathbb{Z}$,

$$\int_0^\pi \sin(2x \sin\theta) \sin n\theta \, d\theta = 0.$$

[4.85] (Watson, 2011) Prove that $\forall (x, n) \in \mathbb{R} \times (2\mathbb{N} - 2)$,

$$\int_0^\pi \cos(2x \cos\theta) \cos n\theta \, d\theta = (-1)^{n/2} \pi J_n(2x)$$

$$= (-1)^{n/2} \pi x^n \sum_{j=0}^\infty (-1)^j \frac{x^{2j}}{j!(n+j)!}$$

$$= (-1)^{n/2} \pi x^n \left(\frac{1}{0!n!} - \frac{x^2}{1!(n+1)!} + \frac{x^4}{2!(n+2)!} - \frac{x^6}{3!(n+3)!} + \cdots \right).$$

[4.86] Prove that $\forall (x, n) \in \mathbb{R} \times (2\mathbb{Z} + 1)$,

$$\int_0^\pi \cos(2x \cos\theta) \cos n\theta \, d\theta = 0.$$

Problems

[4.87] Prove that $\forall (x,n) \in \mathbb{R} \times (2\mathbb{N} - 2)$,

$$\int_0^\pi \cosh(2x \sin\theta) \cos n\theta \, d\theta = (-1)^{n/2} \pi I_n(2x)$$

$$= (-1)^{n/2} \pi x^n \sum_{j=0}^\infty \frac{x^{2j}}{j!(n+j)!}$$

$$= (-1)^{n/2} \pi x^n \left(\frac{1}{0!n!} + \frac{x^2}{1!(n+1)!} + \frac{x^4}{2!(n+2)!} + \frac{x^6}{3!(n+3)!} + \cdots \right).$$

[4.88] Prove that $\forall (x,n) \in \mathbb{R} \times (2\mathbb{Z} + 1)$,

$$\int_0^\pi \cosh(2x \sin\theta) \cos n\theta \, d\theta = 0.$$

[4.89] Prove that $\forall (x,n) \in \mathbb{R} \times (2\mathbb{N} - 2)$,

$$\int_0^\pi \cosh(2x \cos\theta) \cos n\theta \, d\theta = \pi I_n(2x)$$

$$= \pi x^n \sum_{j=0}^\infty \frac{x^{2j}}{j!(n+j)!}$$

$$= \pi x^n \left(\frac{1}{0!n!} + \frac{x^2}{1!(n+1)!} + \frac{x^4}{2!(n+2)!} + \frac{x^6}{3!(n+3)!} + \cdots \right).$$

[4.90] Prove that $\forall (x,n) \in \mathbb{R} \times (2\mathbb{Z} + 1)$,

$$\int_0^\pi \cosh(2x \cos\theta) \cos n\theta \, d\theta = 0.$$

[4.91] (Watson, 2011) Prove that $\forall (x,n) \in \mathbb{R} \times (2\mathbb{N} - 2)$,

$$\int_0^\pi \cos(2x \sin\theta) \cos n\theta \, d\theta = \pi J_n(2x)$$

$$= \pi x^n \sum_{j=0}^\infty (-1)^j \frac{x^{2j}}{j!(n+j)!}$$

$$= \pi x^n \left(\frac{1}{0!n!} - \frac{x^2}{1!(n+1)!} + \frac{x^4}{2!(n+2)!} - \frac{x^6}{3!(n+3)!} + \cdots \right).$$

[4.92] Prove that $\forall (x, n) \in \mathbb{R} \times (2\mathbb{Z} + 1)$,

$$\int_0^\pi \cos(2x \sin\theta) \cos n\theta \, d\theta = 0.$$

[4.93] Prove that $\forall (x, y, m) \in \mathbb{R}^2 \times (\mathbb{N} - 1)$,

$$\int_0^\pi (y + 2x\cos\theta)^m \, d\theta = \pi \sum_{j=0}^{\lfloor m/2 \rfloor} \binom{2j}{j}\binom{m}{2j} y^{m-2j} x^{2j}$$

$$= \pi \left(\binom{0}{0}\binom{m}{0} y^m x^0 + \binom{2}{1}\binom{m}{2} y^{m-2} x^2 + \binom{4}{2}\binom{m}{4} y^{m-4} x^4 \right.$$

$$\left. + \cdots + \binom{2\lfloor m/2 \rfloor}{\lfloor m/2 \rfloor}\binom{m}{2\lfloor m/2 \rfloor} y^{m-2\lfloor m/2 \rfloor} x^{2\lfloor m/2 \rfloor} \right).$$

[4.94] Prove that $\forall (x, y, m, n) \in \mathbb{R}^2 \times (\mathbb{N} - 1) \times [0..m]$,

$$\int_0^\pi (y + 2x\cos\theta)^m \cos n\theta \, d\theta = \pi \sum_{j=0}^{\lfloor (m-n)/2 \rfloor} \binom{n+2j}{j}\binom{m}{n+2j} y^{m-n-2j} x^{n+2j}$$

$$= \pi \left(\binom{n}{0}\binom{m}{n} y^{m-n} x^n + \binom{n+2}{1}\binom{m}{n+2} y^{m-n-2} x^{n+2} \right.$$

$$+ \binom{n+4}{2}\binom{m}{n+4} y^{m-n-4} x^{n+4}$$

$$+ \cdots + \binom{n+2\lfloor (m-n)/2 \rfloor}{\lfloor (m-n)/2 \rfloor}\binom{m}{n+2\lfloor (m-n)/2 \rfloor}$$

$$\left. y^{m-n-2\lfloor (m-n)/2 \rfloor} x^{n+2\lfloor (m-n)/2 \rfloor} \right).$$

[4.95] Prove that $\forall (x, y, m, n) \in \mathbb{R}^2 \times (\mathbb{N} - 1) \times ((\mathbb{N} - 1) - [0..m])$,

$$\int_0^\pi (y + 2x\cos\theta)^m \cos n\theta \, d\theta = 0.$$

[4.96] Prove that $\forall x \in \mathbb{R}$,

$$\int_0^\pi e^{x\cos\theta} \cos(x\sin\theta) \, d\theta = \pi.$$

[4.97] Prove that $\forall (x, n) \in \mathbb{R} \times \mathbb{N}$,
$$\int_0^\pi e^{x\cos\theta} \cos(x\sin\theta) \cos n\theta \, d\theta = \frac{\pi}{2n!} x^n.$$

[4.98] Prove that $\forall (x, n) \in \mathbb{R} \times \mathbb{N}$,
$$\int_0^\pi e^{x\cos\theta} \sin(x\sin\theta) \sin n\theta \, d\theta = \frac{\pi}{2n!} x^n.$$

[4.99] Prove that $\forall (x, n, m) \in \mathbb{R} \times (\mathbb{N} - 1) \times [0..n]$,
$$\int_0^\pi e^{x\cos\theta} \cos((m-n)\theta + x\sin\theta) \, d\theta = \frac{\pi}{(n-m)!} x^{n-m}.$$

[4.100] Prove that $\forall (x, n, m) \in \mathbb{R} \times (\mathbb{N} - 1) \times ((\mathbb{N} - 1) - [0..n])$,
$$\int_0^\pi e^{x\cos\theta} \cos((m-n)\theta + x\sin\theta) \, d\theta = 0.$$

[4.101] Prove that
$$\int_0^\pi \log\left(2\cos\frac{\theta}{2}\right) d\theta = 0.$$

[4.102] Prove that $\forall x \in \mathbb{R} - (-1, 1)$,
$$\int_0^\pi \log\sqrt{1 + \frac{1}{x^2} + \frac{2}{x}\cos\theta} \, d\theta = 0.$$

[4.103] Prove that $\forall n \in \mathbb{N}$,
$$\int_0^\pi \log\left(2\cos\frac{\theta}{2}\right) \cos n\theta \, d\theta = (-1)^{n+1} \frac{\pi}{2n}.$$

[4.104] Prove that $\forall (x, n) \in (\mathbb{R} - (-1, 1)) \times \mathbb{N}$,

$$\int_0^\pi \cos n\theta \log \sqrt{1 + \frac{1}{x^2} + \frac{2}{x} \cos \theta} \, d\theta = (-1)^{n+1} \frac{\pi}{2nx^n}.$$

[4.105] Prove that $\forall (x, n) \in (\mathbb{R} - (-1, 1)) \times \mathbb{N}$,

$$\int_0^\pi \sin n\theta \tan^{-1} \left(\frac{\sin \theta}{x + \cos \theta} \right) d\theta = (-1)^{n+1} \frac{\pi}{2nx^n}.$$

[4.106] Prove that $\forall x \in \left(-\frac{1}{2}, \frac{1}{2} \right)$,

$$\int_0^\pi \log(1 + 2x \cos \theta) \, d\theta = -\pi \sum_{j=1}^\infty \binom{2j}{j} \frac{x^{2j}}{2j}$$

$$= -\pi \left(\frac{\binom{2}{1}}{2} x^2 + \frac{\binom{4}{2}}{4} x^4 + \frac{\binom{6}{3}}{6} x^6 + \frac{\binom{8}{4}}{8} x^8 + \cdots \right).$$

[4.107] Prove that $\forall (x, n) \in \left(-\frac{1}{2}, \frac{1}{2} \right) \times \mathbb{N}$,

$$\int_0^\pi \log(1 + 2x \cos \theta) \cos n\theta \, d\theta = (-1)^{n+1} \pi \sum_{j=0}^\infty \binom{n + 2j}{j} \frac{x^{n+2j}}{n + 2j}$$

$$= (-1)^{n+1} \pi \left(\frac{\binom{n}{0}}{n} x^n + \frac{\binom{n+2}{1}}{n+2} x^{n+2} \right.$$

$$\left. + \frac{\binom{n+4}{2}}{n+4} x^{n+4} + \frac{\binom{n+6}{3}}{n+6} x^{n+6} + \cdots \right).$$

[4.108] Prove that $\forall (x, y) \in \left(-\dfrac{1}{2}, \dfrac{1}{2}\right) \times \mathbb{R}$,

$$\int_0^\pi (1 + 2x\cos\theta)^y \, d\theta = \pi \sum_{j=0}^\infty \binom{2j}{j} \dfrac{\prod_{k=0}^{2j-1}(y-k)}{(2j)!} x^{2j}$$

$$= \pi \left(1 + \dfrac{\binom{2}{1}}{2!} y(y-1)x^2 + \dfrac{\binom{4}{2}}{4!} y(y-1)(y-2)(y-3)x^4 \right.$$

$$\left. + \dfrac{\binom{6}{3}}{6!} y(y-1)(y-2)(y-3)(y-4)(y-5)x^6 + \cdots \right).$$

[4.109] Prove that $\forall (x, y, n) \in \left(-\dfrac{1}{2}, \dfrac{1}{2}\right) \times \mathbb{R} \times \mathbb{N}$,

$$\int_0^\pi (1 + 2x\cos\theta)^y \cos n\theta \, d\theta = \pi \sum_{j=0}^\infty \binom{n+2j}{j} \dfrac{\prod_{k=0}^{n+2j-1}(y-k)}{(n+2j)!} x^{n+2j}$$

$$= \pi \left(\dfrac{\binom{n}{0}}{n!} y(y-1)(y-2)\ldots(y-n+1)x^n \right.$$

$$+ \dfrac{\binom{n+2}{1}}{(n+2)!} y(y-1)(y-2)\ldots(y-n-1)x^{n+2}$$

$$+ \dfrac{\binom{n+4}{2}}{(n+4)!} y(y-1)(y-2)\ldots(y-n-3)x^{n+4}$$

$$\left. + \dfrac{\binom{n+6}{3}}{(n+6)!} y(y-1)(y-2)\ldots(y-n-5)x^{n+6} + \cdots \right).$$

[4.110] Let the function $\phi(x, y, \theta)$ be defined as

$$\phi(x, y, \theta) = \tan^{-1}\left(\frac{x-y}{x+y}\tan\theta\right). \qquad (4.48)$$

Prove that $\forall (x, y, m) \in \mathbb{R} \times (\mathbb{R} - \{-x\}) \times (2\mathbb{N} - 2)$,

$$\int_0^\pi (x^2 + y^2 + 2xy\cos 2\theta)^{m/2} \cos m\phi(x, y, \theta)\, d\theta = \pi \binom{m}{m/2}(xy)^{m/2}.$$

[4.111] Let the function $\phi(x, y, \theta)$ be as defined in (4.48). Prove that $\forall (x, y, m) \in \mathbb{R} \times (\mathbb{R} - \{-x\}) \times (2\mathbb{N} - 1)$,

$$\int_0^\pi (x^2 + y^2 + 2xy\cos 2\theta)^{m/2} \cos m\phi(x, y, \theta)\, d\theta = 0.$$

[4.112] Let the function $\phi(x, y, \theta)$ be as defined in (4.48). Prove that $\forall (x, y, m, n) \in \mathbb{R} \times (\mathbb{R} - \{-x\}) \times (\mathbb{N} - 1)^2$, if $n \leq m$ and $(m - n)/2$ is an integer, then

$$\int_0^\pi (x^2 + y^2 + 2xy\cos 2\theta)^{m/2} \cos m\phi(x, y, \theta) \cos n\theta\, d\theta$$
$$= \frac{\pi(x^n + y^n)}{2}\binom{m}{(m-n)/2}(xy)^{(m-n)/2}.$$

[4.113] Let the function $\phi(x, y, \theta)$ be as defined in (4.48). Prove that $\forall (x, y, m, n) \in \mathbb{R} \times (\mathbb{R} - \{-x\}) \times (\mathbb{N} - 1)^2$, if $n > m$ or $(m - n)/2$ is not an integer, then

$$\int_0^\pi (x^2 + y^2 + 2xy\cos 2\theta)^{m/2} \cos m\phi(x, y, \theta) \cos n\theta\, d\theta = 0.$$

[4.114] Let the function $\phi(x, y, \theta)$ be defined as

$$\phi(x, y, \theta) = \tan^{-1}\left(\frac{y\sin\theta}{x + y\cos\theta}\right). \qquad (4.49)$$

Prove that $\forall (x, y, m) \in (\mathbb{R}^2 - \{(0, 0)\}) \times (\mathbb{N} - 1)$,

$$\int_0^\pi (x^2 + y^2 + 2xy\cos\theta)^{m/2} \cos m\phi(x, y, \theta)\, d\theta = \pi x^m.$$

[4.115] Let the function $\phi(x, y, \theta)$ be as defined in (4.49). Prove that $\forall (x, y, m, n) \in \left(\mathbb{R}^2 - \{ (0, 0) \} \right) \times \mathbb{N} \times [1..m]$,

$$\int_0^\pi (x^2 + y^2 + 2xy \cos \theta)^{m/2} \cos m\phi(x, y, \theta) \cos n\theta \, d\theta = \frac{\pi}{2} \binom{m}{n} x^{m-n} y^n.$$

[4.116] Let the function $\phi(x, y, \theta)$ be as defined in (4.49). Prove that $\forall (x, y, m, n) \in \left(\mathbb{R}^2 - \{ (0, 0) \} \right) \times \mathbb{N} \times (\mathbb{N} - [1..m])$,

$$\int_0^\pi (x^2 + y^2 + 2xy \cos \theta)^{m/2} \cos m\phi(x, y, \theta) \cos n\theta \, d\theta = 0.$$

[4.117] Let the function $\phi(x, y, \theta)$ be defined as

$$\phi(x, y, \theta) = \tan^{-1} \left(\frac{(x - y) \sin \theta}{1 + (x + y) \cos \theta} \right). \qquad (4.50)$$

Prove that $\forall (x, y, m) \in \mathbb{R}^2 \times (\mathbb{N} - 1)$,

$$\int_0^\pi (1 + x^2 + y^2 + 2xy \cos 2\theta + 2(x + y) \cos \theta)^{m/2} \cos m\phi(x, y, \theta) \, d\theta$$

$$= \pi \sum_{j=0}^{\lfloor m/2 \rfloor} \binom{2j}{j} \binom{m}{2j} (xy)^j$$

$$= \pi \left(1 + \binom{2}{1} \binom{m}{2} xy + \binom{4}{2} \binom{m}{4} x^2 y^2 + \cdots \right.$$

$$\left. + \binom{2\lfloor m/2 \rfloor}{\lfloor m/2 \rfloor} \binom{m}{2\lfloor m/2 \rfloor} (xy)^{\lfloor m/2 \rfloor} \right).$$

[4.118] Let the function $\phi(x, y, \theta)$ be as defined in (4.50). Prove that $\forall (x, y, m, n) \in \mathbb{R}^2 \times (\mathbb{N} - 1)^2$,

$$\int_0^\pi \left(1 + x^2 + y^2 + 2xy \cos 2\theta + 2(x+y) \cos \theta\right)^{m/2} \cos m\phi(x, y, \theta) \cos n\theta \, d\theta$$

$$= \frac{\pi (x^n + y^n)}{2} \sum_{j=0}^{\lfloor (m-n)/2 \rfloor} \binom{n+2j}{j} \binom{m}{n+2j} (xy)^j$$

$$= \frac{\pi (x^n + y^n)}{2} \left(\binom{n}{0} \binom{m}{n} + \binom{n+2}{1} \binom{m}{n+2} xy + \binom{n+4}{2} \binom{m}{n+4} x^2 y^2 \right.$$

$$\left. + \cdots + \binom{n + 2\lfloor (m-n)/2 \rfloor}{\lfloor (m-n)/2 \rfloor} \binom{m}{n + 2\lfloor (m-n)/2 \rfloor} (xy)^{\lfloor (m-n)/2 \rfloor} \right).$$

[4.119] Let the function $\phi(x, y, \theta)$ be defined as

$$\phi(x, y, \theta) = \tan^{-1} \left(\frac{y \sin \theta}{1 + x + y \cos \theta} \right). \qquad (4.51)$$

Prove that $\forall (x, y, m) \in \left(\mathbb{R}^2 - \{(-1, 0)\} \right) \times (\mathbb{N} - 1)$,

$$\int_0^\pi \left((1+x)^2 + y^2 + 2y(1+x) \cos \theta\right)^{m/2} \cos m\phi(x, y, \theta) \, d\theta = \pi (1+x)^m.$$

[4.120] Let the function $\phi(x, y, \theta)$ be as defined in (4.51). Prove that $\forall (x, y, m, n) \in \left(\mathbb{R}^2 - \{(-1, 0)\} \right) \times (\mathbb{N} - 1) \times \mathbb{N}$,

$$\int_0^\pi \left((1+x)^2 + y^2 + 2y(1+x) \cos \theta\right)^{m/2} \cos m\phi(x, y, \theta) \cos n\theta \, d\theta$$

$$= \frac{\pi y^n}{2} \sum_{j=0}^{m-n} \binom{n+j}{j} \binom{m}{n+j} x^j$$

$$= \frac{\pi y^n}{2} \left(\binom{n}{0} \binom{m}{n} + \binom{n+1}{1} \binom{m}{n+1} x + \binom{n+2}{2} \binom{m}{n+2} x^2 \right.$$

$$\left. + \cdots + \binom{m}{m-n} \binom{m}{m} x^{m-n} \right)$$

$$= \frac{\pi}{2} \binom{m}{n} (1+x)^{m-n} y^n.$$

Problems

[4.121] Prove that $\forall (x,y) \in \mathbb{R}^2$, such that $|x|+|y| \leq 1$,

$$\int_0^\pi \log\left(1+x^2+y^2+2(x+y)\cos\theta+2xy\cos 2\theta\right) d\theta = -\pi \sum_{j=1}^\infty \binom{2j}{j}\frac{(xy)^j}{j}$$

$$= -\pi\left(\frac{\binom{2}{1}}{1}xy + \frac{\binom{4}{2}}{2}x^2y^2 + \frac{\binom{6}{3}}{3}x^3y^3 + \frac{\binom{8}{4}}{4}x^4y^4 + \cdots\right).$$

[4.122] Prove that $\forall (x,y,n) \in \mathbb{R}^2 \times \mathbb{N}$, such that $|x|+|y| \leq 1$,

$$\int_0^\pi \log\left(1+x^2+y^2+2(x+y)\cos\theta+2xy\cos 2\theta\right)\cos n\theta\, d\theta$$

$$= (-1)^{n+1}\pi(x^n+y^n)\sum_{j=0}^\infty \binom{n+2j}{j}\frac{(xy)^j}{n+2j}$$

$$= (-1)^{n+1}\pi(x^n+y^n)\left(\frac{\binom{n}{0}}{n} + \frac{\binom{n+2}{1}}{n+2}xy + \frac{\binom{n+4}{2}}{n+4}x^2y^2\right.$$

$$\left.+\frac{\binom{n+6}{3}}{n+6}x^3y^3+\cdots\right).$$

[4.123] Prove that $\forall x \in [-1,1]$,

$$\int_0^\pi \log\left(1+x^2+2x\cos\theta\right) d\theta = 0.$$

[4.124] Prove that $\forall (x,n) \in [-1,1] \times \mathbb{N}$,

$$\int_0^\pi \log\left(1+x^2+2x\cos\theta\right)\cos n\theta\, d\theta = (-1)^{n+1}\frac{\pi}{n}x^n.$$

[4.125] Prove that $\forall (x, y) \in \mathbb{R}^2$,

$$\int_0^\pi \sin(x + y\cos\theta)\cosh(y\sin\theta)\,d\theta = \pi\sin x.$$

[4.126] Prove that $\forall (x, y) \in \mathbb{R}^2$,

$$\int_0^\pi \sinh(x + y\cos\theta)\cos(y\sin\theta)\,d\theta = \pi\sinh x.$$

[4.127] Prove that $\forall (x, y) \in \mathbb{R}^2$,

$$\int_0^\pi \cos(x + y\cos\theta)\cosh(y\sin\theta)\,d\theta = \pi\cos x.$$

[4.128] Prove that $\forall (x, y) \in \mathbb{R}^2$,

$$\int_0^\pi \cosh(x + y\cos\theta)\cos(y\sin\theta)\,d\theta = \pi\cosh x.$$

[4.129] Prove that $\forall (x, y, n) \in \mathbb{R}^2 \times (2\mathbb{N} - 1)$,

$$\int_0^\pi \sin(x + y\cos\theta)\cosh(y\sin\theta)\cos n\theta\,d\theta = (-1)^{(n-1)/2}\frac{\pi y^n}{2n!}\cos x.$$

[4.130] Prove that $\forall (x, y, n) \in \mathbb{R}^2 \times 2\mathbb{N}$,

$$\int_0^\pi \sin(x + y\cos\theta)\cosh(y\sin\theta)\cos n\theta\,d\theta = (-1)^{n/2}\frac{\pi y^n}{2n!}\sin x.$$

[4.131] Prove that $\forall (x, y, n) \in \mathbb{R}^2 \times (2\mathbb{N} - 1)$,

$$\int_0^\pi \sinh(x + y\cos\theta)\cos(y\sin\theta)\cos n\theta\,d\theta = \frac{\pi y^n}{2n!}\cosh x.$$

Problems

[4.132] Prove that $\forall (x, y, n) \in \mathbb{R}^2 \times 2\mathbb{N}$,

$$\int_0^\pi \sinh(x + y \cos \theta) \cos(y \sin \theta) \cos n\theta \, d\theta = \frac{\pi y^n}{2n!} \sinh x.$$

[4.133] Prove that $\forall (x, y, n) \in \mathbb{R}^2 \times (2\mathbb{N} - 1)$,

$$\int_0^\pi \cos(x + y \cos \theta) \cosh(y \sin \theta) \cos n\theta \, d\theta = (-1)^{(n+1)/2} \frac{\pi y^n}{2n!} \sin x.$$

[4.134] Prove that $\forall (x, y, n) \in \mathbb{R}^2 \times 2\mathbb{N}$,

$$\int_0^\pi \cos(x + y \cos \theta) \cosh(y \sin \theta) \cos n\theta \, d\theta = (-1)^{n/2} \frac{\pi y^n}{2n!} \cos x.$$

[4.135] Prove that $\forall (x, y, n) \in \mathbb{R}^2 \times (2\mathbb{N} - 1)$,

$$\int_0^\pi \cosh(x + y \cos \theta) \cos(y \sin \theta) \cos n\theta \, d\theta = \frac{\pi y^n}{2n!} \sinh x.$$

[4.136] Prove that $\forall (x, y, n) \in \mathbb{R}^2 \times 2\mathbb{N}$,

$$\int_0^\pi \cosh(x + y \cos \theta) \cos(y \sin \theta) \cos n\theta \, d\theta = \frac{\pi y^n}{2n!} \cosh x.$$

[4.137] Let the function $\phi(x, \theta)$ be given by

$$\phi(x, \theta) = \tan^{-1}\left(\frac{x^2 - 1}{x^2 + 1} \tan \theta\right). \tag{4.52}$$

Prove that $\forall (x, m) \in (\mathbb{R} - \{0\}) \times (2\mathbb{N} - 2)$,

$$\int_0^\pi \left(x^2 + \frac{1}{x^2} + 2\cos 2\theta\right)^{m/2} \cos m\phi(x, \theta) \, d\theta = \pi \binom{m}{m/2}.$$

[4.138] Let the function $\phi(x, \theta)$ be given by (4.52). Prove that $\forall (x, m) \in (\mathbb{R} - \{0\}) \times (2\mathbb{N} - 1)$,

$$\int_0^\pi \left(x^2 + \frac{1}{x^2} + 2\cos 2\theta\right)^{m/2} \cos m\phi(x, \theta)\, d\theta = 0.$$

[4.139] Let the function $\phi(x, \theta)$ be given by

$$\phi(x, \theta) = \tan^{-1}\left(\frac{x \sin \theta}{1 + x \cos \theta}\right).$$

Prove that $\forall (x, m) \in \mathbb{R} \times (\mathbb{N} - 1)$,

$$\int_0^\pi \left(1 + x^2 + 2x \cos \theta\right)^{m/2} \cos m\phi(x, \theta)\, d\theta = \pi.$$

[4.140] Let the function $\phi(x, \theta)$ be given by

$$\phi(x, \theta) = \tan^{-1}\left(\frac{\sin \theta}{x^2 + \cos \theta}\right).$$

Prove that $\forall (x, m) \in (\mathbb{R} - \{0\}) \times (\mathbb{N} - 1)$,

$$\int_0^\pi \left(x^2 + \frac{1}{x^2} + 2\cos \theta\right)^{m/2} \cos m\phi(x, \theta)\, d\theta = \pi x^m.$$

[4.141] Let the function $\phi(\theta)$ be given by

$$\phi(\theta) = \tan^{-1}\left(\frac{\sin \theta}{2 + \cos \theta}\right). \tag{4.53}$$

Prove that $\forall m \in \mathbb{N} - 1$,

$$\int_0^\pi (5 + 4\cos \theta)^{m/2} \cos m\phi(\theta)\, d\theta = 2^m \pi.$$

[4.142] Let the function $\phi(\theta)$ be given by (4.53). Prove that $\forall (m, n) \in (\mathbb{N} - 1) \times \mathbb{N}$,

$$\int_0^\pi (5 + 4\cos\theta)^{m/2} \cos m\phi(\theta) \cos n\theta \, d\theta = \frac{\pi}{2} \sum_{j=0}^{m-n} \binom{n+j}{j}\binom{m}{n+j}$$

$$= \frac{\pi}{2}\left(\binom{n}{0}\binom{m}{n} + \binom{n+1}{1}\binom{m}{n+1} + \binom{n+2}{2}\binom{m}{n+2} + \cdots \right.$$

$$\left. + \binom{m}{m-n}\binom{m}{m}\right).$$

[4.143] (Boros and Moll, 2006) Prove that $\forall m \in 2\mathbb{N} - 2$,

$$\int_0^\pi \cos^m \theta \, d\theta = \frac{\pi}{2^m} \binom{m}{m/2}.$$

[4.144] Prove that $\forall m \in 2\mathbb{N} - 1$,

$$\int_0^\pi \cos^m \theta \, d\theta = 0.$$

[4.145] Prove that $\forall m \in \mathbb{N} - 1$,

$$\int_0^\pi \cos^m \frac{\theta}{2} \cos \frac{m\theta}{2} \, d\theta = \frac{\pi}{2^m}.$$

[4.146] Prove that

$$\int_0^\pi \log(5 + 4\cos\theta) \, d\theta = 2\pi \log 2.$$

[4.147] (Boros and Moll, 2006) Prove that

$$\int_0^\pi \log \sin\theta \, d\theta = -\pi \log 2.$$

[4.148] Prove that $\forall n \in \mathbb{N}$,

$$\int_0^\pi \log\left(\cos\frac{\theta}{2}\right) \cos n\theta \, d\theta = (-1)^{n+1} \frac{\pi}{2n}.$$

[4.149] Prove that
$$\int_0^\pi \log\left(e^2 + \frac{1}{e^2} + 2\cos\theta\right) d\theta = 2\pi.$$

[4.150] Prove that $\forall x \in \left[-1, -\frac{1}{4}\right] \cup \left[\frac{1}{4}, 1\right]$,

$$\int_0^\pi \log\left(1 + x^2 + \frac{1}{16x^2} + 2\left(x - \frac{1}{4x}\right)\cos\theta - \frac{\cos 2\theta}{2}\right) d\theta$$
$$= \pi \sum_{j=1}^\infty \frac{(-1)^{j-1}}{2^{2j} j} \binom{2j}{j}$$
$$= \pi \left(\frac{\binom{2}{1}}{1 \cdot 2^2} - \frac{\binom{4}{2}}{2 \cdot 2^4} + \frac{\binom{6}{3}}{3 \cdot 2^6} - \frac{\binom{8}{4}}{4 \cdot 2^8} + \cdots\right).$$

[4.151] Prove that $\forall (x, n) \in \left(\left[-1, -\frac{1}{4}\right] \cup \left[\frac{1}{4}, 1\right]\right) \times \mathbb{N}$,

$$\int_0^\pi \log\left(1 + x^2 + \frac{1}{16x^2} + 2\left(x - \frac{1}{4x}\right)\cos\theta - \frac{\cos 2\theta}{2}\right) \cos n\theta \, d\theta$$
$$= (-1)^{n+1} \left(x^n + \frac{(-1)^n}{4^n x^n}\right) \pi \sum_{j=0}^\infty \frac{(-1)^j}{2^{2j}(n+2j)} \binom{n+2j}{j}$$
$$= (-1)^{n+1} \left(x^n + \frac{(-1)^n}{4^n x^n}\right) \left(\frac{\binom{n}{0}}{n} - \frac{\binom{n+2}{1}}{2^2(n+2)} + \frac{\binom{n+4}{2}}{2^4(n+4)} - \frac{\binom{n+6}{3}}{2^6(n+6)} + \cdots\right).$$

[4.152] Prove that
$$\int_0^\pi \log\left(1 + \sin^2\theta\right) d\theta = \pi \sum_{j=1}^\infty \frac{(-1)^{j-1}}{2^{2j} j} \binom{2j}{j}$$
$$= \pi \left(\frac{\binom{2}{1}}{1 \cdot 2^2} - \frac{\binom{4}{2}}{2 \cdot 2^4} + \frac{\binom{6}{3}}{3 \cdot 2^6} - \frac{\binom{8}{4}}{4 \cdot 2^8} + \cdots\right).$$

[4.153] Prove that $\forall n \in 2\mathbb{N}$,

$$\int_0^\pi \log\left(1 + \sin^2\theta\right) \cos n\theta \, d\theta = -\frac{\pi}{2^{n-1}} \sum_{j=0}^\infty \frac{(-1)^j}{2^{2j}(n+2j)} \binom{n+2j}{j}$$

$$= -\frac{\pi}{2^{n-1}} \left(\frac{\binom{n}{0}}{n} - \frac{\binom{n+2}{1}}{2^2(n+2)} + \frac{\binom{n+4}{2}}{2^4(n+4)} - \frac{\binom{n+6}{3}}{2^6(n+6)} + \cdots \right).$$

[4.154] Prove that $\forall n \in 2\mathbb{N} - 1$,

$$\int_0^\pi \log\left(1 + \sin^2\theta\right) \cos n\theta \, d\theta = 0.$$

[4.155] Let the function $\phi(x, \theta)$ be given by

$$\phi(x, \theta) = \tan^{-1}\left(\frac{(x^2 - 1) \sin\theta}{x + (x^2 + 1) \cos\theta} \right).$$

Prove that $\forall (x, m) \in (\mathbb{R} - \{0\}) \times (\mathbb{N} - 1)$,

$$\int_0^\pi \left(1 + x^2 + \frac{1}{x^2} + 2\cos 2\theta + 2\left(x + \frac{1}{x}\right) \cos\theta \right)^{m/2} \cos m\phi(x, \theta) \, d\theta$$

$$= \pi \sum_{j=0}^{\lfloor m/2 \rfloor} \binom{2j}{j}\binom{m}{2j}$$

$$= \pi \left(1 + \binom{2}{1}\binom{m}{2} + \binom{4}{2}\binom{m}{4} + \cdots + \binom{2\lfloor m/2 \rfloor}{\lfloor m/2 \rfloor}\binom{m}{2\lfloor m/2 \rfloor} \right).$$

[4.156] Let the function $\phi(x, \theta)$ be given by

$$\phi(x, \theta) = \tan^{-1}\left(\frac{(x^2 + 1) \sin\theta}{x + (x^2 - 1) \cos\theta} \right).$$

Prove that $\forall (x, m) \in (\mathbb{R} - \{0\}) \times (\mathbb{N} - 1)$,

$$\int_0^\pi \left(1 + x^2 + \frac{1}{x^2} - 2\cos 2\theta + 2\left(x - \frac{1}{x}\right)\cos\theta\right)^{m/2} \cos m\phi(x,\theta)\, d\theta$$
$$= \pi \sum_{j=0}^{\lfloor m/2 \rfloor} (-1)^j \binom{2j}{j}\binom{m}{2j}$$
$$= \pi \left(1 - \binom{2}{1}\binom{m}{2} + \binom{4}{2}\binom{m}{4} - \cdots + (-1)^{\lfloor m/2 \rfloor}\binom{2\lfloor m/2 \rfloor}{\lfloor m/2 \rfloor}\binom{m}{2\lfloor m/2 \rfloor}\right).$$

[4.157] Prove that $\forall m \in \mathbb{N} - 1$,

$$\int_0^\pi (3 + 4\cos\theta + 2\cos 2\theta)^{m/2}\, d\theta = \pi \sum_{j=0}^{\lfloor m/2 \rfloor} \binom{2j}{j}\binom{m}{2j}$$
$$= \pi \left(1 + \binom{2}{1}\binom{m}{2} + \binom{4}{2}\binom{m}{4} + \cdots + \binom{2\lfloor m/2 \rfloor}{\lfloor m/2 \rfloor}\binom{m}{2\lfloor m/2 \rfloor}\right).$$

[4.158] Prove that $\forall (m,n) \in (\mathbb{N}-1)^2$,

$$\int_0^\pi (3 + 4\cos\theta + 2\cos 2\theta)^{m/2} \cos n\theta\, d\theta = \pi \sum_{j=0}^{\lfloor (m-n)/2 \rfloor} \binom{n+2j}{j}\binom{m}{n+2j}$$
$$= \pi \left(\binom{n}{0}\binom{m}{n} + \binom{n+2}{1}\binom{m}{n+2} + \binom{n+4}{2}\binom{m}{n+4}\right.$$
$$\left. + \cdots + \binom{n+2\lfloor (m-n)/2 \rfloor}{\lfloor (m-n)/2 \rfloor}\binom{m}{n+2\lfloor (m-n)/2 \rfloor}\right).$$

[4.159] Let the function $\phi(\theta)$ be given by

$$\phi(\theta) = \tan^{-1}(2\sin\theta). \tag{4.54}$$

Prove that $\forall m \in \mathbb{N} - 1$,

$$\int_0^\pi (3 - 2\cos 2\theta)^{m/2} \cos m\phi(\theta)\, d\theta = \pi \sum_{j=0}^{\lfloor m/2 \rfloor} (-1)^j \binom{2j}{j}\binom{m}{2j}$$
$$= \pi \left(1 - \binom{2}{1}\binom{m}{2} + \binom{4}{2}\binom{m}{4} - \cdots + (-1)^{\lfloor m/2 \rfloor}\binom{2\lfloor m/2 \rfloor}{\lfloor m/2 \rfloor}\binom{m}{2\lfloor m/2 \rfloor}\right).$$

[4.160] Let the function $\phi(\theta)$ be given by (4.54). Prove that $\forall (m, n) \in (\mathbb{N} - 1) \times (2\mathbb{N} - 2)$,

$$\int_0^\pi (3 - 2\cos 2\theta)^{m/2} \cos m\phi(\theta) \cos n\theta \, d\theta = \pi \sum_{j=0}^{\lfloor (m-n)/2 \rfloor} (-1)^j \binom{n+2j}{j}\binom{m}{n+2j}$$
$$= \pi \left(\binom{n}{0}\binom{m}{n} - \binom{n+2}{1}\binom{m}{n+2} + \binom{n+4}{2}\binom{m}{n+4} \right.$$
$$\left. - \cdots + (-1)^{\lfloor (m-n)/2 \rfloor} \binom{n+2\lfloor (m-n)/2 \rfloor}{\lfloor (m-n)/2 \rfloor}\binom{m}{n+2\lfloor (m-n)/2 \rfloor} \right).$$

[4.161] Let the function $\phi(\theta)$ be given by (4.54). Prove that $\forall (m, n) \in (\mathbb{N} - 1) \times (2\mathbb{N} - 1)$,

$$\int_0^\pi (3 - 2\cos 2\theta)^{m/2} \cos m\phi(\theta) \cos n\theta \, d\theta = 0.$$

[4.162] Let the function $\phi(\theta)$ be defined as

$$\phi(\theta) = \tan^{-1}\left(\frac{\sin\theta}{2 + \cos\theta} \right). \tag{4.55}$$

Prove that $\forall m \in \mathbb{N} - 1$,

$$\int_0^\pi (5 + 4\cos\theta)^{m/2} \cos m\phi(\theta) \, d\theta = 2^m \pi.$$

[4.163] Let the function $\phi(\theta)$ be defined as in (4.55). Prove that $\forall (m, n) \in (\mathbb{N} - 1) \times \mathbb{N}$,

$$\int_0^\pi (5 + 4\cos\theta)^{m/2} \cos m\phi(\theta) \cos n\theta \, d\theta = \frac{\pi}{2} \sum_{j=0}^{m-n} \binom{n+j}{j}\binom{m}{n+j}$$
$$= \frac{\pi}{2}\left(\binom{n}{0}\binom{m}{n} + \binom{n+1}{1}\binom{m}{n+1} + \binom{n+2}{2}\binom{m}{n+2} \right.$$
$$\left. + \cdots + \binom{m}{m-n}\binom{m}{m} \right).$$

Chapter 5
Generalized Bessel Functions

The most common elementary complex functions are the exponential, circular trigonometrical, and hyperbolic trigonometrical functions. $\forall z \in \mathbb{C}$, the exponential function, e^z, is defined as

$$e^z = \sum_{j=0}^{\infty} \frac{z^j}{j!} = 1 + \frac{z}{1!} + \frac{z^2}{2!} + \frac{z^3}{3!} + \cdots \tag{5.1}$$

$\forall z \in \mathbb{C}$, the circular trigonometrical functions, $\sin z$ and $\cos z$, are defined as

$$\sin z = \sum_{j=0}^{\infty} (-1)^j \frac{z^{2j+1}}{(2j+1)!} = \frac{z}{1!} - \frac{z^3}{3!} + \frac{z^5}{5!} + \frac{z^7}{7!} + \cdots, \tag{5.2}$$

and

$$\cos z = \sum_{j=0}^{\infty} (-1)^j \frac{z^{2j}}{(2j)!} = 1 - \frac{z^2}{2!} + \frac{z^4}{4!} + \frac{z^6}{6!} + \cdots, \tag{5.3}$$

respectively. The circular trigonometrical functions are related to the exponential function by Euler's formula, which is equivalent to

$$\sin z = \frac{e^{iz} - e^{-iz}}{2i},$$

and

$$\cos z = \frac{e^{iz} + e^{-iz}}{2}.$$

© The Author(s), under exclusive license to Springer Nature Singapore Pte Ltd. 2025
A. Mishra, *Special Integrals*, University Texts in the Mathematical Sciences,
https://doi.org/10.1007/978-981-97-7514-9_5

$\forall z \in \mathbb{C}$, the hyperbolic trigonometrical functions, sinh z and cosh z, are defined as

$$\sinh z = \sum_{j=0}^{\infty} \frac{z^{2j+1}}{(2j+1)!} = \frac{z}{1!} + \frac{z^3}{3!} + \frac{z^5}{5!} + \frac{z^7}{7!} + \cdots, \qquad (5.4)$$

and

$$\cosh z = \sum_{j=0}^{\infty} \frac{z^{2j}}{(2j)!} = 1 + \frac{z^2}{2!} + \frac{z^4}{4!} + \frac{z^6}{6!} + \cdots, \qquad (5.5)$$

respectively. We can easily see that the hyperbolic trigonometrical functions are also related to the exponential functions by the equations

$$\sinh z = \frac{e^z - e^{-z}}{2},$$

and

$$\cosh z = \frac{e^z + e^{-z}}{2}.$$

Finally, the circular and hyperbolic trigonometrical functions are also related to each other by the equations

$$\sin(iz) = i \sinh z,$$

and

$$\cos(iz) = \cosh z.$$

Collectively, in this text, the above functions are called the elementary exponential functions.

In this chapter, first we will define the Mittag-Leffler functions $E_\alpha(z)$ (Mittag-Leffler, 1903; Erdélyi, 1955) and their various generalizations: $E_{\alpha,\beta}(z)$ by Wiman (1905), $E_{\alpha,\beta}^{\gamma}(z)$ by Prabhakar (1971), and $E_{\alpha,\beta}^{\gamma,q}(z)$ by Shukla and Prajapati (2007). Then, we will define the generalized Bessel functions $\phi(\rho, \beta; z)$ as defined by Wright (1933). Finally, we will define the *Pseudo-Exponential* functions, which look very similar to the above functions. Using the theorems proved in the previous chapter, we can derive the integral representations of the pseudo-exponential functions. We will also find the derivatives and integrals of the pseudo-exponential functions and see that they are very similar to the elementary exponential functions. The problems listed in this chapter show the relations between the pseudo-exponential functions and the elementary exponential functions. In particular, we will see that the pseudo-exponential functions are the special integrals of the elementary exponential functions. We will also see that many special integrals can be simplified using the pseudo-exponential functions.

5.1 The Mittag-Leffler Functions $E_\alpha(z)$

In this section, we will give a generalization of the exponential function (5.1), called the Mittag-Leffler functions, $E_\alpha(z)$ (Mittag-Leffler, 1903).

Definition 5.1 *(The Mittag-Leffler Functions $E_\alpha(z)$)* (Mittag-Leffler, 1903)
$\forall (z, \alpha) \in \mathbb{C} \times \Re(\mathbb{C})^+$, the Mittag-Leffler functions, $E_\alpha(z)$, are defined as

$$E_\alpha(z) = \sum_{j=0}^{\infty} \frac{z^j}{\Gamma(\alpha j + 1)} \quad (5.6)$$
$$= 1 + \frac{z}{\Gamma(\alpha + 1)} + \frac{z^2}{\Gamma(2\alpha + 1)} + \frac{z^3}{\Gamma(3\alpha + 1)} + \cdots$$

Example 5.2 *(The Exponential Function as a Special Case of $E_\alpha(z)$)* (Erdélyi, 1955)
Prove that $\forall z \in \mathbb{C}$,
$$e^z = E_1(z).$$

Proof Putting $\alpha = 1$ in (5.6), and using (5.1) and [3.6], we get

$$E_1(z) = \sum_{j=0}^{\infty} \frac{z^j}{\Gamma(j+1)}$$
$$= \sum_{j=0}^{\infty} \frac{z^j}{j!}$$
$$= e^z.$$

□

5.2 Wiman's Generalized Mittag-Leffler Functions $E_{\alpha,\beta}(z)$

In this section, we will give a generalization of the Mittag-Leffler functions $E_\alpha(z)$ (5.6), called the Wiman's generalized Mittag-Leffler functions, $E_{\alpha,\beta}(z)$ (Wiman, 1905).

Definition 5.3 *(Wiman's Generalized Mittag-Leffler Functions $E_{\alpha,\beta}(z)$)* (Wiman, 1905)
$\forall (z, \alpha, \beta) \in \mathbb{C} \times \left(\Re(\mathbb{C})^+\right)^2$, the Wiman's Generalized Mittag-Leffler functions, $E_{\alpha,\beta}(z)$, are defined as

$$E_{\alpha,\beta}(z) = \sum_{j=0}^{\infty} \frac{z^j}{\Gamma(\alpha j + \beta)} \tag{5.7}$$
$$= \frac{1}{\Gamma(\beta)} + \frac{z}{\Gamma(\alpha+\beta)} + \frac{z^2}{\Gamma(2\alpha+\beta)} + \frac{z^3}{\Gamma(3\alpha+\beta)} + \cdots$$

From (5.6) and (5.7), it is obvious that (Erdélyi, 1955)

$$E_\alpha(z) = E_{\alpha,1}(z).$$

5.3 Prabhakar's Generalized Mittag-Leffler Functions $E^\gamma_{\alpha,\beta}(z)$

In this section, we will give a generalization of the Wiman's generalized Mittag-Leffler functions $E_{\alpha,\beta}(z)$ (5.7), called the Prabhakar's Generalized Mittag-Leffler functions, $E^\gamma_{\alpha,\beta}(z)$ (Prabhakar, 1971).

Definition 5.4 *(Prabhakar's Generalized Mittag-Leffler Functions $E^\gamma_{\alpha,\beta}(z)$)* (Prabhakar, 1971)
$\forall (z, \alpha, \beta, \gamma) \in \mathbb{C} \times \left(\Re(\mathbb{C})^+\right)^3$, the Prabhakar's Generalized Mittag-Leffler functions, $E^\gamma_{\alpha,\beta}(z)$, are defined as

$$E^\gamma_{\alpha,\beta}(z) = \sum_{j=0}^{\infty} \frac{\prod_{k=0}^{j-1}(\gamma+k)}{j!\Gamma(\alpha j + \beta)} z^j$$
$$= \frac{1}{\Gamma(\beta)} + \frac{\gamma}{1!\Gamma(\alpha+\beta)} z + \frac{\gamma(\gamma+1)}{2!\Gamma(2\alpha+\beta)} z^2 + \frac{\gamma(\gamma+1)(\gamma+2)}{3!\Gamma(3\alpha+\beta)} z^3 + \cdots \tag{5.8}$$

From (5.6), (5.7), and (5.8), it is obvious that (Prabhakar, 1971)

$$E_\alpha(z) = E^1_{\alpha,1}(z),$$

and

$$E_{\alpha,\beta}(z) = E^1_{\alpha,\beta}(z).$$

5.4 The Shukla-Prajapati Generalized Mittag-Leffler Functions $E_{\alpha,\beta}^{\gamma,q}(z)$

In this section, we will give a generalization of Prabhakar's generalized Mittag-Leffler functions $E_{\alpha,\beta}^{\gamma}(z)$ (5.8), called the Shukla-Prajapati generalized Mittag-Leffler functions, $E_{\alpha,\beta}^{\gamma,q}(z)$ (Shukla and Prajapati, 2007).

Definition 5.5 *(The Shukla-Prajapati Generalized Mittag-Leffler Functions $E_{\alpha,\beta}^{\gamma,q}(z)$)* (Shukla and Prajapati, 2007)

$\forall (z, \alpha, \beta, \gamma, q) \in \mathbb{C} \times (\Re(\mathbb{C})^+)^3 \times ((0, 1) \cup \mathbb{N})$, the Shukla-Prajapati Generalized Mittag-Leffler functions, $E_{\alpha,\beta}^{\gamma,q}(z)$, are defined as

$$E_{\alpha,\beta}^{\gamma,q}(z) = \sum_{j=0}^{\infty} \frac{\prod_{k=0}^{qj-1}(\gamma+k)}{j!\Gamma(\alpha j + \beta)} z^j$$

$$= \frac{1}{\Gamma(\beta)} + \frac{\prod_{k=0}^{q-1}(\gamma+k)}{1!\Gamma(\alpha+\beta)} z + \frac{\prod_{k=0}^{2q-1}(\gamma+k)}{2!\Gamma(2\alpha+\beta)} z^2 + \frac{\prod_{k=0}^{3q-1}(\gamma+k)}{3!\Gamma(3\alpha+\beta)} z^3 + \cdots \tag{5.9}$$

From (5.8) and (5.9), it is obvious that

$$E_{\alpha,\beta}^{\gamma}(z) = E_{\alpha,\beta}^{\gamma,1}(z).$$

5.5 Wright's Generalized Bessel Functions $\phi(\alpha, \beta; z)$

In this section, we will give a generalization of Bessel functions $J_v(z)$ (3.27), called the Wright's generalized Bessel functions, $\phi(\alpha, \beta; z)$ (Wright, 1933).

Definition 5.6 *(Wright's Generalized Bessel Functions $\phi(\alpha, \beta; z)$)* (Wright, 1933)
$\forall (z, \alpha, \beta) \in \mathbb{C} \times (\Re(\mathbb{C})^+)^2$, the Wright's Generalized Bessel functions, $\phi(\alpha, \beta; z)$, are defined as

$$\phi(\alpha, \beta; z) = \sum_{j=0}^{\infty} \frac{z^j}{j!\Gamma(\alpha j + \beta)}$$

$$= \frac{1}{\Gamma(\beta)} + \frac{z}{1!\Gamma(\alpha+\beta)} + \frac{z^2}{\Gamma(2\alpha+\beta)} + \frac{z^3}{3!\Gamma(3\alpha+\beta)} + \cdots \tag{5.10}$$

Example 5.7 *(The Bessel Functions $J_v(z)$ as a Special Case of $\phi(\rho, \beta; z)$)* (Erdélyi, 1955)
Prove that $\forall (z, v) \in \mathbb{C} \times (\mathbb{C} - (\mathbb{Z} - \mathbb{N}))$,

$$J_v(z) = \left(\frac{z}{2}\right)^v \phi\left(1, v+1; -\frac{z^2}{4}\right).$$

Proof Putting $\alpha = 1$ and $\beta = v+1$ in (5.10), and using (3.27), we get

$$\left(\frac{z}{2}\right)^v \phi\left(1, v+1; -\frac{z^2}{4}\right) = \left(\frac{z}{2}\right)^v \sum_{j=0}^{\infty} \frac{\left(-\frac{z^2}{4}\right)^j}{j!\Gamma(j+v+1)}$$

$$= \sum_{j=0}^{\infty} (-1)^j \frac{z^{v+2j}}{j!\Gamma(v+j+1)2^{v+2j}}$$

$$= J_v(z).$$

$\square$

5.6 The Pseudo-Exponential Functions

In this section, we will define the Pseudo-Exponential functions as a special case of Wright's generalized Bessel functions $\phi(\alpha, \beta; z)$ (5.10).

Definition 5.8 *(The Pseudo-Exponential Functions)*
$\forall (z, n) \in \mathbb{C} \times \mathbb{Z}$, we define the Pseudo-Exponential function, $e(z, n)$, as

$$e(z, n) = z^{\max\{0, -n\}} \phi(1, |n|+1; z)$$

$$= \sum_{j=\max\{0,-n\}}^{\infty} \frac{z^j}{j!(n+j)!}$$

$$= \begin{cases} \dfrac{1}{0!n!} + \dfrac{z}{1!(n+1)!} + \dfrac{z^2}{2!(n+2)!} + \dfrac{z^3}{3!(n+3)!} + \cdots, & \text{if } n \in \mathbb{N} - 1; \\ \dfrac{z^{-n}}{0!(-n)!} + \dfrac{z^{-n+1}}{1!(-n+1)!} + \dfrac{z^{-n+2}}{2!(-n+2)!} \\ + \dfrac{z^{-n+3}}{3!(-n+3)!} + \cdots, & \text{if } n \in -\mathbb{N}. \end{cases}$$

(5.11)

Example 5.9 *($e(z, n)$ for negative values of n)*
Prove that $\forall (z, n) \in \mathbb{C} \times \mathbb{Z}$,

$$e(z, -n) = z^n e(z, n). \tag{5.12}$$

5.6 The Pseudo-Exponential Functions

Proof The result follows from (5.11). □

Example 5.10 *(Relation Between $e(z,n)$ and $I_n(z)$)*
Prove that $\forall (z,n) \in \mathbb{C} \times \mathbb{Z}$,

$$e(z,n) = z^{-n/2} I_n \left(2\sqrt{z} \right).$$

Proof Using (5.11) and (3.19), $\forall (z,n) \in \mathbb{C} \times (\mathbb{N} - 1)$, we have

$$z^{-n/2} I_n \left(2\sqrt{z} \right) = z^{-n/2} \sum_{j=0}^{\infty} \frac{\left(2\sqrt{z} \right)^{n+2j}}{j!(n+j)!2^{n+2j}}$$

$$= \sum_{j=0}^{\infty} \frac{z^j}{j!(n+j)!} \quad (5.13)$$

$$= e(z,n).$$

Using (5.12), (5.13), and [3.19], $\forall (z,n) \in \mathbb{C} \times (-\mathbb{N})$, we have

$$e(z,n) = z^{-n} e(z,-n)$$
$$= z^{-n} \cdot z^{n/2} I_{-n} \left(2\sqrt{z} \right)$$
$$= z^{-n/2} I_n \left(2\sqrt{z} \right).$$

□

Definition 5.11 *(The Pseudo-Sine and Pseudo-Cosine Functions)*
$\forall (z,n) \in \mathbb{C} \times \mathbb{Z}$, we define the Pseudo-Sine function, $sin(z,n)$, and the Pseudo-Cosine function, $cos(z,n)$, related by the equation

$$e(iz,n) = \cos(z,n) + i \sin(z,n), \quad (5.14)$$

as

$$\sin(z,n) = \sum_{j=\max\{0,\lfloor -n/2 \rfloor\}}^{\infty} (-1)^j \frac{z^{2j+1}}{(2j+1)!(n+2j+1)!}$$

$$= \begin{cases} \dfrac{z}{1!(n+1)!} - \dfrac{z^3}{3!(n+3)!} + \dfrac{z^5}{5!(n+5)!} \\ \quad - \dfrac{z^7}{7!(n+7)!} + \cdots, & \text{if } n \in \mathbb{N} - 1; \\[4pt] (-1)^{-n/2} \left(\dfrac{z^{-n+1}}{1!(-n+1)!} - \dfrac{z^{-n+3}}{3!(-n+3)!} \right. \\ \quad \left. + \dfrac{z^{-n+5}}{5!(-n+5)!} - \dfrac{z^{-n+7}}{7!(-n+7)!} + \cdots \right), & \text{if } n \in -2\mathbb{N}; \\[4pt] (-1)^{(-n-1)/2} \left(\dfrac{z^{-n}}{0!(-n)!} - \dfrac{z^{-n+2}}{2!(-n+2)!} \right. \\ \quad \left. + \dfrac{z^{-n+4}}{4!(-n+4)!} - \dfrac{z^{-n+6}}{6!(-n+6)!} + \cdots \right), & \text{if } n \in -(2\mathbb{N}-1); \end{cases}$$

(5.15)

and

$$\cos(z, n) = \sum_{j=\max\{0, \lceil -n/2 \rceil\}}^{\infty} (-1)^j \frac{z^{2j}}{(2j)!(n+2j)!}$$

$$= \begin{cases} \dfrac{1}{0!n!} - \dfrac{z^2}{2!(n+2)!} + \dfrac{z^4}{4!(n+4)!} \\ \quad - \dfrac{z^6}{6!(n+6)!} + \cdots, & \text{if } n \in \mathbb{N} - 1; \\ (-1)^{-n/2} \left(\dfrac{z^{-n}}{0!(-n)!} - \dfrac{z^{-n+2}}{2!(-n+2)!} \right. \\ \quad \left. + \dfrac{z^{-n+4}}{4!(-n+4)!} - \dfrac{z^{-n+6}}{6!(-n+6)!} + \cdots \right), & \text{if } n \in -2\mathbb{N}; \\ (-1)^{(-n+1)/2} \left(\dfrac{z^{-n+1}}{1!(-n+1)!} - \dfrac{z^{-n+3}}{3!(-n+3)!} \right. \\ \quad \left. + \dfrac{z^{-n+5}}{5!(-n+5)!} - \dfrac{z^{-n+7}}{7!(-n+7)!} + \cdots \right), & \text{if } n \in -(2\mathbb{N}-1); \end{cases}$$

(5.16)

respectively.

Definition 5.12 (*The Pseudo-Hyperbolic Sine and Pseudo-Hyperbolic Cosine Functions*)
$\forall (z, n) \in \mathbb{C} \times \mathbb{Z}$, we define the Pseudo-Hyperbolic Sine function, $sinh(z, n)$, and the Pseudo-Hyperbolic Cosine function, $cosh(z, n)$, defined using the equations

$$\sinh(z, n) = -i \sin(iz, n), \qquad (5.17)$$

and

$$\cosh(z, n) = \cos(iz, n), \qquad (5.18)$$

respectively, as

$$\sinh(z, n) = \sum_{j=\max\{0, \lfloor -n/2 \rfloor\}}^{\infty} \frac{z^{2j+1}}{(2j+1)!(n+2j+1)!}$$

$$= \begin{cases} \dfrac{z}{1!(n+1)!} + \dfrac{z^3}{3!(n+3)!} + \dfrac{z^5}{5!(n+5)!} \\ \quad + \dfrac{z^7}{7!(n+7)!} + \cdots, & \text{if } n \in \mathbb{N} - 1; \\ \dfrac{z^{-n+1}}{1!(-n+1)!} + \dfrac{z^{-n+3}}{3!(-n+3)!} \\ \quad + \dfrac{z^{-n+5}}{5!(-n+5)!} + \dfrac{z^{-n+7}}{7!(-n+7)!} + \cdots, & \text{if } n \in -2\mathbb{N}; \\ \dfrac{z^{-n}}{0!(-n)!} + \dfrac{z^{-n+2}}{2!(-n+2)!} \\ \quad + \dfrac{z^{-n+4}}{4!(-n+4)!} + \dfrac{z^{-n+6}}{6!(-n+6)!} + \cdots, & \text{if } n \in -(2\mathbb{N}-1); \end{cases}$$

(5.19)

5.6 The Pseudo-Exponential Functions

and

$$\cosh(z, n) = \sum_{j=\max\{0, \lceil -n/2 \rceil\}}^{\infty} \frac{z^{2j}}{(2j)!(n+2j)!}$$

$$= \begin{cases} \dfrac{1}{0!n!} + \dfrac{z^2}{2!(n+2)!} + \dfrac{z^4}{4!(n+4)!} \\ \quad + \dfrac{z^6}{6!(n+6)!} + \cdots, & \text{if } n \in \mathbb{N} - 1; \\[6pt] \dfrac{z^{-n}}{0!(-n)!} + \dfrac{z^{-n+2}}{2!(-n+2)!} \\ \quad + \dfrac{z^{-n+4}}{4!(-n+4)!} + \dfrac{z^{-n+6}}{6!(-n+6)!} + \cdots, & \text{if } n \in -2\mathbb{N}; \\[6pt] \dfrac{z^{-n+1}}{1!(-n+1)!} + \dfrac{z^{-n+3}}{3!(-n+3)!} \\ \quad + \dfrac{z^{-n+5}}{5!(-n+5)!} + \dfrac{z^{-n+7}}{7!(-n+7)!} + \cdots, & \text{if } n \in -(2\mathbb{N} - 1); \end{cases}$$

(5.20)

respectively.

Before proceeding to derive the properties of the pseudo-exponential functions, we should plot their graphs and compare them with the elementary exponential functions.

Example 5.13 *(The Plot of $y = e(x, 0)$)*
The plot of $y = e(x, 0)$ in $[-6\pi, \pi]$ is shown in Fig. 5.1. As compared to the function $y = e^x$, we notice that their structures are similar in the $x \geq 0$ region. Both tend to infinity. However, we notice a difference for the $x < 0$ region. $y = e(x, 0)$ is taking both positive and negative values (including 0), but $y = e^x$ takes only the positive values.

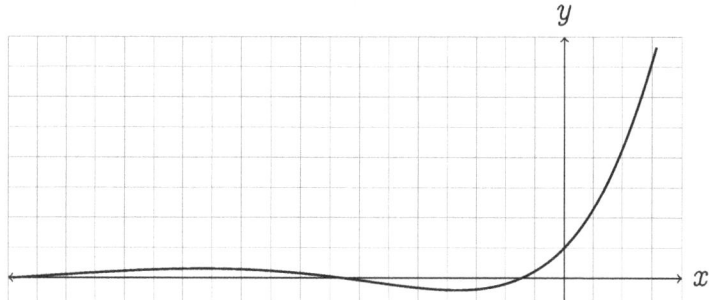

Fig. 5.1 The plot of $y = e(x, 0)$ in $[-6\pi, \pi]$

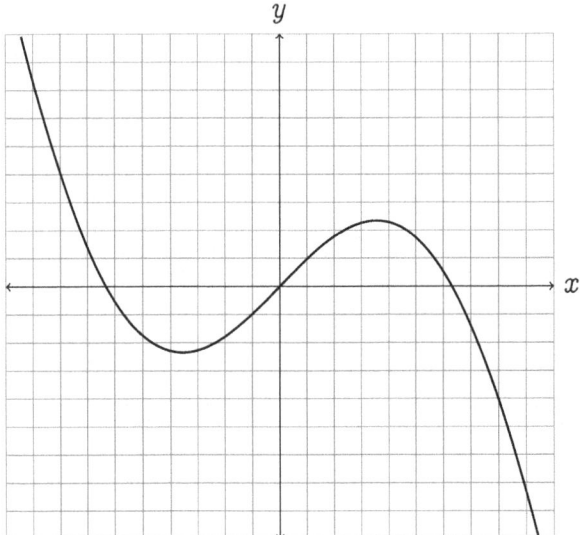

Fig. 5.2 The plot of $y = \sin(x, 0)$ in $[-3\pi, 3\pi]$

Example 5.14 *(The plot of $y = \sin(x, 0)$)*
The plot of $y = \sin(x, 0)$ in $[-3\pi, 3\pi]$ is shown in Fig. 5.2. As before, we notice that both $y = \sin(x, 0)$ and $y = \sin x$ have similar structures in the plotted region. The difference may be in the periodicity of the functions. $y = \sin x$ is periodic with period 2π, while $y = \sin(x, 0)$ is not seen to be periodic. Does $y = \sin(x, 0)$ takes the value 0 infinitely often, similar to $y = \sin x$? We have to find the answer.

Example 5.15 *(The plot of $y = \cos(x, 0)$)*
The plot of $y = \cos(x, 0)$ in $[-9\pi/2, 9\pi/2]$ is shown in Fig. 5.3. We notice that both $y = \cos(x, 0)$ and $y = \cos x$ have similar structures in the plotted region. The difference may be in the periodicity of the functions. $y = \cos x$ is periodic with period 2π, while $y = \cos(x, 0)$ is not seen to be periodic. Does $y = \cos(x, 0)$ takes the value 0 infinitely often, similar to $y = \cos x$? We have to find the answer.

Example 5.16 *(The plot of $y = \sinh(x, 0)$)*
The plot of $y = \sinh(x, 0)$ in $[-2\pi, 2\pi]$ is shown in Fig. 5.4. As before, both $y = \sinh(x, 0)$ and $y = \sinh x$ have similar structures in the plotted region.

5.6 The Pseudo-Exponential Functions

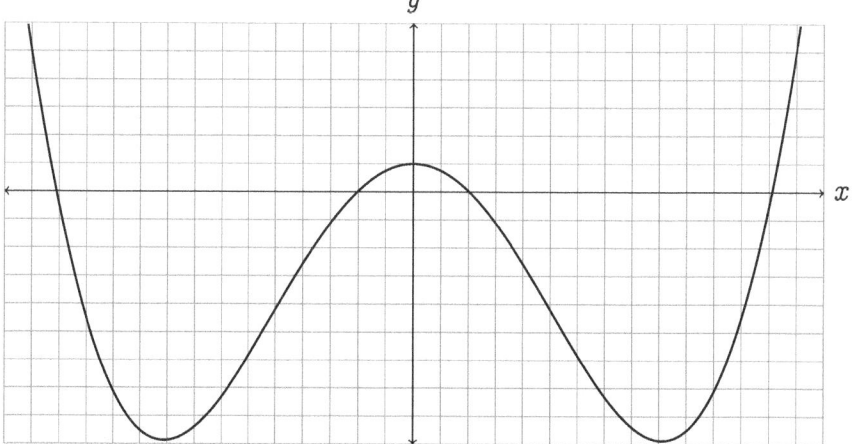

Fig. 5.3 The plot of $y = \cos(x, 0)$ in $[-9\pi/2, 9\pi/2]$

Fig. 5.4 The plot of
$y = \sinh(x, 0)$ in $[-2\pi, 2\pi]$

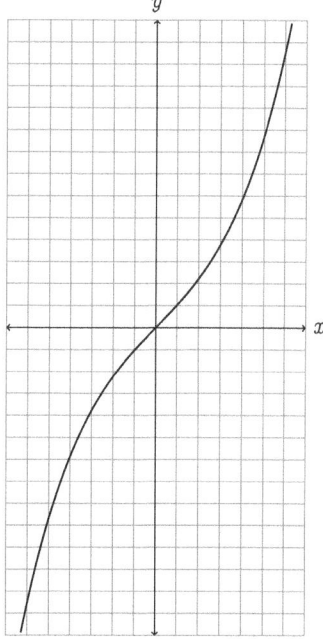

Example 5.17 *(The plot of $y = \cosh(x, 0)$)*
The plot of $y = \cosh(x, 0)$ in $[-2\pi, 2\pi]$ is shown in Fig. 5.5. Both $y = \cosh(x, 0)$ and $y = \cosh x$ have similar structures in the plotted region.

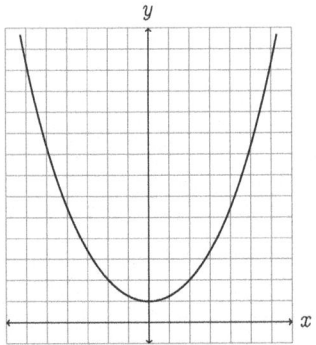

Fig. 5.5 The plot of $y = \cosh(x, 0)$ in $[-2\pi, 2\pi]$

5.7 Derivatives of the Pseudo-Exponential Functions

Recall the derivatives of the elementary exponential functions. We have

$$\frac{de^z}{dz} = e^z,$$
$$\frac{d \sin z}{dz} = \cos z,$$
$$\frac{d \cos z}{dz} = -\sin z,$$
$$\frac{d \sinh z}{dz} = \cosh z,$$

and

$$\frac{d \cosh z}{dz} = \sinh z.$$

We will see that the derivatives of the pseudo-exponential functions also exhibit similar properties. We will also observe that differentiating a pseudo-exponential function increments its level by 1 (the parameter n).

Example 5.18 *(Derivative of $e(z, n)$)*
Prove that $\forall (z, n) \in \mathbb{C} \times \mathbb{Z}$,

$$\frac{de(z, n)}{dz} = e(z, n+1). \tag{5.21}$$

Proof From (5.11), we get

5.7 Derivatives of the Pseudo-Exponential Functions

$$\frac{de(z,n)}{dz} = \frac{d}{dz}\left(\sum_{j=\max\{0,-n\}}^{\infty} \frac{z^j}{j!(n+j)!}\right)$$

$$= \sum_{j=\max\{1,-n\}}^{\infty} \frac{jz^{j-1}}{j!(n+j)!}$$

$$= \sum_{j=\max\{1,-n\}}^{\infty} \frac{z^{j-1}}{(j-1)!(n+j)!}$$

$$= \sum_{j=\max\{0,-(n+1)\}}^{\infty} \frac{z^j}{j!(n+1+j)!}$$

$$= e(z, n+1).$$

□

Example 5.19 *(Derivative of $\sin(z,n)$)*
Prove that $\forall (z,n) \in \mathbb{C} \times \mathbb{Z}$,

$$\frac{d\sin(z,n)}{dz} = \cos(z, n+1). \tag{5.22}$$

Proof From (5.15) and (5.16), we get

$$\frac{d\sin(z,n)}{dz} = \frac{d}{dz}\left(\sum_{j=\max\{0,\lfloor -n/2 \rfloor\}}^{\infty} (-1)^j \frac{z^{2j+1}}{(2j+1)!(n+2j+1)!}\right)$$

$$= \sum_{j=\max\{0,\lfloor -n/2 \rfloor\}}^{\infty} (-1)^j \frac{(2j+1)z^{2j}}{(2j+1)!(n+2j+1)!}$$

$$= \sum_{j=\max\{0,\lceil -(n+1)/2 \rceil\}}^{\infty} (-1)^j \frac{z^{2j}}{(2j)!(2j+n+1)!}$$

$$= \cos(z, n+1).$$

□

Example 5.20 *(Derivative of $\cos(z,n)$)*
Prove that $\forall (z,n) \in \mathbb{C} \times \mathbb{Z}$,

$$\frac{d\cos(z,n)}{dz} = -\sin(z, n+1). \tag{5.23}$$

Proof From (5.16) and (5.15), we get

$$\frac{d\cos(z,n)}{dz} = \frac{d}{dz}\left(\sum_{j=\max\{0,\lceil -n/2\rceil\}}^{\infty} (-1)^j \frac{z^{2j}}{(2j)!(n+2j)!}\right)$$

$$= \sum_{j=\max\{1,\lceil -n/2\rceil\}}^{\infty} (-1)^j \frac{(2j)z^{2j-1}}{(2j)!(n+2j)!}$$

$$= \sum_{j=\max\{1,\lceil -n/2\rceil\}}^{\infty} (-1)^j \frac{z^{2j-1}}{(2j-1)!(n+2j)!}$$

$$= -\sum_{j=\max\{0,\lfloor -(n+1)/2\rfloor\}}^{\infty} (-1)^j \frac{z^{2j+1}}{(2j+1)!(n+1+2j+1)!}$$

$$= -\sin(z, n+1).$$

$\square$

Example 5.21 *(Derivative of* $\sinh(z,n)$*)*
Prove that $\forall (z,n) \in \mathbb{C} \times \mathbb{Z}$,

$$\frac{d\sinh(z,n)}{dz} = \cosh(z, n+1). \tag{5.24}$$

Proof From (5.19) and (5.20), we get

$$\frac{d\sinh(z,n)}{dz} = \frac{d}{dz}\left(\sum_{j=\max\{0,\lfloor -n/2\rfloor\}}^{\infty} \frac{z^{2j+1}}{(2j+1)!(n+2j+1)!}\right)$$

$$= \sum_{j=\max\{0,\lfloor -n/2\rfloor\}}^{\infty} \frac{(2j+1)z^{2j}}{(2j+1)!(n+2j+1)!}$$

$$= \sum_{j=\max\{0,\lceil -(n+1)/2\rceil\}}^{\infty} \frac{z^{2j}}{(2j)!(n+1+2j)!}$$

$$= \cosh(z, n+1).$$

$\square$

Example 5.22 *(Derivative of* $\cosh(z,n)$*)*
Prove that $\forall (z,n) \in \mathbb{C} \times \mathbb{Z}$,

$$\frac{d\cosh(z,n)}{dz} = \sinh(z, n+1). \tag{5.25}$$

Proof From (5.20) and (5.19), we get

$$\frac{d\cosh(z,n)}{dz} = \frac{d}{dz}\left(\sum_{j=\max\{0,\lceil -n/2\rceil\}}^{\infty} \frac{z^{2j}}{(2j)!(n+2j)!}\right)$$

$$= \sum_{j=\max\{1,\lceil -n/2\rceil\}}^{\infty} \frac{2jz^{2j-1}}{(2j)!(n+2j)!}$$

$$= \sum_{j=\max\{1,\lceil -n/2\rceil\}}^{\infty} \frac{z^{2j-1}}{(2j-1)!(n+2j)!}$$

$$= \sum_{j=\max\{0,\lfloor -(n+1)/2\rfloor\}}^{\infty} \frac{z^{2j+1}}{(2j+1)!(n+1+2j+1)!}$$

$$= \sinh(z, n+1).$$

□

5.8 Integrals of the Pseudo-Exponential Functions

As before, we revisit the integrals of the elementary exponential functions. We have

$$\int e^z \, dz = e^z,$$

$$\int \sin z \, dz = -\cos z,$$

$$\int \cos z \, dz = \sin z,$$

$$\int \sinh z \, dz = \cosh z,$$

and

$$\int \cosh z \, dz = \sinh z.$$

We will see that the integrals of the pseudo-exponential functions show similar properties and that integrating a pseudo-exponential function decrements its level by 1 (the parameter n).

Example 5.23 (*Integral of $e(z,n)$*)
Prove that $\forall (z,n) \in \mathbb{C} \times \mathbb{Z}$,

$$\int e(z,n) \, dz = e(z, n-1).$$

Proof From (5.21), we get the above result. □

Example 5.24 *(Integral of* $\sin(z,n)$*)*
Prove that $\forall (z,n) \in \mathbb{C} \times \mathbb{Z}$,

$$\int \sin(z,n)\, dz = -\cos(z, n-1).$$

Proof From (5.23), we get the above result. □

Example 5.25 *(Integral of* $\cos(z,n)$*)*
Prove that $\forall (z,n) \in \mathbb{C} \times \mathbb{Z}$,

$$\int \cos(z,n)\, dz = \sin(z, n-1).$$

Proof From (5.22), we get the above result. □

Example 5.26 *(Integral of* $\sinh(z,n)$*)*
Prove that $\forall (z,n) \in \mathbb{C} \times \mathbb{Z}$,

$$\int \sinh(z,n)\, dz = \cosh(z, n-1).$$

Proof From (5.25), we get the above result. □

Example 5.27 *(Integral of* $\cosh(z,n)$*)*
Prove that $\forall (z,n) \in \mathbb{C} \times \mathbb{Z}$,

$$\int \cosh(z,n)\, dz = \sinh(z, n-1).$$

Proof From (5.24), we get the above result. □

Problems

[5.1] (Erdélyi, 1955) Prove that $\forall z \in \mathbb{C}$,

$$\cosh z = E_2(z^2).$$

[5.2] (Erdélyi, 1955) Prove that $\forall (z, \alpha, \beta) \in \mathbb{C} \times \left(\Re(\mathbb{C})^+\right)^2$,

$$E_{\alpha,\beta}(z) = \frac{1}{\Gamma(\beta)} + z E_{\alpha, \alpha+\beta}(z).$$

[5.3] (Shukla and Prajapati, 2007) Prove that $\forall (z, \alpha, \beta, \gamma) \in \mathbb{C} \times \left(\Re(\mathbb{C})^+\right)^3$,

$$E^{\gamma}_{\alpha, \beta-\alpha}(z) - E^{\gamma-1}_{\alpha, \beta-\alpha}(z) = z E^{\gamma}_{\alpha,\beta}(z).$$

[5.4] (Shukla and Prajapati, 2007) Prove that $\forall (z, \alpha, \beta, \gamma, q) \in \mathbb{C} \times \left(\Re(\mathbb{C})^+\right)^3 \times ((0,1) \cup \mathbb{N})$,

$$E_{\alpha,\beta}^{\gamma,q}(z) = \beta E_{\alpha,\beta+1}^{\gamma,q}(z) + \alpha z \frac{d}{dz}\left(E_{\alpha,\beta+1}^{\gamma,q}(z)\right).$$

[5.5] (Erdélyi, 1955) Prove that $\forall (z, \alpha, \beta) \in \mathbb{C} \times \left(\Re(\mathbb{C})^+\right)^2$,

$$\alpha z \phi(\alpha, \alpha + \beta; z) = \phi(\alpha, \beta - 1; z) + (1 - \beta)\phi(\alpha, \beta; z).$$

[5.6] (Erdélyi, 1955) Prove that $\forall (z, \alpha, \beta) \in \mathbb{C} \times \left(\Re(\mathbb{C})^+\right)^2$,

$$\frac{d\phi(\alpha, \beta; z)}{dz} = \phi(\alpha, \alpha + \beta; z).$$

[5.7] (Erdélyi, 1955) Prove that $\forall (z, \alpha, \beta) \in \mathbb{C} \times \left(\Re(\mathbb{C})^+\right)^2$,

$$\alpha z \frac{d\phi(\alpha, \beta; z)}{dz} = \phi(\alpha, \beta - 1; z) + (1 - \beta)\phi(\alpha, \beta; z).$$

[5.8] Prove that
$$e(1, 0) = \frac{1}{\pi} \int_0^\pi e^{2\cos\theta} d\theta.$$

[5.9] Prove that $\forall n \in \mathbb{Z}$,
$$e(1, n) = \frac{1}{\pi} \int_0^\pi e^{2\cos\theta} \cos n\theta \, d\theta.$$

[5.10] Prove that $\forall x \in \mathbb{R}$,
$$e\left(x^2, 0\right) = \frac{1}{\pi} \int_0^\pi e^{2x\cos\theta} d\theta.$$

[5.11] Prove that $\forall (x, n) \in (\mathbb{R} - \{0\}) \times \mathbb{Z}$,
$$e\left(x^2, n\right) = \frac{1}{\pi x^n} \int_0^\pi e^{2x\cos\theta} \cos n\theta \, d\theta.$$

[5.12] (Watson, 2011) Prove that $\forall (x, n) \in (\mathbb{R} - \{0\}) \times (2\mathbb{Z} + 1)$,
$$e\left(-x^2, n\right) = \frac{J_n(2x)}{x^n} = \frac{(-1)^{(n-1)/2}}{\pi x^n} \int_0^\pi \sin(2x\cos\theta) \cos n\theta \, d\theta.$$

[5.13] (Watson, 2011) Prove that $\forall (x, n) \in (\mathbb{R} - \{0\}) \times 2\mathbb{Z}$,

$$e\left(-x^2, n\right) = \frac{J_n(2x)}{x^n} = \frac{(-1)^{n/2}}{\pi x^n} \int_0^\pi \cos(2x \cos\theta) \cos n\theta \, d\theta.$$

[5.14] Prove that $\forall (x, n) \in (\mathbb{R} - \{0\}) \times (2\mathbb{Z} + 1)$,

$$e\left(x^2, n\right) = \frac{I_n(2x)}{x^n} = \frac{(-1)^{(n-1)/2}}{\pi x^n} \int_0^\pi \sinh(2x \sin\theta) \sin n\theta \, d\theta.$$

[5.15] Prove that $\forall (x, n) \in (\mathbb{R} - \{0\}) \times 2\mathbb{Z}$,

$$e\left(x^2, n\right) = \frac{I_n(2x)}{x^n} = \frac{(-1)^{n/2}}{\pi x^n} \int_0^\pi \cosh(2x \sin\theta) \cos n\theta \, d\theta.$$

[5.16] Prove that $\forall (x, n) \in (\mathbb{R} - \{0\}) \times (2\mathbb{Z} + 1)$,

$$e\left(x^2, n\right) = \frac{I_n(2x)}{x^n} = \frac{1}{\pi x^n} \int_0^\pi \sinh(2x \cos\theta) \cos n\theta \, d\theta.$$

[5.17] Prove that $\forall (x, n) \in (\mathbb{R} - \{0\}) \times 2\mathbb{Z}$,

$$e\left(x^2, n\right) = \frac{I_n(2x)}{x^n} = \frac{1}{\pi x^n} \int_0^\pi \cosh(2x \cos\theta) \cos n\theta \, d\theta.$$

[5.18] (Watson, 2011) Prove that $\forall (x, n) \in (\mathbb{R} - \{0\}) \times (2\mathbb{Z} + 1)$,

$$e\left(-x^2, n\right) = \frac{J_n(2x)}{x^n} = \frac{1}{\pi x^n} \int_0^\pi \sin(2x \sin\theta) \sin n\theta \, d\theta.$$

[5.19] (Watson, 2011) Prove that $\forall (x, n) \in (\mathbb{R} - \{0\}) \times 2\mathbb{Z}$,

$$e\left(-x^2, n\right) = \frac{J_n(2x)}{x^n} = \frac{1}{\pi x^n} \int_0^\pi \cos(2x \sin\theta) \cos n\theta \, d\theta.$$

[5.20] Prove that $\forall x \in \mathbb{R}$,

$$e(x, 0) = \frac{1}{\pi} \int_0^\pi e^{(x+1)\cos\theta} \cos((x-1)\sin\theta) \, d\theta.$$

[5.21] Prove that $\forall (x, n) \in \mathbb{R} \times \mathbb{Z} - \{-1\} \times (2\mathbb{Z} + 1)$,

$$e(x, n) = \frac{2}{\pi(1 + x^n)} \int_0^\pi e^{(x+1)\cos\theta} \cos((x-1)\sin\theta) \cos n\theta \, d\theta.$$

Problems

[5.22] Prove that $\forall (x, n) \in \mathbb{R} \times 2\mathbb{Z}$,

$$e(x, n) = \frac{2}{\pi(1+x^n)} \int_0^\pi \cosh\left((x+1)\cos\theta\right) \cos\left((x-1)\sin\theta\right) \cos n\theta \, d\theta.$$

[5.23] Prove that $\forall (x, n) \in (\mathbb{R} - \{-1\}) \times (2\mathbb{Z}+1)$,

$$e(x, n) = \frac{2}{\pi(1+x^n)} \int_0^\pi \sinh\left((x+1)\cos\theta\right) \cos\left((x-1)\sin\theta\right) \cos n\theta \, d\theta.$$

[5.24] Prove that $\forall x \in \mathbb{R}$,

$$e(x, 0) = -1 + \frac{2}{\pi} \int_0^\pi e^{(x+1)\cos\theta} \cos(\sin\theta) \cos(x\sin\theta) \, d\theta.$$

[5.25] Prove that $\forall x \in \mathbb{R}$,

$$e(x, 0) = 1 + \frac{2}{\pi} \int_0^\pi e^{(x+1)\cos\theta} \sin(\sin\theta) \sin(x\sin\theta) \, d\theta.$$

[5.26] Prove that $\forall x \in \mathbb{R}$,

$$\int_0^\pi e^{(x+1)\cos\theta} \cos\left((x+1)\sin\theta\right) d\theta = \pi.$$

[5.27] Prove that $\forall x \in \mathbb{R}$,

$$\sinh\left(x^2, 0\right) = \frac{1}{\pi} \int_0^\pi \left(\sinh^2(x\cos\theta)\cos^2(x\sin\theta)\right.$$
$$\left. + \cosh^2(x\cos\theta)\sin^2(x\sin\theta)\right) d\theta.$$

[5.28] Prove that $\forall x \in \mathbb{R}$,

$$\sinh\left(x^2, 0\right) = \frac{1}{\pi} \int_0^\pi \left(\sin^2(x\cos\theta)\cosh^2(x\sin\theta)\right.$$
$$\left. + \cos^2(x\cos\theta)\sinh^2(x\sin\theta)\right) d\theta.$$

[5.29] Prove that $\forall (x, n) \in (\mathbb{R} - \{0\}) \times 2\mathbb{Z}$,

$$\sinh\left(x^2, n\right) = \frac{(-1)^{n/2}}{\pi x^n} \int_0^\pi \left(\sin^2(x\cos\theta)\cosh^2(x\sin\theta)\right.$$
$$\left. + \cos^2(x\cos\theta)\sinh^2(x\sin\theta)\right) \cos n\theta \, d\theta.$$

[5.30] Prove that $\forall (x, n) \in (\mathbb{R} - \{0\}) \times 2\mathbb{Z}$,

$$\sinh(x^2, n) = \frac{1}{\pi x^n} \int_0^\pi \left(\sinh^2(x\cos\theta)\cos^2(x\sin\theta) + \cosh^2(x\cos\theta)\sin^2(x\sin\theta)\right)\cos n\theta \, d\theta.$$

[5.31] Prove that $\forall x \in \mathbb{R}$,

$$\cosh(x^2, 0) = \frac{1}{\pi} \int_0^\pi \left(\cos^2(x\cos\theta)\cosh^2(x\sin\theta) + \sin^2(x\cos\theta)\sinh^2(x\sin\theta)\right) d\theta.$$

[5.32] Prove that $\forall x \in \mathbb{R}$,

$$\cosh(x^2, 0) = \frac{1}{\pi} \int_0^\pi \left(\cosh^2(x\cos\theta)\cos^2(x\sin\theta) + \sinh^2(x\cos\theta)\sin^2(x\sin\theta)\right) d\theta.$$

[5.33] Prove that $\forall (x, n) \in (\mathbb{R} - \{0\}) \times 2\mathbb{Z}$,

$$\cosh(x^2, n) = \frac{(-1)^{n/2}}{\pi x^n} \int_0^\pi \left(\cos^2(x\cos\theta)\cosh^2(x\sin\theta) + \sin^2(x\cos\theta)\sinh^2(x\sin\theta)\right)\cos n\theta \, d\theta.$$

[5.34] Prove that $\forall (x, n) \in (\mathbb{R} - \{0\}) \times 2\mathbb{Z}$,

$$\cosh(x^2, n) = \frac{1}{\pi x^n} \int_0^\pi \left(\cosh^2(x\cos\theta)\cos^2(x\sin\theta) + \sinh^2(x\cos\theta)\sin^2(x\sin\theta)\right)\cos n\theta \, d\theta.$$

[5.35] Let $S_1(x, \theta)$ be given by

$$S_1(x, \theta) = \sinh(x\sin\theta)\cos(x\cos\theta)\sin(\sin\theta) + \cosh(x\sin\theta)\sin(x\cos\theta)\cos(\sin\theta). \tag{5.26}$$

Prove that $\forall (x, n) \in \mathbb{R} \times 2\mathbb{Z} - \{-1, 1\} \times 2(2\mathbb{Z} + 1)$,

$$\sin(x, n) = \frac{2}{\pi(1 + (-1)^{n/2} x^n)} \int_0^\pi S_1(x, \theta) e^{\cos\theta} \cos n\theta \, d\theta.$$

Problems

[5.36] Let $S_1(x, \theta)$ be given by (5.26). Prove that $\forall (x, n) \in \{-1, 1\} \times 2(2\mathbb{Z}+1)$,

$$\int_0^\pi S_1(x, \theta) e^{\cos \theta} \cos n\theta \, d\theta = 0.$$

[5.37] Let $S_1(x, \theta)$ be given by (5.26), and $C_1(x, \theta)$ be given by

$$C_1(x, \theta) = \cosh(x \sin \theta) \cos(x \cos \theta) \cos(\sin \theta) \\ - \sinh(x \sin \theta) \sin(x \cos \theta) \sin(\sin \theta). \quad (5.27)$$

Prove that $\forall (x, n) \in \mathbb{R} \times (2\mathbb{Z}+1)$,

$$\sin(x, n) = \frac{2}{\pi (1 + x^{2n})} \int_0^\pi S_1(x, \theta) e^{\cos \theta} \cos n\theta \, d\theta \\ + (-1)^{(n+1)/2} \frac{2x^n}{\pi (1 + x^{2n})} \int_0^\pi C_1(x, \theta) e^{\cos \theta} \cos n\theta \, d\theta.$$

[5.38] Let $C_1(x, \theta)$ be given by (5.27). Prove that $\forall (x, n) \in \mathbb{R} \times 2\mathbb{Z} - \{-1, 1\} \times 2(2\mathbb{Z}+1)$,

$$\cos(x, n) = \frac{2}{\pi (1 + (-1)^{n/2} x^n)} \int_0^\pi C_1(x, \theta) e^{\cos \theta} \cos n\theta \, d\theta.$$

[5.39] Let $C_1(x, \theta)$ be given by (5.27). Prove that $\forall (x, n) \in \{-1, 1\} \times 2(2\mathbb{Z}+1)$,

$$\int_0^\pi C_1(x, \theta) e^{\cos \theta} \cos n\theta \, d\theta = 0.$$

[5.40] Let $C_1(x, \theta)$ be given by (5.27), and $S_1(x, \theta)$ be given by (5.26). Prove that $\forall (x, n) \in \mathbb{R} \times (2\mathbb{Z}+1)$,

$$\cos(x, n) = \frac{2}{\pi (1 + x^{2n})} \int_0^\pi C_1(x, \theta) e^{\cos \theta} \cos n\theta \, d\theta \\ + (-1)^{(n-1)/2} \frac{2x^n}{\pi (1 + x^{2n})} \int_0^\pi S_1(x, \theta) e^{\cos \theta} \cos n\theta \, d\theta.$$

[5.41] Let $S_2(x, \theta)$ be given by

$$S_2(x, \theta) = \cosh(x \cos \theta) \sin(x \sin \theta) \sin(\sin \theta) \\ + \sinh(x \cos \theta) \cos(x \sin \theta) \cos(\sin \theta). \quad (5.28)$$

Prove that $\forall (x, n) \in \mathbb{R} \times 2\mathbb{Z}$,

$$\sinh(x,n) = \frac{2}{\pi(1+x^n)} \int_0^\pi S_2(x,\theta) e^{\cos\theta} \cos n\theta \, d\theta.$$

[5.42] Let $S_2(x,\theta)$ be given by (5.28), and $C_2(x,\theta)$ be given by

$$\begin{aligned}C_2(x,\theta) &= \cosh(x\cos\theta)\cos(x\sin\theta)\cos(\sin\theta) \\ &\quad + \sinh(x\cos\theta)\sin(x\sin\theta)\sin(\sin\theta).\end{aligned} \quad (5.29)$$

Prove that $\forall (x,n) \in (\mathbb{R} - \{-1, 1\}) \times (2\mathbb{Z}+1)$,

$$\begin{aligned}\sinh(x,n) &= \frac{2}{\pi(1-x^{2n})} \int_0^\pi S_2(x,\theta) e^{\cos\theta} \cos n\theta \, d\theta \\ &\quad - \frac{2x^n}{\pi(1-x^{2n})} \int_0^\pi C_2(x,\theta) e^{\cos\theta} \cos n\theta \, d\theta.\end{aligned}$$

[5.43] Let $S_2(x,\theta)$ be given by (5.28). Prove that $\forall (x,n) \in \{-1,1\} \times (2\mathbb{Z}+1)$,

$$\int_0^\pi S_2(x,\theta) e^{\cos\theta} \cos n\theta \, d\theta = x^n \int_0^\pi C_2(x,\theta) e^{\cos\theta} \cos n\theta \, d\theta.$$

[5.44] Let $C_2(x,\theta)$ be given by (5.29). Prove that $\forall (x,n) \in \mathbb{R} \times 2\mathbb{Z}$,

$$\cosh(x,n) = \frac{2}{\pi(1+x^n)} \int_0^\pi C_2(x,\theta) e^{\cos\theta} \cos n\theta \, d\theta.$$

[5.45] Let $C_2(x,\theta)$ be given by (5.29), and $S_2(x,\theta)$ be given by (5.28). Prove that $\forall (x,n) \in (\mathbb{R} - \{-1,1\}) \times (2\mathbb{Z}+1)$,

$$\begin{aligned}\cosh(x,n) &= \frac{2}{\pi(1-x^{2n})} \int_0^\pi C_2(x,\theta) e^{\cos\theta} \cos n\theta \, d\theta \\ &\quad - \frac{2x^n}{\pi(1-x^{2n})} \int_0^\pi S_2(x,\theta) e^{\cos\theta} \cos n\theta \, d\theta.\end{aligned}$$

[5.46] Let $C_2(x,\theta)$ be given by (5.29), and $S_2(x,\theta)$ be given by (5.28). Prove that $\forall (x,n) \in \{-1,1\} \times (2\mathbb{Z}+1)$,

$$\int_0^\pi C_2(x,\theta) e^{\cos\theta} \cos n\theta \, d\theta = x^n \int_0^\pi S_2(x,\theta) e^{\cos\theta} \cos n\theta \, d\theta.$$

[5.47] Prove that $\forall (x,y,n) \in \mathbb{R}^2 \times \mathbb{Z}$,

$$\int_0^\pi e^{(x+y)\cos\theta} \cos((x-y)\sin\theta) \cos n\theta \, d\theta = \frac{\pi}{2}(x^n + y^n) e(xy, n).$$

[5.48] Prove that $\forall (x, y, n) \in \mathbb{R}^2 \times (2\mathbb{Z}+1)$,

$$\int_0^\pi \sin((x+y)\cos\theta)\cosh((x-y)\sin\theta)\cos n\theta\, d\theta$$
$$= (-1)^{(n-1)/2}\frac{\pi}{2}\left(x^n + y^n\right)e(-xy, n).$$

[5.49] Prove that $\forall (x, y, n) \in \mathbb{R}^2 \times 2\mathbb{Z}$,

$$\int_0^\pi \sin((x+y)\cos\theta)\cosh((x-y)\sin\theta)\cos n\theta\, d\theta = 0.$$

[5.50] Prove that $\forall (x, y, n) \in \mathbb{R}^2 \times 2\mathbb{Z}$,

$$\int_0^\pi \cos((x+y)\cos\theta)\cosh((x-y)\sin\theta)\cos n\theta\, d\theta$$
$$= (-1)^{n/2}\frac{\pi}{2}\left(x^n + y^n\right)e(-xy, n).$$

[5.51] Prove that $\forall (x, y, n) \in \mathbb{R}^2 \times (2\mathbb{Z}+1)$,

$$\int_0^\pi \cos((x+y)\cos\theta)\cosh((x-y)\sin\theta)\cos n\theta\, d\theta = 0.$$

[5.52] Prove that $\forall (x, y, n) \in \mathbb{R}^2 \times (2\mathbb{Z}+1)$,

$$\int_0^\pi \sinh((x+y)\cos\theta)\cos((x-y)\sin\theta)\cos n\theta\, d\theta = \frac{\pi}{2}\left(x^n + y^n\right)e(xy, n).$$

[5.53] Prove that $\forall (x, y, n) \in \mathbb{R}^2 \times 2\mathbb{Z}$,

$$\int_0^\pi \sinh((x+y)\cos\theta)\cos((x-y)\sin\theta)\cos n\theta\, d\theta = 0.$$

[5.54] Prove that $\forall (x, y, n) \in \mathbb{R}^2 \times 2\mathbb{Z}$,

$$\int_0^\pi \cosh((x+y)\cos\theta)\cos((x-y)\sin\theta)\cos n\theta\, d\theta = \frac{\pi}{2}\left(x^n + y^n\right)e(xy, n).$$

[5.55] Prove that $\forall (x, y, n) \in \mathbb{R}^2 \times (2\mathbb{Z}+1)$,

$$\int_0^\pi \cosh((x+y)\cos\theta)\cos((x-y)\sin\theta)\cos n\theta\, d\theta = 0.$$

[5.56] Prove that $\forall (x, n) \in (\mathbb{R} - \{0\}) \times \mathbb{Z}$,

$$\int_0^\pi e^{(x+1/x)\cos\theta} \cos\left(\left(x - \frac{1}{x}\right)\sin\theta\right) \cos n\theta \, d\theta = \frac{\pi}{2}\left(x^n + \frac{1}{x^n}\right) e(1, n).$$

[5.57] Prove that $\forall (x, n) \in (\mathbb{R} - \{0\}) \times (2\mathbb{Z} + 1)$,

$$\int_0^\pi \sin\left(\left(x + \frac{1}{x}\right)\cos\theta\right) \cosh\left(\left(x - \frac{1}{x}\right)\sin\theta\right) \cos n\theta \, d\theta$$
$$= (-1)^{(n-1)/2} \frac{\pi}{2}\left(x^n + \frac{1}{x^n}\right) e(-1, n).$$

[5.58] Prove that $\forall (x, n) \in (\mathbb{R} - \{0\}) \times 2\mathbb{Z}$,

$$\int_0^\pi \sin\left(\left(x + \frac{1}{x}\right)\cos\theta\right) \cosh\left(\left(x - \frac{1}{x}\right)\sin\theta\right) \cos n\theta \, d\theta = 0.$$

[5.59] Prove that $\forall (x, n) \in (\mathbb{R} - \{0\}) \times 2\mathbb{Z}$,

$$\int_0^\pi \cos\left(\left(x + \frac{1}{x}\right)\cos\theta\right) \cosh\left(\left(x - \frac{1}{x}\right)\sin\theta\right) \cos n\theta \, d\theta$$
$$= (-1)^{n/2} \frac{\pi}{2}\left(x^n + \frac{1}{x^n}\right) e(-1, n).$$

[5.60] Prove that $\forall (x, n) \in (\mathbb{R} - \{0\}) \times (2\mathbb{Z} + 1)$,

$$\int_0^\pi \cos\left(\left(x + \frac{1}{x}\right)\cos\theta\right) \cosh\left(\left(x - \frac{1}{x}\right)\sin\theta\right) \cos n\theta \, d\theta = 0.$$

[5.61] Prove that $\forall (x, n) \in (\mathbb{R} - \{0\}) \times (2\mathbb{Z} + 1)$,

$$\int_0^\pi \sinh\left(\left(x + \frac{1}{x}\right)\cos\theta\right) \cos\left(\left(x - \frac{1}{x}\right)\sin\theta\right) \cos n\theta \, d\theta$$
$$= \frac{\pi}{2}\left(x^n + \frac{1}{x^n}\right) e(1, n).$$

[5.62] Prove that $\forall (x, n) \in (\mathbb{R} - \{0\}) \times 2\mathbb{Z}$,

$$\int_0^\pi \sinh\left(\left(x + \frac{1}{x}\right)\cos\theta\right) \cos\left(\left(x - \frac{1}{x}\right)\sin\theta\right) \cos n\theta \, d\theta = 0.$$

[5.63] Prove that $\forall (x, n) \in (\mathbb{R} - \{0\}) \times 2\mathbb{Z}$,

$$\int_0^\pi \cosh\left(\left(x + \frac{1}{x}\right)\cos\theta\right) \cos\left(\left(x - \frac{1}{x}\right)\sin\theta\right) \cos n\theta \, d\theta$$
$$= \frac{\pi}{2}\left(x^n + \frac{1}{x^n}\right) e(1, n).$$

[5.64] Prove that $\forall (x, n) \in (\mathbb{R} - \{0\}) \times (2\mathbb{Z} + 1)$,

$$\int_0^\pi \cosh\left(\left(x + \frac{1}{x}\right)\cos\theta\right) \cos\left(\left(x - \frac{1}{x}\right)\sin\theta\right) \cos n\theta \, d\theta = 0.$$

[5.65] Prove that $\forall x \in \mathbb{R} - \{0\}$,

$$\int_0^\pi e^{(x+1/x)\cos\theta} \cos\left(\left(x - \frac{1}{x}\right)\sin\theta\right) d\theta = \pi e(1, 0).$$

[5.66] Prove that $\forall x \in \mathbb{R} - \{0\}$,

$$\int_0^\pi \cos\left(\left(x - \frac{1}{x}\right)\cos\theta\right) \cosh\left(\left(x + \frac{1}{x}\right)\sin\theta\right) d\theta = \pi e(1, 0).$$

[5.67] Prove that $\forall x \in \mathbb{R} - \{0\}$,

$$\int_0^\pi \cosh\left(\left(x + \frac{1}{x}\right)\cos\theta\right) \cos\left(\left(x - \frac{1}{x}\right)\sin\theta\right) d\theta = \pi e(1, 0).$$

[5.68] Prove that $\forall n \in \mathbb{Z}$,

$$\int_0^\pi e^{2\cos\theta} \cos n\theta \, d\theta = \pi e(1, n).$$

[5.69] Prove that $\forall n \in 2\mathbb{Z} + 1$,

$$\int_0^\pi \sin(2\cos\theta) \cos n\theta \, d\theta = (-1)^{(n-1)/2} \pi e(-1, n).$$

[5.70] Prove that $\forall n \in 2\mathbb{Z}$,

$$\int_0^\pi \sin(2\cos\theta) \cos n\theta \, d\theta = 0.$$

[5.71] Prove that $\forall n \in 2\mathbb{Z}$,

$$\int_0^\pi \cos(2\cos\theta)\cos n\theta\, d\theta = (-1)^{n/2}\pi e(-1, n).$$

[5.72] Prove that $\forall n \in 2\mathbb{Z}+1$,

$$\int_0^\pi \cos(2\cos\theta)\cos n\theta\, d\theta = 0.$$

[5.73] Prove that $\forall n \in 2\mathbb{Z}+1$,

$$\int_0^\pi \sinh(2\cos\theta)\cos n\theta\, d\theta = \pi e(1, n).$$

[5.74] Prove that $\forall n \in 2\mathbb{Z}$,

$$\int_0^\pi \sinh(2\cos\theta)\cos n\theta\, d\theta = 0.$$

[5.75] Prove that $\forall n \in 2\mathbb{Z}$,

$$\int_0^\pi \cosh(2\cos\theta)\cos n\theta\, d\theta = \pi e(1, n).$$

[5.76] Prove that $\forall n \in 2\mathbb{Z}+1$,

$$\int_0^\pi \cosh(2\cos\theta)\cos n\theta\, d\theta = 0.$$

[5.77] Prove that $\forall n \in 2\mathbb{Z}$,

$$\int_0^\pi \left(\sin^2(\cos\theta)\cosh^2(\sin\theta) + \cos^2(\cos\theta)\sinh^2(\sin\theta)\right)\cos n\theta\, d\theta$$
$$= (-1)^{n/2}\pi\sinh(1, n).$$

[5.78] Prove that $\forall n \in 2\mathbb{Z}+1$,

$$\int_0^\pi \left(\sin^2(\cos\theta)\cosh^2(\sin\theta) + \cos^2(\cos\theta)\sinh^2(\sin\theta)\right)\cos n\theta\, d\theta = 0.$$

[5.79] Prove that $\forall n \in 2\mathbb{Z}$,

$$\int_0^\pi \left(\sinh^2(\cos\theta)\cos^2(\sin\theta) + \cosh^2(\cos\theta)\sin^2(\sin\theta)\right)\cos n\theta\, d\theta$$
$$= \pi\sinh(1, n).$$

Problems

[5.80] Prove that $\forall n \in 2\mathbb{Z}+1$,
$$\int_0^\pi \left(\sinh^2(\cos\theta)\cos^2(\sin\theta)+\cosh^2(\cos\theta)\sin^2(\sin\theta)\right)\cos n\theta\,d\theta = 0.$$

[5.81] Prove that $\forall n \in 2\mathbb{Z}$,
$$\int_0^\pi \left(\cos^2(\cos\theta)\cosh^2(\sin\theta)+\sin^2(\cos\theta)\sinh^2(\sin\theta)\right)\cos n\theta\,d\theta$$
$$= (-1)^{n/2}\pi\cosh(1,n).$$

[5.82] Prove that $\forall n \in 2\mathbb{Z}+1$,
$$\int_0^\pi \left(\cos^2(\cos\theta)\cosh^2(\sin\theta)+\sin^2(\cos\theta)\sinh^2(\sin\theta)\right)\cos n\theta\,d\theta = 0.$$

[5.83] Prove that $\forall n \in 2\mathbb{Z}$,

$$\int_0^\pi \left(\cosh^2(\cos\theta)\cos^2(\sin\theta)+\sinh^2(\cos\theta)\sin^2(\sin\theta)\right)\cos n\theta\,d\theta$$
$$= \pi\cosh(1,n).$$

[5.84] Prove that $\forall n \in 2\mathbb{Z}+1$,
$$\int_0^\pi \left(\cosh^2(\cos\theta)\cos^2(\sin\theta)+\sinh^2(\cos\theta)\sin^2(\sin\theta)\right)\cos n\theta\,d\theta = 0.$$

[5.85] Prove that
$$\int_0^\pi e^{\cos\theta}\cos(\sin\theta)\,d\theta = \pi.$$

[5.86] Prove that $\forall n \in \mathbb{N}$,
$$\int_0^\pi e^{\cos\theta}\cos(\sin\theta)\cos n\theta\,d\theta = \frac{\pi}{2n!}.$$

[5.87] Prove that $\forall n \in 2\mathbb{N}$,
$$\int_0^\pi \cosh(\cos\theta)\cos(\sin\theta)\cos n\theta\,d\theta = \frac{\pi}{2n!}.$$

[5.88] Prove that $\forall n \in 2\mathbb{Z}+1$,
$$\int_0^\pi \cosh(\cos\theta)\cos(\sin\theta)\cos n\theta\,d\theta = 0.$$

[5.89] Prove that $\forall n \in 2\mathbb{N} - 1$,
$$\int_0^\pi \sinh(\cos\theta) \cos(\sin\theta) \cos n\theta \, d\theta = \frac{\pi}{2n!}.$$

[5.90] Prove that $\forall n \in 2\mathbb{Z}$,
$$\int_0^\pi \sinh(\cos\theta) \cos(\sin\theta) \cos n\theta \, d\theta = 0.$$

[5.91] Prove that $\forall n \in 2\mathbb{Z}$,
$$\int_0^\pi \cos(2\sin\theta) \cos n\theta \, d\theta = \pi e(-1, n).$$

[5.92] Prove that $\forall n \in 2\mathbb{Z} + 1$,
$$\int_0^\pi \cos(2\sin\theta) \cos n\theta \, d\theta = 0.$$

[5.93] Prove that $\forall (z, n) \in \mathbb{C} \times \mathbb{Z}$,
$$\sin(z, n) = \frac{e(iz, n) - e(-iz, n)}{2i}.$$

[5.94] Prove that $\forall (z, n) \in \mathbb{C} \times \mathbb{Z}$,
$$\cos(z, n) = \frac{e(iz, n) + e(-iz, n)}{2}.$$

[5.95] Prove that $\forall (z, n) \in \mathbb{C} \times \mathbb{Z}$,
$$\sinh(z, n) = \frac{e(z, n) - e(-z, n)}{2}.$$

[5.96] Prove that $\forall (z, n) \in \mathbb{C} \times \mathbb{Z}$,
$$\cosh(z, n) = \frac{e(z, n) + e(-z, n)}{2}.$$

Chapter 6
Series Sums Using Special Integrals

In this chapter, we will explore some more interesting properties of the pseudo-exponential functions. First, we will prove that the pseudo-exponential functions are expressible as a finite-polynomial combination of each other. Next, we will derive the differential equations satisfied by the pseudo-exponential functions. Finally, we will define the *star product* of series and find its integral representation. In the problems listed in this chapter, we will see that several interesting series can be summed using the special integral of [5.21]. We will also evaluate special integrals and series sums derived using the star product identity.

6.1 Polynomial Dependence in the Pseudo-Exponential Functions

Recall the concept of linear dependence in *Linear Algebra* (Hoffman and Kunje, 2015). We say that the functions $(f_j(x))_{j=1}^n$ are linearly dependent if there exist constants $(\alpha_j)_{j=1}^n$ such that $\sum_{j=1}^n \alpha_j f_j(x) = 0$. We generalize the concept of linear dependence of functions into polynomial dependence, in which the constants $(\alpha_j)_{j=1}^n$ are replaced by finite-degree polynomials $(\alpha_j(x))_{j=1}^n$.

Definition 6.1 *(Polynomially Dependent Functions)*

We say that the functions $(f_j(x))_{j=1}^n$ are polynomially dependent if there exist finite-degree polynomials $(\alpha_j(x))_{j=1}^n$ such that

$$\sum_{j=1}^n \alpha_j(x) f_j(x) = 0.$$

Now, we will prove polynomial dependence among the pseudo-exponential functions in the examples that follow.

Example 6.2 *(Polynomial Dependence of $e(z,n)$)*
Prove that $\forall (z,n) \in \mathbb{C} \times \mathbb{Z}$,

$$e(z,n) = (n+1)e(z,n+1) + ze(z,n+2). \tag{6.1}$$

Proof Using (5.11), we get

$(n+1)e(z, n+1) + ze(z, n+2)$

$$= (n+1) \sum_{j=\max\{0,-(n+1)\}}^{\infty} \frac{z^j}{j!(n+1+j)!} + z \sum_{j=\max\{0,-(n+2)\}}^{\infty} \frac{z^j}{j!(n+2+j)!}$$

$$= (n+1) \sum_{j=\max\{0,-(n+1)\}}^{\infty} \frac{z^j}{j!(n+1+j)!} + \sum_{j=\max\{0,-(n+2)\}}^{\infty} \frac{z^{j+1}}{j!(n+2+j)!}$$

$$= (n+1) \sum_{j=\max\{0,-(n+1)\}}^{\infty} \frac{z^j}{j!(n+1+j)!} + \sum_{j=\max\{1,-(n+1)\}}^{\infty} \frac{z^j}{(j-1)!(n+1+j)!}$$

$$= \frac{[n \geq 0](n+1)}{0!(n+1)!} + \sum_{j=\max\{1,-(n+1)\}}^{\infty} \left(\frac{n+1}{j!(n+1+j)!} + \frac{1}{(j-1)!(n+1+j)!} \right) z^j$$

$$= \frac{[n \geq 0]}{0!n!} + \sum_{j=\max\{1,-(n+1)\}}^{\infty} \left(\frac{n+1}{j} + 1 \right) \frac{z^j}{(j-1)!(n+1+j)!}$$

$$= \frac{[n \geq 0]}{0!n!} + \sum_{j=\max\{1,-n\}}^{\infty} \frac{z^j}{j!(n+j)!}$$

$$= \sum_{j=\max\{0,-n\}}^{\infty} \frac{z^j}{j!(n+j)!}$$

$$= e(z,n).$$

$\square$

Example 6.3 *(Polynomial Dependence of $\sin(z,n)$)*
Prove that $\forall (z,n) \in \mathbb{C} \times \mathbb{Z}$,

$$\sin(z,n) = (n+1)\sin(z,n+1) + z\cos(z,n+2). \tag{6.2}$$

Proof Using (5.15) and (5.16), we get

$(n+1)\sin(z, n+1) + z\cos(z, n+2)$

$$= (n+1) \sum_{j=\max\{0,\lfloor -(n+1)/2 \rfloor\}}^{\infty} (-1)^j \frac{z^{2j+1}}{(2j+1)!(n+1+2j+1)!}$$

6.1 Polynomial Dependence in the Pseudo-Exponential Functions

$$+z \sum_{j=\max\{0,\lceil -(n+2)/2\rceil\}}^{\infty} (-1)^j \frac{z^{2j}}{(2j)!(n+2+2j)!}$$

$$= (n+1) \sum_{j=\max\{0,\lfloor -(n+1)/2\rfloor\}}^{\infty} (-1)^j \frac{z^{2j+1}}{(2j+1)!(n+1+2j+1)!}$$

$$+ \sum_{j=\max\{0,\lceil -(n+2)/2\rceil\}}^{\infty} (-1)^j \frac{z^{2j+1}}{(2j)!(n+2+2j)!}$$

$$= \sum_{j=\max\{0,\lfloor -(n+1)/2\rfloor\}}^{\infty} (-1)^j \left(\frac{n+1}{(2j+1)!(n+1+2j+1)!} \right.$$

$$\left. + \frac{1}{(2j)!(n+2+2j)!} \right) z^{2j+1}$$

$$= \sum_{j=\max\{0,\lfloor -(n+1)/2\rfloor\}}^{\infty} (-1)^j \left(\frac{n+1}{2j+1} + 1 \right) \frac{z^{2j+1}}{(2j)!(n+2+2j)!}$$

$$= \sum_{j=\max\{0,\lfloor -(n+1)/2\rfloor\}}^{j=\max\{0,\lfloor -(n+1)/2\rfloor\}} (-1)^j \left(\frac{n+2j+2}{2j+1} \right) \frac{z^{2j+1}}{(2j)!(n+2+2j)!}$$

$$+ \sum_{j=\max\{1,\lfloor -(n-1)/2\rfloor\}}^{\infty} (-1)^j \frac{z^{2j+1}}{(2j+1)!(n+2j+1)!}$$

$$= \sum_{j=\max\{0,\lfloor -n/2\rfloor\}}^{\infty} (-1)^j \frac{z^{2j+1}}{(2j+1)!(n+2j+1)!}$$

$$= \sin(z, n).$$

$\square$

Example 6.4 *(Polynomial Dependence of* $\cos(z, n)$*)*
Prove that $\forall (z, n) \in \mathbb{C} \times \mathbb{Z}$,

$$\cos(z, n) = (n+1)\cos(z, n+1) - z\sin(z, n+2). \tag{6.3}$$

Proof Using (5.16) and (5.15), we get

$$(n+1)\cos(z, n+1) - z\sin(z, n+2)$$

$$= (n+1) \sum_{j=\max\{0,\lceil -(n+1)/2\rceil\}}^{\infty} (-1)^j \frac{z^{2j}}{(2j)!(n+1+2j)!}$$

$$-z \sum_{j=\max\{0,\lfloor-(n+2)/2\rfloor\}}^{\infty} (-1)^j \frac{z^{2j+1}}{(2j+1)!(n+2+2j+1)!}$$

$$= (n+1) \sum_{j=\max\{0,\lceil-(n+1)/2\rceil\}}^{\infty} (-1)^j \frac{z^{2j}}{(2j)!(n+1+2j)!}$$

$$- \sum_{j=\max\{0,\lfloor-(n+2)/2\rfloor\}}^{\infty} (-1)^j \frac{z^{2j+2}}{(2j+1)!(n+2+2j+1)!}$$

$$= (n+1) \sum_{j=\max\{0,\lceil-(n+1)/2\rceil\}}^{\infty} (-1)^j \frac{z^{2j}}{(2j)!(n+1+2j)!}$$

$$+ \sum_{j=\max\{1,\lfloor-n/2\rfloor\}}^{\infty} (-1)^j \frac{z^{2j}}{(2j-1)!(n+1+2j)!}$$

$$= [n \geq 0]\frac{n+1}{0!(n+1)!} + \sum_{j=\max\{1,\lfloor-n/2\rfloor\}}^{\infty} (-1)^j \left(\frac{n+1}{2j}+1\right) \frac{z^{2j}}{(2j-1)!(n+1+2j)!}$$

$$= [n \geq 0]\frac{n+1}{0!(n+1)!} + \sum_{j=\max\{1,\lceil-n/2\rceil\}}^{\infty} (-1)^j \frac{z^{2j}}{(2j)!(n+2j)!}$$

$$= \sum_{j=\max\{0,\lceil-n/2\rceil\}}^{\infty} (-1)^j \frac{z^{2j}}{(2j)!(n+2j)!}$$

$$= \cos(z, n).$$

$\square$

Example 6.5 *(Polynomial Dependence of* $\sinh(z, n)$*)*
Prove that $\forall (z, n) \in \mathbb{C} \times \mathbb{Z}$,

$$\sinh(z, n) = (n+1)\sinh(z, n+1) + z\cosh(z, n+2). \qquad (6.4)$$

Proof Replacing z with iz in (6.2) and using (5.17) and (5.18), we get

$$\sin(iz, n) = (n+1)\sin(iz, n+1) + iz\cos(iz, n+2)$$
$$\iff i\sinh(z, n) = (n+1)i\sinh(z, n+1) + iz\cosh(z, n+2)$$
$$\iff \sinh(z, n) = (n+1)\sinh(z, n+1) + z\cosh(z, n+2).$$

$\square$

Example 6.6 *(Polynomial Dependence of* $\cosh(z, n)$*)*
Prove that $\forall (z, n) \in \mathbb{C} \times \mathbb{Z}$,

$$\cosh(z, n) = (n+1)\cosh(z, n+1) + z\sinh(z, n+2). \qquad (6.5)$$

6.2 Differential Equations Satisfied by the Pseudo-Exponential Functions

Proof Replacing z with iz in (6.3) and using (5.18) and (5.17), we get

$$\cos(iz, n) = (n+1)\cos(iz, n+1) - iz\sin(iz, n+2)$$
$$\iff \cosh(z, n) = (n+1)\cosh(z, n+1) + z\sinh(z, n+2).$$

$\square$

6.2 Differential Equations Satisfied by the Pseudo-Exponential Functions

We know that the elementary exponential functions satisfy simple differential equations.

$$\frac{dy}{dz} - y = 0 \implies y = e^z,$$

$$\frac{d^2y}{dz^2} + y = 0 \implies y = \sin z, \quad y = \cos z,$$

$$\frac{d^2y}{dz^2} - y = 0 \implies y = \sinh z, \quad y = \cosh z.$$

From the above, we can see that the elementary exponential functions satisfy up to second-order differential equations. Now, we will derive the differential equations satisfied by the pseudo-exponential functions and see that they satisfy up to fourth-order differential equations.

Example 6.7 (*Differential Equation Satisfied by $e(z, n)$*)
Prove that $\forall (z, n) \in \mathbb{C} \times \mathbb{Z}$, $y = e(z, n)$ satisfies the differential equation

$$z\frac{d^2y}{dz^2} + (n+1)\frac{dy}{dz} - y = 0.$$

Proof From (5.21), we have

$$y = e(z, n)$$
$$\implies \frac{dy}{dz} = e(z, n+1) \tag{6.6}$$
$$\implies \frac{d^2y}{dz^2} = e(z, n+2).$$

From (6.6) and (6.1), we get

$$y = (n+1)\frac{dy}{dz} + z\frac{d^2y}{dz^2}$$

$$\iff z\frac{d^2y}{dz^2} + (n+1)\frac{dy}{dz} - y = 0.$$

□

Example 6.8 *(Differential Equation Satisfied by* $\sin(z, n)$*)*
Prove that $\forall (z, n) \in \mathbb{C} \times \mathbb{Z}$, $y = \sin(z, n)$ satisfies the differential equation

$$z^2\frac{d^4y}{dz^4} + 2(n+2)z\frac{d^3y}{dz^3} + (n+1)(n+2)\frac{d^2y}{dz^2} + y = 0.$$

Proof From (5.22) and (5.23), we have

$$y = \sin(z, n)$$
$$\implies \frac{dy}{dz} = \cos(z, n+1) \qquad (6.7)$$
$$\implies \frac{d^2y}{dz^2} = -\sin(z, n+2).$$

From (6.7) and (6.3), we get

$$\cos(z, n) = (n+1)\frac{dy}{dz} + z\frac{d^2y}{dz^2}. \qquad (6.8)$$

Differentiating (6.8) with respect to z, and using (5.23), we get

$$-\sin(z, n+1) = (n+2)\frac{d^2y}{dz^2} + z\frac{d^3y}{dz^3}. \qquad (6.9)$$

Differentiating (6.9) with respect to z, and using (5.22), we get

$$-\cos(z, n+2) = (n+3)\frac{d^3y}{dz^3} + z\frac{d^4y}{dz^4}. \qquad (6.10)$$

From (6.2), (6.7), (6.9), and (6.10), we get

$$z\left((n+3)\frac{d^3y}{dz^3} + z\frac{d^4y}{dz^4}\right) + (n+1)\left((n+2)\frac{d^2y}{dz^2} + z\frac{d^3y}{dz^3}\right) + y = 0$$
$$\iff z^2\frac{d^4y}{dz^4} + 2(n+2)z\frac{d^3y}{dz^3} + (n+1)(n+2)\frac{d^2y}{dz^2} + y = 0.$$

□

Example 6.9 *(Differential Equation Satisfied by* $\cos(z, n)$*)*
Prove that $\forall (z, n) \in \mathbb{C} \times \mathbb{Z}$, $y = \cos(z, n)$ satisfies the differential equation

$$z^2\frac{d^4y}{dz^4} + 2(n+2)z\frac{d^3y}{dz^3} + (n+1)(n+2)\frac{d^2y}{dz^2} + y = 0.$$

6.2 Differential Equations Satisfied by the Pseudo-Exponential Functions

Proof From (5.23) and (5.22), we have

$$y = \cos(z, n)$$
$$\Longrightarrow \frac{dy}{dz} = -\sin(z, n+1) \quad (6.11)$$
$$\Longrightarrow \frac{d^2y}{dz^2} = -\cos(z, n+2).$$

From (6.11) and (6.2), we get

$$\sin(z, n) = -(n+1)\frac{dy}{dz} - z\frac{d^2y}{dz^2}. \quad (6.12)$$

Differentiating (6.12) with respect to z, and using (5.22), we get

$$\cos(z, n+1) = -(n+2)\frac{d^2y}{dz^2} - z\frac{d^3y}{dz^3}. \quad (6.13)$$

Differentiating (6.13) with respect to z, and using (5.23), we get

$$-\sin(z, n+2) = -(n+3)\frac{d^3y}{dz^3} - z\frac{d^4y}{dz^4}$$
$$\Longleftrightarrow \sin(z, n+2) = (n+3)\frac{d^3y}{dz^3} + z\frac{d^4y}{dz^4}. \quad (6.14)$$

From (6.3), (6.11), (6.13), and (6.14), we get

$$z\left((n+3)\frac{d^3y}{dz^3} + z\frac{d^4y}{dz^4}\right) + (n+1)\left((n+2)\frac{d^2y}{dz^2} + z\frac{d^3y}{dz^3}\right) + y = 0$$
$$\Longleftrightarrow z^2\frac{d^4y}{dz^4} + 2(n+2)z\frac{d^3y}{dz^3} + (n+1)(n+2)\frac{d^2y}{dz^2} + y = 0.$$

$\square$

Example 6.10 *(Differential Equation Satisfied by $\sinh(z, n)$)*
Prove that $\forall (z, n) \in \mathbb{C} \times \mathbb{Z}$, $y = \sinh(z, n)$ satisfies the differential equation

$$z^2\frac{d^4y}{dz^4} + 2(n+2)z\frac{d^3y}{dz^3} + (n+1)(n+2)\frac{d^2y}{dz^2} - y = 0.$$

Proof From (5.24) and (5.25), we have

$$y = \sinh(z, n)$$
$$\implies \frac{dy}{dz} = \cosh(z, n+1) \tag{6.15}$$
$$\implies \frac{d^2y}{dz^2} = \sinh(z, n+2).$$

From (6.15) and (6.5), we get

$$\cosh(z, n) = (n+1)\frac{dy}{dz} + z\frac{d^2y}{dz^2}. \tag{6.16}$$

Differentiating (6.16) with respect to z, and using (5.25), we get

$$\sinh(z, n+1) = (n+2)\frac{d^2y}{dz^2} + z\frac{d^3y}{dz^3}. \tag{6.17}$$

Differentiating (6.17) with respect to z, and using (5.24), we get

$$\cosh(z, n+2) = (n+3)\frac{d^3y}{dz^3} + z\frac{d^4y}{dz^4}. \tag{6.18}$$

From (6.4), (6.15), (6.17), and (6.18), we get

$$z\left((n+3)\frac{d^3y}{dz^3} + z\frac{d^4y}{dz^4}\right) + (n+1)\left((n+2)\frac{d^2y}{dz^2} + z\frac{d^3y}{dz^3}\right) - y = 0$$
$$\iff z^2\frac{d^4y}{dz^4} + 2(n+2)z\frac{d^3y}{dz^3} + (n+1)(n+2)\frac{d^2y}{dz^2} - y = 0.$$

□

Example 6.11 *(Differential Equation Satisfied by $\cosh(z, n)$)*
Prove that $\forall (z, n) \in \mathbb{C} \times \mathbb{Z}$, $y = \cosh(z, n)$ satisfies the differential equation

$$z^2\frac{d^4y}{dz^4} + 2(n+2)z\frac{d^3y}{dz^3} + (n+1)(n+2)\frac{d^2y}{dz^2} - y = 0.$$

Proof From (5.25) and (5.24), we have

$$y = \cosh(z, n)$$
$$\implies \frac{dy}{dz} = \sinh(z, n+1) \tag{6.19}$$
$$\implies \frac{d^2y}{dz^2} = \cosh(z, n+2).$$

6.3 Series Transformations

From (6.19) and (6.4), we get

$$\sinh(z, n) = (n+1)\frac{dy}{dz} + z\frac{d^2y}{dz^2}. \tag{6.20}$$

Differentiating (6.20) with respect to z, and using (5.24), we get

$$\cosh(z, n+1) = (n+2)\frac{d^2y}{dz^2} + z\frac{d^3y}{dz^3}. \tag{6.21}$$

Differentiating (6.21) with respect to z, and using (5.25), we get

$$\sinh(z, n+2) = (n+3)\frac{d^3y}{dz^3} + z\frac{d^4y}{dz^4}. \tag{6.22}$$

From (6.5), (6.19), (6.21), and (6.22), we get

$$z\left((n+3)\frac{d^3y}{dz^3} + z\frac{d^4y}{dz^4}\right) + (n+1)\left((n+2)\frac{d^2y}{dz^2} + z\frac{d^3y}{dz^3}\right) - y = 0$$

$$\iff z^2\frac{d^4y}{dz^4} + 2(n+2)z\frac{d^3y}{dz^3} + (n+1)(n+2)\frac{d^2y}{dz^2} - y = 0.$$

$\square$

6.3 Series Transformations

In Theorem 4.5, we computed the sum of squares of terms in a series and generalized it to find a quadratic sum of terms in Theorem 4.6. Similarly, we computed several series sums involving the binomial coefficients in Theorems 4.7, 4.8, 4.10, 4.11, and 4.13. We computed some more series sums mainly involving the binomial coefficients in Theorems 4.17, 4.28, and 4.41. We can view the above series sums as an instance of the *star product* of two series that we will define below. In this section, we will find a special integral to compute the star product of two functions. The above-mentioned theorems can easily be proved using the star product special integral.

Definition 6.12 (*Star Product of Series*)
Let

$$f(z) = \sum_{j=0}^{\infty} a_j z^j = a_0 + a_1 z + a_2 z^2 + a_3 z^3 + \cdots \tag{6.23}$$

and

$$g(z) = \sum_{j=0}^{\infty} b_j z^j = b_0 + b_1 z + b_2 z^2 + b_3 z^3 + \cdots \qquad (6.24)$$

be functions that can be expanded by Maclaurin's series. Then, the star product, $f \star g$, of f and g is defined by

$$(f \star g)(z) = \sum_{j=0}^{\infty} a_j b_j z^j = a_0 b_0 + a_1 b_1 z + a_2 b_2 z^2 + a_3 b_3 z^3 + \cdots \qquad (6.25)$$

We also write $(f \star g)(xy)$ as $f(x) \star g(y)$.

Theorem 6.13 (A Star Product Identity)
Let the functions f and g be defined as in Definition 6.12. Assuming that $f\left(xe^{i\theta}\right)$, $f\left(xe^{-i\theta}\right)$, $g\left(ye^{i\theta}\right)$, and $g\left(ye^{-i\theta}\right)$ are all convergent, the star product, $f \star g$, of f and g is given by

$$(f \star g)(z) = -f(0)g(0) + \frac{1}{2\pi} \int_0^{\pi} \left(f\left(xe^{i\theta}\right) + f\left(xe^{-i\theta}\right)\right) \left(g\left(ye^{i\theta}\right) + g\left(ye^{-i\theta}\right)\right) d\theta,$$

where $z = xy$.

Proof Using (6.23) and (6.24), we get

$$\left(f\left(xe^{i\theta}\right) + f\left(xe^{-i\theta}\right)\right) \left(g\left(ye^{i\theta}\right) + g\left(ye^{-i\theta}\right)\right)$$

$$= \left(\sum_{k=0}^{\infty} a_k \left(e^{ik\theta} + e^{-ik\theta}\right) x^k\right) \left(\sum_{j=0}^{\infty} b_j \left(e^{ij\theta} + e^{-ij\theta}\right) y^j\right)$$

$$= 4 \left(\sum_{k=0}^{\infty} a_k x^k \cos k\theta\right) \left(\sum_{j=0}^{\infty} b_j y^j \cos j\theta\right)$$

$$= 4 \sum_{k=0}^{\infty} a_k b_k x^k y^k \cos^2 k\theta + 4 \sum_{k=0}^{\infty} \sum_{j=0, j \neq k}^{\infty} a_k b_j x^k y^j \cos k\theta \cos j\theta$$

$$= 2 \sum_{k=0}^{\infty} a_k b_k (1 + \cos 2k\theta) z^k + 2 \sum_{k=0}^{\infty} \sum_{j=0, j \neq k}^{\infty} a_k b_j (\cos(k+j)\theta + \cos(k-j)\theta) x^k y^j.$$

$$(6.26)$$

Integrating (6.26) with respect to θ from 0 to π, dividing the result by 2π, subtracting by $f(0)g(0)$, and using (6.25), we get

$$-f(0)g(0) + \frac{1}{2\pi} \int_0^{\pi} \left(f\left(xe^{i\theta}\right) + f\left(xe^{-i\theta}\right)\right) \left(g\left(ye^{i\theta}\right) + g\left(ye^{-i\theta}\right)\right) d\theta$$

6.3 Series Transformations

$$= -a_0 b_0 + \frac{1}{\pi} \int_0^\pi \left(\sum_{k=0}^\infty a_k b_k (1 + \cos 2k\theta) z^k \right) d\theta$$

$$+ \frac{1}{\pi} \int_0^\pi \left(\sum_{k=0}^\infty \sum_{j=0, j \neq k}^\infty a_k b_j (\cos(k+j)\theta + \cos(k-j)\theta) x^k y^j \right) d\theta$$

$$= -a_0 b_0 + \frac{1}{\pi} \sum_{k=0}^\infty a_k b_k z^k \int_0^\pi (1 + \cos 2k\theta) d\theta$$

$$+ \frac{1}{\pi} \sum_{k=0}^\infty \sum_{j=0, j \neq k}^\infty a_k b_j x^k y^j \int_0^\pi (\cos(k+j)\theta + \cos(k-j)\theta) \, d\theta$$

$$= -a_0 b_0 + \frac{a_0 b_0}{\pi} (2\theta) \Big|_0^\pi + \frac{1}{\pi} \sum_{k=1}^\infty a_k b_k z^k \left(\theta + \frac{\sin 2k\theta}{2k} \right) \Big|_0^\pi$$

$$+ \frac{1}{\pi} \sum_{k=0}^\infty \sum_{j=0, j \neq k}^\infty a_k b_j x^k y^j \left(\frac{\sin(k+j)\theta}{k+j} + \frac{\sin(k-j)\theta}{k-j} \right) \Big|_0^\pi$$

$$= \sum_{k=0}^\infty a_k b_k z^k$$

$$= (f \star g)(z).$$

$\square$

Theorem 6.14 (Another Star Product Identity)
Let the functions f and g be defined as in Definition 6.12. Assuming that $f\left(xe^{i\theta}\right)$, $f\left(xe^{-i\theta}\right)$, $g\left(ye^{i\theta}\right)$, and $g\left(ye^{-i\theta}\right)$ are all convergent, the star product, $f \star g$, of f and g is given by

$$(f \star g)(z) = f(0)g(0) - \frac{1}{2\pi} \int_0^\pi \left(f\left(xe^{i\theta}\right) - f\left(xe^{-i\theta}\right) \right) \left(g\left(ye^{i\theta}\right) - g\left(ye^{-i\theta}\right) \right) d\theta,$$

where $z = xy$.

Proof Using (6.23) and (6.24), we get

$$\left(f\left(xe^{i\theta}\right) - f\left(xe^{-i\theta}\right)\right)\left(g\left(ye^{i\theta}\right) - g\left(ye^{-i\theta}\right)\right)$$

$$= \left(\sum_{k=0}^{\infty} a_k \left(e^{ik\theta} - e^{-ik\theta}\right) x^k\right) \left(\sum_{j=0}^{\infty} b_j \left(e^{ij\theta} - e^{-ij\theta}\right) y^j\right)$$

$$= -4 \left(\sum_{k=0}^{\infty} a_k x^k \sin k\theta\right) \left(\sum_{j=0}^{\infty} b_j y^j \sin j\theta\right)$$

$$= -4 \sum_{k=0}^{\infty} a_k b_k x^k y^k \sin^2 k\theta - 4 \sum_{k=0}^{\infty} \sum_{j=0, j\neq k}^{\infty} a_k b_j x^k y^j \sin k\theta \sin j\theta \quad (6.27)$$

$$= -2 \sum_{k=0}^{\infty} a_k b_k (1 - \cos 2k\theta) z^k$$

$$+ 2 \sum_{k=0}^{\infty} \sum_{j=0, j\neq k}^{\infty} a_k b_j (\cos(k+j)\theta - \cos(k-j)\theta) x^k y^j.$$

Integrating (6.27) with respect to θ from 0 to π, dividing the result by 2π, subtracting from $f(0)g(0)$, and using (6.25), we get

$$f(0)g(0) - \frac{1}{2\pi} \int_0^{\pi} \left(f\left(xe^{i\theta}\right) - f\left(xe^{-i\theta}\right)\right)\left(g\left(ye^{i\theta}\right) - g\left(ye^{-i\theta}\right)\right) d\theta$$

$$= a_0 b_0 + \frac{1}{\pi} \int_0^{\pi} \left(\sum_{k=0}^{\infty} a_k b_k (1 - \cos 2k\theta) z^k\right) d\theta$$

$$- \frac{1}{\pi} \int_0^{\pi} \left(\sum_{k=0}^{\infty} \sum_{j=0, j\neq k}^{\infty} a_k b_j (\cos(k+j)\theta - \cos(k-j)\theta) x^k y^j\right) d\theta$$

$$= a_0 b_0 + \frac{1}{\pi} \sum_{k=0}^{\infty} a_k b_k z^k \int_0^{\pi} (1 - \cos 2k\theta) d\theta$$

$$- \frac{1}{\pi} \sum_{k=0}^{\infty} \sum_{j=0, j\neq k}^{\infty} a_k b_j x^k y^j \int_0^{\pi} (\cos(k+j)\theta - \cos(k-j)\theta) d\theta$$

$$= a_0 b_0 + \frac{1}{\pi} \sum_{k=1}^{\infty} a_k b_k z^k \left(\theta - \frac{\sin 2k\theta}{2k}\right)\bigg|_0^{\pi}$$

$$- \frac{1}{\pi} \sum_{k=0}^{\infty} \sum_{j=0, j\neq k}^{\infty} a_k b_j x^k y^j \left(\frac{\sin(k+j)\theta}{k+j} - \frac{\sin(k-j)\theta}{k-j}\right)\bigg|_0^{\pi}$$

$$= \sum_{k=0}^{\infty} a_k b_k z^k = (f \star g)(z).$$

$\square$

Problems

[6.1] Let the function $C_3(x, \theta)$ be defined as

$$C_3(x, \theta) = \cosh(x \sin \theta) \cos(x \cos \theta) \cos(2 \sin \theta) \\ - \sinh(x \sin \theta) \sin(x \cos \theta) \sin(2 \sin \theta), \quad (6.28)$$

and the function $S_3(x, \theta)$ be defined as

$$S_3(x, \theta) = \sinh(x \sin \theta) \cos(x \cos \theta) \sin(2 \sin \theta) \\ + \cosh(x \sin \theta) \sin(x \cos \theta) \cos(2 \sin \theta). \quad (6.29)$$

Let the function $\phi(x)$ be defined as $\phi(x) = \tan^{-1} x$. Prove that $\forall (x, n) \in \mathbb{R} \times (\mathbb{N} - 1) - \{0\} \times (2\mathbb{N} - 1)$,

$$\sum_{j=0}^{\infty} \frac{(1+x^2)^j \cos 2j\phi(x)}{(2j)!(n+2j)!} - \sqrt{1+x^2} \sum_{j=0}^{\infty} \frac{(1+x^2)^j \cos(2j+1)\phi(x)}{(2j+1)!(n+2j+1)!}$$

$$= \left(\frac{1}{0!n!} + \frac{(1+x^2) \cos 2\phi(x)}{2!(n+2)!} + \frac{(1+x^2)^2 \cos 4\phi(x)}{4!(n+4)!} \right.$$
$$\left. + \frac{(1+x^2)^3 \cos 6\phi(x)}{6!(n+6)!} + \cdots \right) - \sqrt{1+x^2} \left(\frac{\cos \phi(x)}{1!(n+1)!} \right.$$
$$\left. + \frac{(1+x^2) \cos 3\phi(x)}{3!(n+3)!} + \frac{(1+x^2)^2 \cos 5\phi(x)}{5!(n+5)!} + \frac{(1+x^2)^3 \cos 7\phi(x)}{7!(n+7)!} + \cdots \right)$$

$$= \frac{2}{\pi} \cdot \frac{1 + (1+x^2)^{n/2} \cos n(\pi - \phi(x))}{1 + (1+x^2)^n + 2(1+x^2)^{n/2} \cos n(\pi - \phi(x))} \int_0^\pi C_3(x, \theta) \cos n\theta \, d\theta$$

$$+ \frac{2}{\pi} \cdot \frac{(1+x^2)^{n/2} \sin n(\pi - \phi(x))}{1 + (1+x^2)^n + 2(1+x^2)^{n/2} \cos n(\pi - \phi(x))} \int_0^\pi S_3(x, \theta) \cos n\theta \, d\theta.$$

[6.2] Let the function $\phi(x)$ be defined as $\phi(x) = \tan^{-1} x$ and $C_3(x, \theta)$ be given by (6.28). Prove that $\forall x \in \mathbb{R}$,

$$\sum_{j=0}^{\infty} \frac{(1+x^2)^j \cos 2j\phi(x)}{((2j)!)^2} - \sqrt{1+x^2} \sum_{j=0}^{\infty} \frac{(1+x^2)^j \cos(2j+1)\phi(x)}{((2j+1)!)^2}$$

$$= \left(\frac{1}{(0!)^2} + \frac{(1+x^2) \cos 2\phi(x)}{(2!)^2} + \frac{(1+x^2)^2 \cos 4\phi(x)}{(4!)^2} \right.$$

$$\left. + \frac{(1+x^2)^3 \cos 6\phi(x)}{(6!)^2} + \cdots \right)$$

$$- \sqrt{1+x^2} \left(\frac{\cos \phi(x)}{(1!)^2} + \frac{(1+x^2) \cos 3\phi(x)}{(3!)^2} + \frac{(1+x^2)^2 \cos 5\phi(x)}{(5!)^2} \right.$$

$$\left. + \frac{(1+x^2)^3 \cos 7\phi(x)}{(7!)^2} + \cdots \right)$$

$$= \frac{1}{\pi} \int_0^{\pi} C_3(x, \theta) \, d\theta.$$

[6.3] Let the function $\phi(x)$ be defined as $\phi(x) = \tan^{-1} x$, $S_3(x, \theta)$ be given by (6.29), and $C_3(x, \theta)$ be given by (6.28). Prove that $\forall (x, n) \in \mathbb{R} \times (\mathbb{N} - 1) - \{0\} \times (2\mathbb{N} - 1)$,

$$\sqrt{1+x^2} \sum_{j=0}^{\infty} \frac{(1+x^2)^j \sin(2j+1)\phi(x)}{(2j+1)!(n+2j+1)!} - \sum_{j=1}^{\infty} \frac{(1+x^2)^j \sin 2j\phi(x)}{(2j)!(n+2j)!}$$

$$= \sqrt{1+x^2} \left(\frac{\sin \phi(x)}{1!(n+1)!} + \frac{(1+x^2) \sin 3\phi(x)}{3!(n+3)!} + \frac{(1+x^2)^2 \sin 5\phi(x)}{5!(n+5)!} \right.$$

$$\left. + \frac{(1+x^2)^3 \sin 7\phi(x)}{7!(n+7)!} + \cdots \right)$$

$$- \left(\frac{(1+x^2) \sin 2\phi(x)}{2!(n+2)!} + \frac{(1+x^2)^2 \sin 4\phi(x)}{4!(n+4)!} + \frac{(1+x^2)^3 \sin 6\phi(x)}{6!(n+6)!} \right.$$

$$\left. + \frac{(1+x^2)^4 \sin 8\phi(x)}{8!(n+8)!} + \cdots \right)$$

$$= \frac{2}{\pi} \cdot \frac{1 + (1+x^2)^{n/2} \cos n(\pi - \phi(x))}{1 + (1+x^2)^n + 2(1+x^2)^{n/2} \cos n(\pi - \phi(x))} \int_0^{\pi} S_3(x, \theta) \cos n\theta \, d\theta$$

$$- \frac{2}{\pi} \cdot \frac{(1+x^2)^{n/2} \sin n(\pi - \phi(x))}{1 + (1+x^2)^n + 2(1+x^2)^{n/2} \cos n(\pi - \phi(x))} \int_0^{\pi} C_3(x, \theta) \cos n\theta \, d\theta.$$

Problems

[6.4] Let the function $\phi(x)$ be defined as $\phi(x) = \tan^{-1} x$ and $S_3(x, \theta)$ be given by (6.29). Prove that $\forall x \in \mathbb{R}$,

$$\sqrt{1+x^2} \sum_{j=0}^{\infty} \frac{(1+x^2)^j \sin(2j+1)\phi(x)}{((2j+1)!)^2} - \sum_{j=1}^{\infty} \frac{(1+x^2)^j \sin 2j\phi(x)}{((2j)!)^2}$$

$$= \sqrt{1+x^2} \left(\frac{\sin \phi(x)}{(1!)^2} + \frac{(1+x^2) \sin 3\phi(x)}{(3!)^2} + \frac{(1+x^2)^2 \sin 5\phi(x)}{(5!)^2} \right.$$

$$\left. + \frac{(1+x^2)^3 \sin 7\phi(x)}{(7!)^2} + \cdots \right)$$

$$- \left(\frac{(1+x^2) \sin 2\phi(x)}{(2!)^2} + \frac{(1+x^2)^2 \sin 4\phi(x)}{(4!)^2} + \frac{(1+x^2)^3 \sin 6\phi(x)}{(6!)^2} \right.$$

$$\left. + \frac{(1+x^2)^4 \sin 8\phi(x)}{(8!)^2} + \cdots \right)$$

$$= \frac{1}{\pi} \int_0^{\pi} S_3(x, \theta) \, d\theta.$$

[6.5] Let $S_3(1, \theta)$ be given by (6.29). Prove that

$$\sum_{j=0}^{\infty} (-1)^j \left(\frac{2^{2j}}{((4j+1)!)^2} + \frac{2^{2j+1}}{((4j+3)!)^2} \right) - \sum_{j=0}^{\infty} (-1)^j \frac{2^{2j+1}}{((4j+2)!)^2}$$

$$= \left(\frac{1}{(1!)^2} + \frac{2}{(3!)^2} - \frac{2^2}{(5!)^2} - \frac{2^3}{(7!)^2} + \cdots \right)$$

$$- \left(\frac{2}{(2!)^2} - \frac{2^3}{(6!)^2} + \frac{2^5}{(10!)^2} - \frac{2^7}{(14!)^2} + \cdots \right)$$

$$= \frac{1}{\pi} \int_0^{\pi} S_3(1, \theta) \, d\theta.$$

[6.6] Let $C_3(1, \theta)$ be given by (6.28). Prove that

$$\sum_{j=0}^{\infty} (-1)^j \frac{2^{2j}}{((4j)!)^2} - \sum_{j=0}^{\infty} (-1)^j \left(\frac{2^{2j}}{((4j+1)!)^2} - \frac{2^{2j+1}}{((4j+3)!)^2} \right)$$

$$= \left(\frac{1}{(0!)^2} - \frac{2^2}{(4!)^2} + \frac{2^4}{(8!)^2} - \frac{2^6}{(12!)^2} + \cdots \right)$$

$$- \left(\frac{1}{(1!)^2} - \frac{2^1}{(3!)^2} - \frac{2^2}{(5!)^2} + \frac{2^3}{(7!)^2} + \cdots \right)$$

$$= \frac{1}{\pi} \int_0^{\pi} C_3(1, \theta) \, d\theta.$$

[6.7] Let the function $C_4(x, \theta)$ be defined as

$$C_4(x, \theta) = \cosh(x\cos\theta)\cos(x\sin\theta)\cos(2\sin\theta) \\ + \sinh(x\cos\theta)\sin(x\sin\theta)\sin(2\sin\theta), \quad (6.30)$$

and the function $S_4(x, \theta)$ be defined as

$$S_4(x, \theta) = \cosh(x\cos\theta)\sin(x\sin\theta)\sin(2\sin\theta) \\ + \sinh(x\cos\theta)\cos(x\sin\theta)\cos(2\sin\theta). \quad (6.31)$$

Let the function $\phi(x)$ be defined as $\phi(x) = \tanh^{-1} x$. Prove that $\forall(x, n) \in (-1, 1) \times (\mathbb{N} - 1) - \{0\} \times (2\mathbb{N} - 1)$,

$$\sum_{j=0}^{\infty} \frac{(1-x^2)^j \cosh 2j\phi(x)}{(2j)!(n+2j)!} - \sqrt{1-x^2} \sum_{j=0}^{\infty} \frac{(1-x^2)^j \cosh(2j+1)\phi(x)}{(2j+1)!(n+2j+1)!}$$

$$= \left(\frac{1}{0!n!} + \frac{(1-x^2)\cosh 2\phi(x)}{2!(n+2)!} + \frac{(1-x^2)^2 \cosh 4\phi(x)}{4!(n+4)!} \right.$$

$$\left. + \frac{(1-x^2)^3 \cosh 6\phi(x)}{6!(n+6)!} + \cdots \right)$$

$$- \sqrt{1-x^2} \left(\frac{\cosh\phi(x)}{1!(n+1)!} + \frac{(1-x^2)\cosh 3\phi(x)}{3!(n+3)!} + \frac{(1-x^2)^2 \cosh 5\phi(x)}{5!(n+5)!} \right.$$

$$\left. + \frac{(1-x^2)^3 \cosh 7\phi(x)}{7!(n+7)!} + \cdots \right)$$

$$= \frac{2}{\pi} \cdot \frac{1 + (-1)^n (1-x^2)^{n/2} \cosh n\phi(x)}{1 + (1-x^2)^n + (-1)^n 2(1-x^2)^{n/2} \cosh n\phi(x)} \int_0^\pi C_4(x, \theta) \cos n\theta \, d\theta$$

$$+ \frac{2}{\pi} \cdot \frac{(-1)^n (1-x^2)^{n/2} \sinh n\phi(x)}{1 + (1-x^2)^n + (-1)^n 2(1-x^2)^{n/2} \cosh n\phi(x)} \int_0^\pi S_4(x, \theta) \cos n\theta \, d\theta.$$

[6.8] Let the function $\phi(x)$ be defined as $\phi(x) = \tanh^{-1} x$ and $C_4(x, \theta)$ be given by (6.30). Prove that $\forall x \in (-1, 1)$,

$$\sum_{j=0}^{\infty} \frac{(1-x^2)^j \cosh 2j\phi(x)}{((2j)!)^2} - \sqrt{1-x^2} \sum_{j=0}^{\infty} \frac{(1-x^2)^j \cosh(2j+1)\phi(x)}{((2j+1)!)^2}$$

$$= \left(\frac{1}{(0!)^2} + \frac{(1-x^2)\cosh 2\phi(x)}{(2!)^2} + \frac{(1-x^2)^2 \cosh 4\phi(x)}{(4!)^2} \right.$$

$$\left. + \frac{(1-x^2)^3 \cosh 6\phi(x)}{(6!)^2} + \cdots \right)$$

$$-\sqrt{1-x^2}\left(\frac{\cosh\phi(x)}{(1!)^2} + \frac{(1-x^2)\cosh 3\phi(x)}{(3!)^2} + \frac{(1-x^2)^2\cosh 5\phi(x)}{(5!)^2}\right.$$
$$\left.+\frac{(1-x^2)^3\cosh 7\phi(x)}{(7!)^2} + \cdots\right)$$
$$= \frac{1}{\pi}\int_0^\pi C_4(x,\theta)\,d\theta.$$

[6.9] Let the function $\phi(x)$ be defined as $\phi(x) = \tanh^{-1} x$, $S_4(x,\theta)$ be given by (6.31), and $C_4(x,\theta)$ be given by (6.30). Prove that $\forall (x,n) \in (-1,1) \times (\mathbb{N} - 1) - \{0\} \times (2\mathbb{N} - 1)$,

$$\sqrt{1-x^2}\sum_{j=0}^\infty \frac{(1-x^2)^j \sinh(2j+1)\phi(x)}{(2j+1)!(n+2j+1)!} - \sum_{j=1}^\infty \frac{(1-x^2)^j \sinh 2j\phi(x)}{(2j)!(n+2j)!}$$
$$= \sqrt{1-x^2}\left(\frac{\sinh\phi(x)}{1!(n+1)!} + \frac{(1-x^2)\sinh 3\phi(x)}{3!(n+3)!} + \frac{(1-x^2)^2 \sinh 5\phi(x)}{5!(n+5)!}\right.$$
$$\left.+\frac{(1-x^2)^3 \sinh 7\phi(x)}{7!(n+7)!} + \cdots\right)$$
$$-\left(\frac{(1-x^2)\sinh 2\phi(x)}{2!(n+2)!} + \frac{(1-x^2)^2 \sinh 4\phi(x)}{4!(n+4)!} + \frac{(1-x^2)^3 \sinh 6\phi(x)}{6!(n+6)!}\right.$$
$$\left.+\frac{(1-x^2)^4 \sinh 8\phi(x)}{8!(n+8)!} + \cdots\right)$$
$$= \frac{2}{\pi}\cdot \frac{1+(-1)^n(1-x^2)^{n/2}\cosh n\phi(x)}{1+(1-x^2)^n + (-1)^n 2(1-x^2)^{n/2}\cosh n\phi(x)}\int_0^\pi S_4(x,\theta)\cos n\theta\,d\theta$$
$$+ \frac{2}{\pi}\cdot \frac{(-1)^n(1-x^2)^{n/2}\sinh n\phi(x)}{1+(1-x^2)^n + (-1)^n 2(1-x^2)^{n/2}\cosh n\phi(x)}\int_0^\pi C_4(x,\theta)\cos n\theta\,d\theta.$$

[6.10] Let the function $\phi(x)$ be defined as $\phi(x) = \tanh^{-1} x$ and $S_4(x,\theta)$ be given by (6.31). Prove that $\forall x \in (-1,1)$,

$$\sqrt{1-x^2}\sum_{j=0}^\infty \frac{(1-x^2)^j \sinh(2j+1)\phi(x)}{((2j+1)!)^2} - \sum_{j=1}^\infty \frac{(1-x^2)^j \sinh 2j\phi(x)}{((2j)!)^2}$$
$$= \sqrt{1-x^2}\left(\frac{\sinh\phi(x)}{(1!)^2} + \frac{(1-x^2)\sinh 3\phi(x)}{(3!)^2} + \frac{(1-x^2)^2 \sinh 5\phi(x)}{(5!)^2}\right.$$
$$\left.+\frac{(1-x^2)^3 \sinh 7\phi(x)}{(7!)^2} + \cdots\right)$$
$$-\left(\frac{(1-x^2)\sinh 2\phi(x)}{(2!)^2} + \frac{(1-x^2)^2 \sinh 4\phi(x)}{(4!)^2} + \frac{(1-x^2)^3 \sinh 6\phi(x)}{(6!)^2}\right.$$

$$+\frac{(1-x^2)^4 \sinh 8\phi(x)}{(8!)^2}+\cdots\bigg)$$
$$=\frac{1}{\pi}\int_0^\pi S_4(x,\theta)\,d\theta.$$

[6.11] Let the function $\phi(x)$ be defined as $\phi(x)=\tan^{-1}x$. Prove that $\forall(x,n)\in \mathbb{R}\times(\mathbb{N}-1)$,

$$\sum_{j=0}^\infty \frac{(1+x^2)^j \cos 2j\phi(x)}{(2j)!(n+2j)!} + \sqrt{1+x^2}\sum_{j=0}^\infty \frac{(1+x^2)^j \cos(2j+1)\phi(x)}{(2j+1)!(n+2j+1)!}$$
$$=\bigg(\frac{1}{0!n!}+\frac{(1+x^2)\cos 2\phi(x)}{2!(n+2)!}+\frac{(1+x^2)^2\cos 4\phi(x)}{4!(n+4)!}$$
$$+\frac{(1+x^2)^3 \cos 6\phi(x)}{6!(n+6)!}+\cdots\bigg)+\sqrt{1+x^2}\bigg(\frac{\cos\phi(x)}{1!(n+1)!}$$
$$+\frac{(1+x^2)\cos 3\phi(x)}{3!(n+3)!}+\frac{(1+x^2)^2\cos 5\phi(x)}{5!(n+5)!}+\frac{(1+x^2)^3\cos 7\phi(x)}{7!(n+7)!}+\cdots\bigg)$$
$$=\frac{2}{\pi}\cdot\frac{1+(1+x^2)^{n/2}\cos n\phi(x)}{1+(1+x^2)^n+2(1+x^2)^{n/2}\cos n\phi(x)}$$
$$\int_0^\pi e^{2\cos\theta}\cos(x\cos\theta)\cosh(x\sin\theta)\cos n\theta\,d\theta$$
$$+\frac{2}{\pi}\cdot\frac{(1+x^2)^{n/2}\sin n\phi(x)}{1+(1+x^2)^n+2(1+x^2)^{n/2}\cos n\phi(x)}$$
$$\int_0^\pi e^{2\cos\theta}\sin(x\cos\theta)\cosh(x\sin\theta)\cos n\theta\,d\theta.$$

[6.12] Let the function $\phi(x)$ be defined as $\phi(x)=\tan^{-1}x$. Prove that $\forall x\in\mathbb{R}$,

$$\sum_{j=0}^\infty \frac{(1+x^2)^j \cos 2j\phi(x)}{((2j)!)^2}+\sqrt{1+x^2}\sum_{j=0}^\infty \frac{(1+x^2)^j\cos(2j+1)\phi(x)}{((2j+1)!)^2}$$
$$=\bigg(\frac{1}{(0!)^2}+\frac{(1+x^2)\cos 2\phi(x)}{(2!)^2}+\frac{(1+x^2)^2\cos 4\phi(x)}{(4!)^2}$$
$$+\frac{(1+x^2)^3\cos 6\phi(x)}{(6!)^2}+\cdots\bigg)+\sqrt{1+x^2}\bigg(\frac{\cos\phi(x)}{(1!)^2}$$
$$+\frac{(1+x^2)\cos 3\phi(x)}{(3!)^2}+\frac{(1+x^2)^2\cos 5\phi(x)}{(5!)^2}$$
$$+\frac{(1+x^2)^3\cos 7\phi(x)}{(7!)^2}+\cdots\bigg)$$

$$= \frac{1}{\pi} \int_0^\pi e^{2\cos\theta} \cos(x\cos\theta) \cosh(x\sin\theta) \, d\theta.$$

[6.13] Let the function $\phi(x)$ be defined as $\phi(x) = \tan^{-1} x$. Prove that $\forall (x, n) \in \mathbb{R} \times (\mathbb{N} - 1)$,

$$\sum_{j=1}^\infty \frac{(1+x^2)^j \sin 2j\phi(x)}{(2j)!(n+2j)!} + \sqrt{1+x^2} \sum_{j=0}^\infty \frac{(1+x^2)^j \sin(2j+1)\phi(x)}{(2j+1)!(n+2j+1)!}$$

$$= \left(\frac{(1+x^2) \sin 2\phi(x)}{2!(n+2)!} + \frac{(1+x^2)^2 \sin 4\phi(x)}{4!(n+4)!} + \frac{(1+x^2)^3 \sin 6\phi(x)}{6!(n+6)!} \right.$$

$$\left. + \frac{(1+x^2)^4 \sin 8\phi(x)}{8!(n+8)!} + \cdots \right)$$

$$+ \sqrt{1+x^2} \left(\frac{\sin \phi(x)}{1!(n+1)!} + \frac{(1+x^2) \sin 3\phi(x)}{3!(n+3)!} + \frac{(1+x^2)^2 \sin 5\phi(x)}{5!(n+5)!} \right.$$

$$\left. + \frac{(1+x^2)^3 \sin 7\phi(x)}{7!(n+7)!} + \cdots \right)$$

$$= \frac{2}{\pi} \cdot \frac{1 + (1+x^2)^{n/2} \cos n\phi(x)}{1 + (1+x^2)^n + 2(1+x^2)^{n/2} \cos n\phi(x)}$$

$$\int_0^\pi e^{2\cos\theta} \sin(x\cos\theta) \cosh(x\sin\theta) \cos n\theta \, d\theta$$

$$- \frac{2}{\pi} \cdot \frac{(1+x^2)^{n/2} \sin n\phi(x)}{1 + (1+x^2)^n + 2(1+x^2)^{n/2} \cos n\phi(x)}$$

$$\int_0^\pi e^{2\cos\theta} \cos(x\cos\theta) \cosh(x\sin\theta) \cos n\theta \, d\theta.$$

[6.14] Let the function $\phi(x)$ be defined as $\phi(x) = \tan^{-1} x$. Prove that $\forall x \in \mathbb{R}$,

$$\sum_{j=1}^\infty \frac{(1+x^2)^j \sin 2j\phi(x)}{((2j)!)^2} + \sqrt{1+x^2} \sum_{j=0}^\infty \frac{(1+x^2)^j \sin(2j+1)\phi(x)}{((2j+1)!)^2}$$

$$= \left(\frac{(1+x^2) \sin 2\phi(x)}{(2!)^2} + \frac{(1+x^2)^2 \sin 4\phi(x)}{(4!)^2} + \frac{(1+x^2)^3 \sin 6\phi(x)}{(6!)^2} \right.$$

$$\left. + \frac{(1+x^2)^4 \sin 8\phi(x)}{(8!)^2} + \cdots \right)$$

$$+ \sqrt{1+x^2} \left(\frac{\sin \phi(x)}{(1!)^2} + \frac{(1+x^2) \sin 3\phi(x)}{(3!)^2} + \frac{(1+x^2)^2 \sin 5\phi(x)}{(5!)^2} \right.$$

$$\left. + \frac{(1+x^2)^3 \sin 7\phi(x)}{(7!)^2} + \cdots \right)$$

$$= \frac{1}{\pi} \int_0^\pi e^{2\cos\theta} \sin(x\cos\theta) \cosh(x\sin\theta) \, d\theta.$$

[6.15] Prove that

$$\sum_{j=0}^{\infty}(-1)^j \frac{2^{2j}}{((4j)!)^2} + \sum_{j=0}^{\infty}(-1)^j \left(\frac{2^{2j}}{((4j+1)!)^2} - \frac{2^{2j+1}}{((4j+3)!)^2}\right)$$

$$= \left(\frac{1}{(0!)^2} - \frac{2^2}{(4!)^2} + \frac{2^4}{(8!)^2} - \frac{2^6}{(12!)^2} + \cdots\right)$$

$$+ \left(\frac{1}{(1!)^2} - \frac{2}{(3!)^2} - \frac{2^2}{(5!)^2} + \frac{2^3}{(7!)^2} + \cdots\right)$$

$$= \frac{1}{\pi}\int_0^{\pi} e^{2\cos\theta}\cos(\cos\theta)\cosh(\sin\theta)\,d\theta.$$

[6.16] Prove that

$$\sum_{j=0}^{\infty}(-1)^j \frac{2^{2j+1}}{((4j+2)!)^2} + \sum_{j=0}^{\infty}(-1)^j \left(\frac{2^{2j}}{((4j+1)!)^2} + \frac{2^{2j+1}}{((4j+3)!)^2}\right)$$

$$= \left(\frac{2}{(2!)^2} - \frac{2^3}{(6!)^2} + \frac{2^5}{(10!)^2} - \frac{2^7}{(14!)^2} + \cdots\right)$$

$$+ \left(\frac{1}{(1!)^2} + \frac{2}{(3!)^2} - \frac{2^2}{(5!)^2} - \frac{2^3}{(7!)^2} + \cdots\right)$$

$$= \frac{1}{\pi}\int_0^{\pi} e^{2\cos\theta}\sin(\cos\theta)\cosh(\sin\theta)\,d\theta.$$

[6.17] Let the function $\phi(x)$ be defined as $\phi(x) = \tanh^{-1} x$. Prove that $\forall (x, n) \in (-1, 1) \times (\mathbb{N} - 1)$,

$$\sum_{j=0}^{\infty} \frac{(1-x^2)^j \cosh 2j\phi(x)}{(2j)!(n+2j)!} + \sqrt{1-x^2}\sum_{j=0}^{\infty} \frac{(1-x^2)^j \cosh(2j+1)\phi(x)}{(2j+1)!(n+2j+1)!}$$

$$= \left(\frac{1}{0!n!} + \frac{(1-x^2)\cosh 2\phi(x)}{2!(n+2)!} + \frac{(1-x^2)^2 \cosh 4\phi(x)}{4!(n+4)!}\right.$$

$$\left.+ \frac{(1-x^2)^3 \cosh 6\phi(x)}{6!(n+6)!} + \cdots\right)$$

$$+ \sqrt{1-x^2}\left(\frac{\cosh\phi(x)}{1!(n+1)!} + \frac{(1-x^2)\cosh 3\phi(x)}{3!(n+3)!} + \frac{(1-x^2)^2 \cosh 5\phi(x)}{5!(n+5)!}\right.$$

$$\left.+ \frac{(1-x^2)^3 \cosh 7\phi(x)}{7!(n+7)!} + \cdots\right)$$

$$= \frac{2}{\pi} \cdot \frac{1 + (1-x^2)^{n/2} \cosh n\phi(x)}{1 + (1-x^2)^n + 2(1-x^2)^{n/2} \cosh n\phi(x)}$$
$$\int_0^\pi e^{2\cos\theta} \cosh(x\cos\theta)\cos(x\sin\theta)\cos n\theta\, d\theta$$
$$- \frac{2}{\pi} \cdot \frac{(1-x^2)^{n/2} \sinh n\phi(x)}{1 + (1-x^2)^n + 2(1-x^2)^{n/2} \cosh n\phi(x)}$$
$$\int_0^\pi e^{2\cos\theta} \sinh(x\cos\theta)\cos(x\sin\theta)\cos n\theta\, d\theta.$$

[6.18] Let the function $\phi(x)$ be defined as $\phi(x) = \tanh^{-1} x$. Prove that $\forall x \in (-1, 1)$,

$$\sum_{j=0}^\infty \frac{(1-x^2)^j \cosh 2j\phi(x)}{((2j)!)^2} + \sqrt{1-x^2} \sum_{j=0}^\infty \frac{(1-x^2)^j \cosh(2j+1)\phi(x)}{((2j+1)!)^2}$$
$$= \left(\frac{1}{(0!)^2} + \frac{(1-x^2)\cosh 2\phi(x)}{(2!)^2} + \frac{(1-x^2)^2 \cosh 4\phi(x)}{(4!)^2} \right.$$
$$\left. + \frac{(1-x^2)^3 \cosh 6\phi(x)}{(6!)^2} + \cdots \right)$$
$$+ \sqrt{1-x^2} \left(\frac{\cosh\phi(x)}{(1!)^2} + \frac{(1-x^2)\cosh 3\phi(x)}{(3!)^2} + \frac{(1-x^2)^2 \cosh 5\phi(x)}{(5!)^2} \right.$$
$$\left. + \frac{(1-x^2)^3 \cosh 7\phi(x)}{(7!)^2} + \cdots \right)$$
$$= \frac{1}{\pi} \int_0^\pi e^{2\cos\theta} \cosh(x\cos\theta)\cos(x\sin\theta)\, d\theta.$$

[6.19] Let the function $\phi(x)$ be defined as $\phi(x) = \tanh^{-1} x$. Prove that $\forall (x, n) \in (-1, 1) \times (\mathbb{N} - 1)$,

$$\sum_{j=1}^\infty \frac{(1-x^2)^j \sinh 2j\phi(x)}{(2j)!(n+2j)!} + \sqrt{1-x^2} \sum_{j=0}^\infty \frac{(1-x^2)^j \sinh(2j+1)\phi(x)}{(2j+1)!(n+2j+1)!}$$
$$= \left(\frac{(1-x^2)\sinh 2\phi(x)}{2!(n+2)!} + \frac{(1-x^2)^2 \sinh 4\phi(x)}{4!(n+4)!} + \frac{(1-x^2)^3 \sinh 6\phi(x)}{6!(n+6)!} \right.$$
$$\left. + \frac{(1-x^2)^4 \sinh 8\phi(x)}{8!(n+8)!} + \cdots \right)$$
$$+ \sqrt{1-x^2} \left(\frac{\sinh\phi(x)}{1!(n+1)!} + \frac{(1-x^2)\sinh 3\phi(x)}{3!(n+3)!} + \frac{(1-x^2)^2 \sinh 5\phi(x)}{5!(n+5)!} \right.$$
$$\left. + \frac{(1-x^2)^3 \sinh 7\phi(x)}{7!(n+7)!} + \cdots \right)$$

$$= \frac{2}{\pi} \cdot \frac{1 + (1 - x^2)^{n/2} \cosh n\phi(x)}{1 + (1 - x^2)^n + 2(1 - x^2)^{n/2} \cosh n\phi(x)}$$

$$\int_0^\pi e^{2\cos\theta} \sinh(x\cos\theta) \cos(x\sin\theta) \cos n\theta \, d\theta$$

$$- \frac{2}{\pi} \cdot \frac{(1 - x^2)^{n/2} \sinh n\phi(x)}{1 + (1 - x^2)^n + 2(1 - x^2)^{n/2} \cosh n\phi(x)}$$

$$\int_0^\pi e^{2\cos\theta} \cosh(x\cos\theta) \cos(x\sin\theta) \cos n\theta \, d\theta.$$

[6.20] Let the function $\phi(x)$ be defined as $\phi(x) = \tanh^{-1} x$. Prove that $\forall x \in (-1, 1)$,

$$\sum_{j=1}^\infty \frac{(1-x^2)^j \sinh 2j\phi(x)}{((2j)!)^2} + \sqrt{1-x^2} \sum_{j=0}^\infty \frac{(1-x^2)^j \sinh(2j+1)\phi(x)}{((2j+1)!)^2}$$

$$= \left(\frac{(1-x^2)\sinh 2\phi(x)}{(2!)^2} + \frac{(1-x^2)^2 \sinh 4\phi(x)}{(4!)^2} + \frac{(1-x^2)^3 \sinh 6\phi(x)}{(6!)^2} \right.$$

$$\left. + \frac{(1-x^2)^4 \sinh 8\phi(x)}{(8!)^2} + \cdots \right)$$

$$+ \sqrt{1-x^2} \left(\frac{\sinh \phi(x)}{(1!)^2} + \frac{(1-x^2)\sinh 3\phi(x)}{(3!)^2} + \frac{(1-x^2)^2 \sinh 5\phi(x)}{(5!)^2} \right.$$

$$\left. + \frac{(1-x^2)^3 \sinh 7\phi(x)}{(7!)^2} + \cdots \right)$$

$$= \frac{1}{\pi} \int_0^\pi e^{2\cos\theta} \sinh(x\cos\theta) \cos(x\sin\theta) \, d\theta.$$

[6.21] Let the function $C_5(x, \phi, \theta)$ be defined as

$$C_5(x, \phi, \theta)$$
$$= \cosh(x \sin\phi \sin\theta) \cos(x \sin\phi \cos\theta) \cos((x \cos\phi - 1)\sin\theta) \quad (6.32)$$
$$+ \sinh(x \sin\phi \sin\theta) \sin(x \sin\phi \cos\theta) \sin((x \cos\phi - 1)\sin\theta),$$

and the function $S_5(x, \phi, \theta)$ be defined as

$$S_5(x, \phi, \theta)$$
$$= \cosh(x \sin\phi \sin\theta) \sin(x \sin\phi \cos\theta) \cos((x \cos\phi - 1)\sin\theta) \quad (6.33)$$
$$- \sinh(x \sin\phi \sin\theta) \cos(x \sin\phi \cos\theta) \sin((x \cos\phi - 1)\sin\theta).$$

Prove that $\forall (x, \phi, n) \in \mathbb{R}^2 \times (\mathbb{N} - 1)$,

$$\sum_{j=0}^{\infty} \frac{x^j \cos j\phi}{j!(n+j)!} = \frac{1}{0!n!} + \frac{x\cos\phi}{1!(n+1)!} + \frac{x^2\cos 2\phi}{2!(n+2)!} + \frac{x^3\cos 3\phi}{3!(n+3)!} + \cdots$$

$$= \frac{2}{\pi} \cdot \frac{1+x^n\cos n\phi}{1+x^{2n}+2x^n\cos n\phi} \int_0^\pi C_5(x,\phi,\theta) e^{(1+x\cos\phi)\cos\theta} \cos n\theta\, d\theta$$

$$+ \frac{2}{\pi} \cdot \frac{x^n \sin n\phi}{1+x^{2n}+2x^n\cos n\phi} \int_0^\pi S_5(x,\phi,\theta) e^{(1+x\cos\phi)\cos\theta} \cos n\theta\, d\theta.$$

[6.22] Let $C_5(x,\phi,\theta)$ be given by (6.32). Prove that $\forall (x,\phi) \in \mathbb{R}^2$,

$$\sum_{j=0}^{\infty} \frac{x^j \cos j\phi}{(j!)^2} = \frac{1}{(0!)^2} + \frac{x\cos\phi}{(1!)^2} + \frac{x^2\cos 2\phi}{(2!)^2} + \frac{x^3\cos 3\phi}{(3!)^2} + \cdots$$

$$= \frac{1}{\pi} \int_0^\pi C_5(x,\phi,\theta) e^{(1+x\cos\phi)\cos\theta}\, d\theta.$$

[6.23] Let $S_5(x,\phi,\theta)$ be given by (6.33) and $C_5(x,\phi,\theta)$ be given by (6.32). Prove that $\forall (x,\phi,n) \in \mathbb{R}^2 \times (\mathbb{N}-1)$,

$$\sum_{j=1}^{\infty} \frac{x^j \sin j\phi}{j!(n+j)!} = \frac{x\sin\phi}{1!(n+1)!} + \frac{x^2 \sin 2\phi}{2!(n+2)!} + \frac{x^3 \sin 3\phi}{3!(n+3)!} + \frac{x^4 \sin 4\phi}{4!(n+4)!} + \cdots$$

$$= \frac{2}{\pi} \cdot \frac{1+x^n\cos n\phi}{1+x^{2n}+2x^n\cos n\phi} \int_0^\pi S_5(x,\phi,\theta) e^{(1+x\cos\phi)\cos\theta} \cos n\theta\, d\theta$$

$$- \frac{2}{\pi} \cdot \frac{x^n \sin n\phi}{1+x^{2n}+2x^n\cos n\phi} \int_0^\pi C_5(x,\phi,\theta) e^{(1+x\cos\phi)\cos\theta} \cos n\theta\, d\theta.$$

[6.24] Let $S_5(x,\phi,\theta)$ be given by (6.33). Prove that $\forall (x,\phi) \in \mathbb{R}^2$,

$$\sum_{j=1}^{\infty} \frac{x^j \sin j\phi}{(j!)^2} = \frac{x\sin\phi}{(1!)^2} + \frac{x^2 \sin 2\phi}{(2!)^2} + \frac{x^3 \sin 3\phi}{(3!)^2} + \frac{x^4 \sin 4\phi}{(4!)^2} + \cdots$$

$$= \frac{1}{\pi} \int_0^\pi S_5(x,\phi,\theta) e^{(1+x\cos\phi)\cos\theta}\, d\theta.$$

[6.25] Let the function $C_6(x,\phi,\theta)$ be defined as

$$C_6(x,\phi,\theta)$$
$$= \cos(x\sinh\phi\sin\theta)\cosh(x\sinh\phi\cos\theta)\cos((x\cosh\phi - 1)\sin\theta)$$
$$- \sin(x\sinh\phi\sin\theta)\sinh(x\sinh\phi\cos\theta)\sin((x\cosh\phi - 1)\sin\theta),$$
(6.34)

and the function $S_6(x,\phi,\theta)$ be defined as

$S_6(x, \phi, \theta)$
$= \cos(x \sinh \phi \sin \theta) \sinh(x \sinh \phi \cos \theta) \cos((x \cosh \phi - 1) \sin \theta)$
$- \sin(x \sinh \phi \sin \theta) \cosh(x \sinh \phi \cos \theta) \sin((x \cosh \phi - 1) \sin \theta).$
(6.35)

Prove that $\forall (x, \phi, n) \in \mathbb{R}^2 \times (\mathbb{N} - 1)$,

$$\sum_{j=0}^{\infty} \frac{x^j \cosh j\phi}{j!(n+j)!} = \frac{1}{0!n!} + \frac{x \cosh \phi}{1!(n+1)!} + \frac{x^2 \cosh 2\phi}{2!(n+2)!} + \frac{x^3 \cosh 3\phi}{3!(n+3)!} + \cdots$$
$$= \frac{2}{\pi} \cdot \frac{1 + x^n \cosh n\phi}{1 + x^{2n} + 2x^n \cosh n\phi} \int_0^\pi C_6(x, \phi, \theta) e^{(1+x \cosh \phi) \cos \theta} \cos n\theta \, d\theta$$
$$- \frac{2}{\pi} \cdot \frac{x^n \sinh n\phi}{1 + x^{2n} + 2x^n \cosh n\phi} \int_0^\pi S_6(x, \phi, \theta) e^{(1+x \cosh \phi) \cos \theta} \cos n\theta \, d\theta.$$

[6.26] Let $C_6(x, \phi, \theta)$ be given by (6.34). Prove that $\forall (x, \phi) \in \mathbb{R}^2$,

$$\sum_{j=0}^{\infty} \frac{x^j \cosh j\phi}{(j!)^2} = \frac{1}{(0!)^2} + \frac{x \cosh \phi}{(1!)^2} + \frac{x^2 \cosh 2\phi}{(2!)^2} + \frac{x^3 \cosh 3\phi}{(3!)^2}$$
$$= \frac{1}{\pi} \int_0^\pi C_6(x, \phi, \theta) e^{(1+x \cosh \phi) \cos \theta} \, d\theta.$$

[6.27] Let $S_6(x, \phi, \theta)$ be given by (6.35) and $C_6(x, \phi, \theta)$ be given by (6.34). Prove that $\forall (x, \phi, n) \in \mathbb{R}^2 \times (\mathbb{N} - 1)$,

$$\sum_{j=1}^{\infty} \frac{x^j \sinh j\phi}{j!(n+j)!} = \frac{x \sinh \phi}{1!(n+1)!} + \frac{x^2 \sinh 2\phi}{2!(n+2)!} + \frac{x^3 \sinh 3\phi}{3!(n+3)!} + \frac{x^4 \sinh 4\phi}{4!(n+4)!} + \cdots$$
$$= \frac{2}{\pi} \cdot \frac{1 + x^n \cosh n\phi}{1 + x^{2n} + 2x^n \cosh n\phi} \int_0^\pi S_6(x, \phi, \theta) e^{(1+x \cosh \phi) \cos \theta} \cos n\theta \, d\theta$$
$$- \frac{2}{\pi} \cdot \frac{x^n \sinh n\phi}{1 + x^{2n} + 2x^n \cosh n\phi} \int_0^\pi C_6(x, \phi, \theta) e^{(1+x \cosh \phi) \cos \theta} \cos n\theta \, d\theta.$$

[6.28] Let $S_6(x, \phi, \theta)$ is given by (6.35). Prove that $\forall (x, \phi) \in \mathbb{R}^2$,

$$\sum_{j=1}^{\infty} \frac{x^j \sinh j\phi}{(j!)^2} = \frac{x \sinh \phi}{(1!)^2} + \frac{x^2 \sinh 2\phi}{(2!)^2} + \frac{x^3 \sinh 3\phi}{(3!)^2} + \frac{x^4 \sinh 4\phi}{(4!)^2} + \cdots$$
$$= \frac{1}{\pi} \int_0^\pi S_6(x, \phi, \theta) e^{(1+x \cosh \phi) \cos \theta} \, d\theta.$$

Problems

[6.29] Prove that $\forall \phi \in \mathbb{R} - (2\mathbb{Z}+1)\dfrac{\pi}{2}$,

$$\sum_{j=0}^{\infty} \frac{\cos j\phi}{(j!)^2 \cos^j \phi} = \frac{1}{(0!)^2} + \frac{\cos \phi}{(1!)^2 \cos \phi} + \frac{\cos 2\phi}{(2!)^2 \cos^2 \phi} + \frac{\cos 3\phi}{(3!)^2 \cos^3 \phi} + \cdots$$

$$= \frac{1}{\pi} \int_0^{\pi} \cosh(\tan \phi \sin \theta) \cos(\tan \phi \cos \theta) \, e^{2\cos \theta} \, d\theta.$$

[6.30] Prove that $\forall \phi \in \mathbb{R} - (2\mathbb{Z}+1)\dfrac{\pi}{2}$,

$$\sum_{j=0}^{\infty} (-1)^j \frac{\cos j\phi}{(j!)^2 \cos^j \phi} = \frac{1}{(0!)^2} - \frac{\cos \phi}{(1!)^2 \cos \phi} + \frac{\cos 2\phi}{(2!)^2 \cos^2 \phi}$$

$$- \frac{\cos 3\phi}{(3!)^2 \cos^3 \phi} + \cdots$$

$$= \frac{1}{\pi} \int_0^{\pi} (\cosh(\tan \phi \sin \theta) \cos(\tan \phi \cos \theta) \cos(2\sin \theta)$$

$$- \sinh(\tan \phi \sin \theta) \sin(\tan \phi \cos \theta) \sin(2\sin \theta)) \, d\theta.$$

[6.31] Prove that $\forall \phi \in \mathbb{R} - (2\mathbb{Z}+1)\dfrac{\pi}{2}$,

$$\sum_{j=1}^{\infty} \frac{\sin j\phi}{(j!)^2 \cos^j \phi} = \frac{\sin \phi}{(1!)^2 \cos \phi} + \frac{\sin 2\phi}{(2!)^2 \cos^2 \phi} + \frac{\sin 3\phi}{(3!)^2 \cos^3 \phi}$$

$$+ \frac{\sin 4\phi}{(4!)^2 \cos^4 \phi} + \cdots$$

$$= \frac{1}{\pi} \int_0^{\pi} \cosh(\tan \phi \sin \theta) \sin(\tan \phi \cos \theta) \, e^{2\cos \theta} \, d\theta.$$

[6.32] Prove that $\forall \phi \in \mathbb{R} - (2\mathbb{Z}+1)\dfrac{\pi}{2}$,

$$\sum_{j=1}^{\infty} (-1)^{j-1} \frac{\sin j\phi}{(j!)^2 \cos^j \phi}$$

$$= \frac{\sin \phi}{(1!)^2 \cos \phi} - \frac{\sin 2\phi}{(2!)^2 \cos^2 \phi} + \frac{\sin 3\phi}{(3!)^2 \cos^3 \phi} - \frac{\sin 4\phi}{(4!)^2 \cos^4 \phi} + \cdots$$

$$= \frac{1}{\pi} \int_0^{\pi} (\cosh(\tan \phi \sin \theta) \sin(\tan \phi \cos \theta) \cos(2\sin \theta)$$

$$+ \sinh(\tan \phi \sin \theta) \cos(\tan \phi \cos \theta) \sin(2\sin \theta)) \, d\theta.$$

[6.33] Prove that $\forall \phi \in \mathbb{R}$,

$$\sum_{j=0}^{\infty} \frac{\cosh j\phi}{(j!)^2 \cosh^j \phi} = \frac{1}{(0!)^2} + \frac{\cosh \phi}{(1!)^2 \cosh \phi} + \frac{\cosh 2\phi}{(2!)^2 \cosh^2 \phi}$$
$$+ \frac{\cosh 3\phi}{(3!)^2 \cosh^3 \phi} + \cdots$$
$$= \frac{1}{\pi} \int_0^{\pi} \cos(\tanh \phi \sin \theta) \cosh(\tanh \phi \cos \theta) e^{2\cos \theta} \, d\theta.$$

[6.34] Prove that $\forall \phi \in \mathbb{R}$,

$$\sum_{j=0}^{\infty} (-1)^j \frac{\cosh j\phi}{(j!)^2 \cosh^j \phi} = \frac{1}{(0!)^2} - \frac{\cosh \phi}{(1!)^2 \cosh \phi} + \frac{\cosh 2\phi}{(2!)^2 \cosh^2 \phi}$$
$$- \frac{\cosh 3\phi}{(3!)^2 \cosh^3 \phi} + \cdots$$
$$= \frac{1}{\pi} \int_0^{\pi} (\cos(\tanh \phi \sin \theta) \cosh(\tanh \phi \cos \theta) \cos(2 \sin \theta)$$
$$+ \sin(\tanh \phi \sin \theta) \sinh(\tanh \phi \cos \theta) \sin(2 \sin \theta)) \, d\theta.$$

[6.35] Prove that $\forall \phi \in \mathbb{R}$,

$$\sum_{j=1}^{\infty} \frac{\sinh j\phi}{(j!)^2 \cosh^j \phi} = \frac{\sinh \phi}{(1!)^2 \cosh \phi} + \frac{\sinh 2\phi}{(2!)^2 \cosh^2 \phi}$$
$$+ \frac{\sinh 3\phi}{(3!)^2 \cosh^3 \phi} + \frac{\sinh 4\phi}{(4!)^2 \cosh^4 \phi} + \cdots$$
$$= \frac{1}{\pi} \int_0^{\pi} \cos(\tanh \phi \sin \theta) \sinh(\tanh \phi \cos \theta) e^{2\cos \theta} \, d\theta.$$

[6.36] Prove that $\forall \phi \in \mathbb{R}$,

$$\sum_{j=1}^{\infty} (-1)^{j-1} \frac{\sinh j\phi}{(j!)^2 \cosh^j \phi}$$
$$= \frac{\sinh \phi}{(1!)^2 \cosh \phi} - \frac{\sinh 2\phi}{(2!)^2 \cosh^2 \phi} + \frac{\sinh 3\phi}{(3!)^2 \cosh^3 \phi} - \frac{\sinh 4\phi}{(4!)^2 \cosh^4 \phi} + \cdots$$
$$= \frac{1}{\pi} \int_0^{\pi} (\cos(\tanh \phi \sin \theta) \sinh(\tanh \phi \cos \theta) \cos(2 \sin \theta)$$
$$+ \sin(\tanh \phi \sin \theta) \cosh(\tanh \phi \cos \theta) \sin(2 \sin \theta)) \, d\theta.$$

Problems

[6.37] Assuming the convergence conditions of Theorem 6.13, prove that $\forall x \in \mathbb{R}$,

$$f(1) \star e^x = \sum_{j=0}^{\infty} \frac{a_j}{j!} x^j$$

$$= a_0 + a_1 x + \frac{a_2}{2!} x^2 + \frac{a_3}{3!} x^3 + \cdots$$

$$= -f(0) + \frac{1}{\pi} \int_0^{\pi} \left(f\left(e^{i\theta}\right) + f\left(e^{-i\theta}\right) \right) \cos(x \sin \theta) e^{x \cos \theta} \, d\theta.$$

[6.38] Assuming the convergence conditions of Theorem 6.13, prove that $\forall x \in (-1, 1)$,

$$f(1) \star \frac{1}{1-x} = \sum_{j=0}^{\infty} a_j x^j$$

$$= a_0 + a_1 x + a_2 x^2 + a_3 x^3 + \cdots$$

$$= f(x)$$

$$= -f(0) + \frac{1}{\pi} \int_0^{\pi} \left(f\left(e^{i\theta}\right) + f\left(e^{-i\theta}\right) \right) \frac{1 - x \cos \theta}{1 + x^2 - 2x \cos \theta} \, d\theta.$$

[6.39] Assuming the convergence conditions of Theorem 6.13, prove that $\forall x \in (-1, 1)$,

$$f(1) \star (-\log(1-x)) = \sum_{j=1}^{\infty} \frac{a_j}{j} x^j$$

$$= a_1 x + \frac{a_2}{2} x^2 + \frac{a_3}{3} x^3 + \frac{a_4}{4} x^4 + \cdots$$

$$= \int_0^x \frac{f(x) - f(0)}{x} \, dx$$

$$= -\frac{1}{2\pi} \int_0^{\pi} \left(f\left(e^{i\theta}\right) + f\left(e^{-i\theta}\right) \right) \log(1 + x^2 - 2x \cos \theta) \, d\theta.$$

[6.40] Assuming the convergence conditions of Theorem 6.13, prove that $\forall x \in (-1, 1)$,

$$f(1) \star \frac{x}{(1-x)^2} = \sum_{j=1}^{\infty} j a_j x^j$$

$$= a_1 x + 2 a_2 x^2 + 3 a_3 x^3 + 4 a_4 x^4 + \cdots$$

$$= x f'(x)$$

$$= \frac{1}{\pi} \int_0^{\pi} x \left(f\left(e^{i\theta}\right) + f\left(e^{-i\theta}\right) \right) \frac{(1 + x^2) \cos \theta - 2x}{(1 + x^2 - 2x \cos \theta)^2} \, d\theta.$$

[6.41] Prove that $\forall x \in \mathbb{R}$,
$$e(x,0) = -1 + \frac{2}{\pi} \int_0^\pi e^{(x+1)\cos\theta} \cos(\sin\theta) \cos(x\sin\theta)\, d\theta.$$

[6.42] Prove that $\forall x \in \mathbb{R}$,
$$e(x,0) = 1 + \frac{2}{\pi} \int_0^\pi e^{(x+1)\cos\theta} \sin(\sin\theta) \sin(x\sin\theta)\, d\theta.$$

[6.43] Prove that $\forall x \in \mathbb{R}$,
$$\int_0^\pi e^{(x+1)\cos\theta} \cos((x+1)\sin\theta)\, d\theta = \pi.$$

[6.44] Assuming that $f(z)$ is convergent for $|z|=1$, prove that
$$\sum_{j=0}^\infty a_j^2 = a_0^2 + a_1^2 + a_2^2 + a_3^2 + \cdots$$
$$= -f(0)^2 + \frac{1}{2\pi} \int_0^\pi \left(f\left(e^{i\theta}\right) + f\left(e^{-i\theta}\right)\right)^2 d\theta.$$

[6.45] Prove that $\forall (x,y) \in \mathbb{R} \times (-1,1)$,
$$\int_0^\pi \frac{1 - y\cos\theta}{1 + y^2 - 2y\cos\theta} e^{x\cos\theta} \cos(x\sin\theta)\, d\theta = \frac{\pi}{2}(e^{xy} + 1).$$

[6.46] Prove that if $f(0) = 0$ and $x \in (-1,1)$, then
$$\int_0^x \frac{f(x)}{x}\, dx = -\frac{1}{2\pi} \int_0^\pi \left(f\left(e^{i\theta}\right) + f\left(e^{-i\theta}\right)\right) \log(1 + x^2 - 2x\cos\theta)\, d\theta.$$

[6.47] Prove that $\forall x \in (-1,1)$,
$$\int_0^x \frac{\log(1+x)}{x}\, dx = -\frac{1}{\pi} \int_0^\pi \log\left(2\cos\frac{\theta}{2}\right) \log(1 + x^2 - 2x\cos\theta)\, d\theta.$$

[6.48] Prove that $\forall x \in (-1,1)$,
$$\int_0^x \frac{\sin x}{x}\, dx = -\frac{1}{\pi} \int_0^\pi \sin(\cos\theta) \cosh(\sin\theta) \log(1 + x^2 - 2x\cos\theta)\, d\theta.$$

[6.49] Prove that $\forall x \in (-1, 1)$,

$$\int_0^x \frac{\sinh x}{x} dx = -\frac{1}{\pi} \int_0^\pi \sinh(\cos\theta)\cos(\sin\theta) \log(1 + x^2 - 2x\cos\theta)\, d\theta.$$

[6.50] Prove that $\forall (x, y) \in (-1, 1)^2$,

$$\log(1 - xy) = \frac{1}{\pi} \int_0^\pi \frac{1 - y\cos\theta}{1 + y^2 - 2y\cos\theta} \log(1 + x^2 - 2x\cos\theta)\, d\theta.$$

[6.51] Prove that $\forall x \in (-1, 1)$,

$$\log(1 + x) = \frac{2}{\pi} \int_0^\pi \log\left(2\cos\frac{\theta}{2}\right) \frac{1 - x\cos\theta}{1 + x^2 - 2x\cos\theta}\, d\theta.$$

[6.52] Prove that $\forall x \in (-1, 1)$,

$$\sin x = \frac{2}{\pi} \int_0^\pi \sin(\cos\theta)\cosh(\sin\theta) \frac{1 - x\cos\theta}{1 + x^2 - 2x\cos\theta}\, d\theta.$$

[6.53] Prove that $\forall x \in (-1, 1)$,

$$\sinh x = \frac{2}{\pi} \int_0^\pi \sinh(\cos\theta)\cos(\sin\theta) \frac{1 - x\cos\theta}{1 + x^2 - 2x\cos\theta}\, d\theta.$$

[6.54] Prove that $\forall (x, y) \in (-1, 1)^2$,

$$\int_0^\pi \frac{1 - x\cos\theta}{1 + x^2 - 2x\cos\theta} \cdot \frac{1 - y\cos\theta}{1 + y^2 - 2y\cos\theta}\, d\theta = \frac{\pi}{2} \cdot \frac{2 - xy}{1 - xy}.$$

[6.55] Prove that $\forall (x, y) \in (-1, 1)^2$,

$$\int_0^\pi x \log\left(1 + y^2 - 2y\cos\theta\right) \frac{(1 + x^2)\cos\theta - 2x}{(1 + x^2 - 2x\cos\theta)^2}\, d\theta = -\frac{\pi xy}{1 - xy}.$$

[6.56] Prove that $\forall x \in (-1, 1)$,

$$2a_0 + \sum_{j=1}^n a_j x^j = 2a_0 + a_1 x + a_2 x^2 + \cdots + a_n x^n$$

$$= \frac{2}{\pi} \int_0^\pi \left(\sum_{j=0}^n a_j \cos j\theta\right) \frac{1 - x\cos\theta}{1 + x^2 - 2x\cos\theta}\, d\theta.$$

[6.57] Prove that

$$\sum_{j=0}^{\infty} \frac{1}{2^{8j}} \binom{2j}{j}^2 = 1 + \frac{\binom{2}{1}^2}{2^8} + \frac{\binom{4}{2}^2}{2^{16}} + \frac{\binom{6}{3}^2}{2^{24}} + \cdots$$

$$= -\frac{1}{2} + \frac{1}{2\pi} \int_0^\pi \frac{15 + 8\sqrt{9 + 16\sin^2 \frac{\theta}{2}}}{9 + 16\sin^2 \frac{\theta}{2}} \, d\theta.$$

[6.58] Prove that

$$\int_0^\pi \cos^2(\sin\theta) e^{2\cos\theta} \, d\theta = \frac{\pi}{2}(e(1,0) + 1).$$

[6.59] Prove that

$$\int_0^\pi \sin^2(\sin\theta) e^{2\cos\theta} \, d\theta = \frac{\pi}{2}(e(1,0) - 1).$$

Part II
Solutions

Chapter 7
Solutions to Binomial Series

[1.1] Put $\theta = \pi/6$ in [2.14].

[1.2] Put $\theta = \log\left(\left(\sqrt{5}+1\right)/2\right)$ in [2.40].

[1.3] Put $\theta = \pi/6$ in [2.9].

[1.4] Put $\theta = \log\left(\left(\sqrt{5}+1\right)/2\right)$ in [2.35].

[1.5] Put $\theta = \pi/6$ in [2.10].

[1.6] Put $\theta = \log\left(\left(\sqrt{5}+1\right)/2\right)$ in [2.36].

[1.7] Put $\theta = \pi/6$ in [2.12].

[1.8] Put $\theta = \log\left(\left(\sqrt{5}+1\right)/2\right)$ in [2.38].

[1.9] Put $\theta = \pi/6$ in [2.19].

[1.10] Put $\theta = \log\left(\left(\sqrt{5}+1\right)/2\right)$ in [2.45].

[1.11] Put $\theta = \pi/6$ in [2.18].

[1.12] Put $\theta = \log\left(\left(\sqrt{5}+1\right)/2\right)$ in [2.44].

[1.13] Put $\theta = \pi/6$ in [2.13].

[1.14] Put $\theta = \log\left(\left(\sqrt{5}+1\right)/2\right)$ in [2.39].

© The Author(s), under exclusive license to Springer Nature Singapore Pte Ltd. 2025
A. Mishra, *Special Integrals*, University Texts in the Mathematical Sciences,
https://doi.org/10.1007/978-981-97-7514-9_7

[1.15] Put $\theta = \pi/6$ in [2.8].

[1.16] Put $\theta = \log\left(\left(\sqrt{5}+1\right)/2\right)$ in [2.34].

[1.17] Put $\theta = \pi/6$ in [2.11].

[1.18] Put $\theta = \log\left(\left(\sqrt{5}+1\right)/2\right)$ in [2.37].

[1.19] Put $\theta = \pi/6$ in [2.7].

[1.20] Put $\theta = \log\left(\left(\sqrt{5}+1\right)/2\right)$ in [2.33].

[1.21] Put $\theta = \pi/6$ in [2.6].

[1.22] Put $\theta = \log\left(\left(\sqrt{5}+1\right)/2\right)$ in [2.32].

[1.23] Put $\theta = \pi/6$ in [2.15].

[1.24] Put $\theta = \log\left(\left(\sqrt{5}+1\right)/2\right)$ in [2.41].

[1.25] Put $\theta = \pi/6$ in [2.16].

[1.26] Put $\theta = \log\left(\left(\sqrt{5}+1\right)/2\right)$ in [2.42].

[1.27] Put $\theta = \pi/6$ in [2.17].

[1.28] Put $\theta = \log\left(\left(\sqrt{5}+1\right)/2\right)$ in [2.43].

[1.29] Put $x = 6$ in [2.5].

[1.30] Put $x = 1$ in [2.31].

[1.31] Put $x = \pi$ in [2.31].

[1.32] Put $x = \pi$ in [2.5].

[1.33] We use the *partial fractions* method to simplify the sum as follows. Let

$$\frac{4j+1}{\binom{2j+2}{4}} = \frac{4!(4j+1)}{(2j+2)(2j+1)(2j)(2j-1)}$$

$$= \frac{A}{2j+2} + \frac{B}{2j+1} + \frac{C}{2j} + \frac{D}{2j-1} \qquad (7.1)$$

$$\implies 96j + 24 = A(2j+1)(2j)(2j-1) + B(2j+2)(2j)(2j-1)$$

7 Solutions to Binomial Series

$$+C(2j+2)(2j+1)(2j-1) + D(2j+2)(2j+1)(2j).$$

Equating the constant term in (7.1), we get

$$-2C = 24 \implies C = -12. \tag{7.2}$$

Equating the coefficient of j in (7.1) and using (7.2), we get

$$-2A - 4B - 2C + 4D = 96$$
$$\implies -A - 2B + 2D = 36. \tag{7.3}$$

Equating the coefficient of j^2 in (7.1) and using (7.2), we get

$$4B + 8C + 12D = 0$$
$$\implies B = 24 - 3D. \tag{7.4}$$

Equating the coefficient of j^3 in (7.1) and using (7.2), we get

$$8A + 8B + 8C + 8D = 0$$
$$\implies A + B + D = 12. \tag{7.5}$$

Putting the value of B from (7.4) into (7.3), we get

$$-A - 48 + 6D + 2D = 36$$
$$\implies -A + 8D = 84. \tag{7.6}$$

Putting the value of B from (7.4) into (7.5), we get

$$A + 24 - 3D + D = 12$$
$$\implies A - 2D = -12. \tag{7.7}$$

Adding (7.6) and (7.7), we get

$$6D = 72 \implies D = 12. \tag{7.8}$$

Putting the value of D from (7.8) into (7.7), we get

$$A = 2D - 12 = 12. \tag{7.9}$$

Putting the value of D from (7.8) into (7.4), we get

$$B = 24 - 3D = -12. \tag{7.10}$$

Using (7.1), (7.9), (7.10), (7.2), and (7.8), we get

$$\sum_{j=1}^{\infty} \frac{4j+1}{\binom{2j+2}{4}} = \sum_{j=1}^{\infty} \left(\frac{12}{2j+2} - \frac{12}{2j+1} - \frac{12}{2j} + \frac{12}{2j-1} \right)$$

$$= 12 \left(\sum_{j=1}^{\infty} \frac{1}{2j+2} - \sum_{j=1}^{\infty} \frac{1}{2j} + \sum_{j=1}^{\infty} \frac{1}{2j-1} - \sum_{j=1}^{\infty} \frac{1}{2j+1} \right)$$

$$= 12 \left(\sum_{j=2}^{\infty} \frac{1}{2j} - \sum_{j=1}^{\infty} \frac{1}{2j} + \sum_{j=1}^{\infty} \frac{1}{2j-1} - \sum_{j=2}^{\infty} \frac{1}{2j-1} \right)$$

$$= 12 \left(-\frac{1}{2} + 1 \right)$$

$$= 6.$$

[1.34] Using the binomial theorem, we have

$$(1-x)^n = \sum_{j=0}^{n} (-1)^j \binom{n}{j} x^j. \qquad (7.11)$$

Integrating (7.11) with respect to x from 0 to 1, we get

$$-\frac{(1-x)^{n+1}}{n+1} \bigg|_0^1 = \sum_{j=0}^{n} (-1)^j \binom{n}{j} \frac{x^{j+1}}{j+1} \bigg|_0^1$$

$$\iff \frac{1}{n+1} = \sum_{j=0}^{n} \frac{(-1)^j}{j+1} \binom{n}{j}. \qquad (7.12)$$

[1.35] Integrating (7.11) with respect to x from 0 to x, and using the binomial theorem, we get

$$-\frac{(1-x)^{n+1}}{n+1} \bigg|_0^x = \sum_{j=0}^{n} (-1)^j \binom{n}{j} \frac{x^{j+1}}{j+1} \bigg|_0^x$$

$$\iff \frac{1}{n+1} - \frac{(1-x)^{n+1}}{n+1} = \sum_{j=0}^{n} \frac{(-1)^j x^{j+1}}{j+1} \binom{n}{j} \qquad (7.13)$$

$$\iff \frac{1}{n+1} \sum_{j=1}^{n+1} (-1)^{j-1} \binom{n+1}{j} x^{j-1} = \sum_{j=0}^{n} \frac{(-1)^j x^j}{j+1} \binom{n}{j}.$$

7 Solutions to Binomial Series

Integrating (7.13) with respect to x from 0 to 1, we get

$$\frac{1}{n+1}\sum_{j=1}^{n+1}\frac{(-1)^{j-1}}{j}\binom{n+1}{j}x^j\bigg|_0^1 = \sum_{j=0}^{n}\frac{(-1)^j}{(j+1)^2}\binom{n}{j}x^{j+1}\bigg|_0^1 \quad (7.14)$$

$$\iff \frac{1}{n+1}\sum_{j=1}^{n+1}\frac{(-1)^{j-1}}{j}\binom{n+1}{j} = \sum_{j=0}^{n}\frac{(-1)^j}{(j+1)^2}\binom{n}{j}.$$

Using (1.11), (7.12), and (7.14), we get

$$\frac{1}{n+1}\sum_{j=1}^{n+1}\frac{1}{j} = \sum_{j=0}^{n}\frac{(-1)^j}{(j+1)^2}\binom{n}{j}$$

$$\iff \frac{1}{n+1} + \frac{1}{n+1}\sum_{j=1}^{n}\frac{1}{j+1} = \sum_{j=0}^{n}\frac{(-1)^j}{(j+1)^2}\binom{n}{j}$$

$$\iff \frac{1}{n+1}\sum_{j=1}^{n}\frac{1}{j+1} = \sum_{j=0}^{n}\frac{(-1)^j}{(j+1)^2}\binom{n}{j} - \sum_{j=0}^{n}\frac{(-1)^j}{j+1}\binom{n}{j}$$

$$= \sum_{j=1}^{n}(-1)^{j-1}\frac{j}{(j+1)^2}\binom{n}{j}.$$

[1.36] Differentiating (7.11) with respect to x, we get

$$-n(1-x)^{n-1} = \sum_{j=1}^{n}(-1)^j j\binom{n}{j}x^{j-1}$$

$$\iff n(1-x)^{n-1} = \sum_{j=1}^{n}(-1)^{j-1} j\binom{n}{j}x^{j-1}. \quad (7.15)$$

Putting $x = 1$ in (7.15), we get

$$0 = \sum_{j=1}^{n}(-1)^{j-1} j\binom{n}{j}.$$

[1.37] Putting $x = 2$ in [1.70], we get

$$\sum_{j=0}^{n} \frac{(-1)^j}{2j+1} \binom{n}{j} = \frac{n! 2^n}{\prod_{j=0}^{n}(1+2j)}$$

$$= \frac{2^n n!}{2n+1} \prod_{j=0}^{n-1} \frac{1}{2j+1}$$

$$= \frac{2^{2n}(n!)^2}{2n+1} \prod_{j=0}^{n-1} \frac{1}{(2j+1)(2j+2)}$$

$$= \frac{2^{2n}(n!)^2}{(2n+1)(2n)!}$$

$$= \frac{2^{2n}}{(2n+1)} \binom{2n}{n}^{-1}.$$

[1.38] Put $\theta = 0$ in (2.16).

[1.39] Find the limit as $\theta \to 0$ in (2.12).

[1.40] Find the limit as $\theta \to 0$ in [2.1].

[1.41] Put $\theta = 0$ in (2.19).

[1.42] Find the limit as $\theta \to 0$ in (2.15).

[1.43] Find the limit as $\theta \to 0$ in [2.2].

[1.44] Using the geometric series, $\forall x \in \mathbb{R} - [-1, 1]$, we have

$$\frac{1}{1 - \frac{1}{x}} = \sum_{j=0}^{\infty} \frac{1}{x^j}$$

$$\iff \sum_{j=0}^{\infty} \frac{1}{x^j} = \frac{x}{x-1} \qquad (7.16)$$

$$= 1 + \frac{1}{x-1}.$$

7 Solutions to Binomial Series

Differentiating (7.16) with respect to x, we get

$$-\frac{1}{(x-1)^2} = -\sum_{j=1}^{\infty} \frac{j}{x^{j+1}}$$

$$\iff \sum_{j=1}^{\infty} \frac{j}{x^j} = \frac{x}{(x-1)^2} \qquad (7.17)$$

$$= \frac{1}{x-1} + \frac{1}{(x-1)^2}.$$

Differentiating (7.17) with respect to x, we get

$$-\frac{1}{(x-1)^2} - \frac{2}{(x-1)^3} = -\sum_{j=1}^{\infty} \frac{j^2}{x^{j+1}}$$

$$\iff \frac{x(x+1)}{(x-1)^3} = \sum_{j=1}^{\infty} \frac{j^2}{x^j}. \qquad (7.18)$$

[1.45] Replace x with $-x$ in (7.18).

[1.46] We simplify (7.18) using partial fractions to get

$$\frac{((x-1)+1)((x-1)+2)}{(x-1)^3} = \frac{(x-1)^2 + 3(x-1) + 2}{(x-1)^3}$$

$$= \frac{1}{x-1} + \frac{3}{(x-1)^2} + \frac{2}{(x-1)^3} \qquad (7.19)$$

$$= \sum_{j=1}^{\infty} \frac{j^2}{x^j}.$$

Differentiating (7.19) with respect to x, we get

$$-\frac{1}{(x-1)^2} - \frac{6}{(x-1)^3} - \frac{6}{(x-1)^4} = -\sum_{j=1}^{\infty} \frac{j^3}{x^{j+1}}$$

$$\iff \sum_{j=1}^{\infty} \frac{j^3}{x^j} = \frac{x\left((x-1)^2 + 6(x-1) + 6\right)}{(x-1)^4} \qquad (7.20)$$

$$= \frac{x\left(x^2 + 4x + 1\right)}{(x-1)^4}.$$

[1.47] Replace x with $-x$ in (7.20).

[1.48] We simplify (7.20) using partial fractions to get

$$\frac{((x-1)+1)\left(((x-1)+1)^2+4((x-1)+1)+1\right)}{(x-1)^4}$$
$$=\frac{((x-1)+1)\left((x-1)^2+6(x-1)+6\right)}{(x-1)^4}$$
$$=\frac{(x-1)^3+7(x-1)^2+12(x-1)+6}{(x-1)^4} \quad (7.21)$$
$$=\frac{1}{x-1}+\frac{7}{(x-1)^2}+\frac{12}{(x-1)^3}+\frac{6}{(x-1)^4}$$
$$=\sum_{j=1}^{\infty}\frac{j^3}{x^j}.$$

Differentiating (7.21) with respect to x, we get

$$-\frac{1}{(x-1)^2}-\frac{14}{(x-1)^3}-\frac{36}{(x-1)^4}-\frac{24}{(x-1)^5}=-\sum_{j=1}^{\infty}\frac{j^4}{x^{j+1}}$$
$$\iff \sum_{j=1}^{\infty}\frac{j^4}{x^j}=\frac{x\left((x-1)^3+14(x-1)^2+36(x-1)+24\right)}{(x-1)^5}$$
$$=\frac{x\left(x^3+11x^2+11x+1\right)}{(x-1)^5}$$
$$=\frac{x(x+1)\left(x^2+10x+1\right)}{(x-1)^5}. \quad (7.22)$$

[1.49] Replace x with $-x$ in (7.22).

[1.50] Put $\theta = \pi$ in [2.46].

[1.51] Put $\theta = \pi/2$ in [2.47].

[1.52] Put $\theta = \pi$ in [2.48].

[1.53] Put $\theta = \pi/2$ in [2.49].

[1.54] Put $\theta = \pi/6$ in [2.20].

7 Solutions to Binomial Series

[1.55] Put $\theta = \pi/6$ in [2.21].

[1.56] Put $\theta = \pi/6$ in [2.22].

[1.57] Put $\theta = \pi/6$ in [2.23].

[1.58] Put $\theta = 1$ in [2.46].

[1.59] Put $\theta = 1/2$ in [2.47].

[1.60] Put $\theta = 1$ in [2.48].

[1.61] Put $\theta = 1/2$ in [2.49].

[1.62] Put $\theta = 1$ in [2.20].

[1.63] Put $\theta = 1/2$ in [2.21].

[1.64] Put $\theta = 1$ in [2.22].

[1.65] Put $\theta = 1/2$ in [2.23].

[1.66] Using the infinite series for e, we get

$$\begin{aligned}\frac{\lfloor n!e \rfloor}{n!} &= \frac{1}{n!}\left\lfloor n!\sum_{j=0}^{\infty}\frac{1}{j!}\right\rfloor \\ &= \frac{1}{n!}\left\lfloor \sum_{j=0}^{n}\frac{n!}{j!} + \sum_{j=n+1}^{\infty}\frac{n!}{j!}\right\rfloor \\ &= \frac{1}{n!}\sum_{j=0}^{n}\frac{n!}{j!} + \frac{1}{n!}\left\lfloor \sum_{j=n+1}^{\infty}\frac{n!}{j!}\right\rfloor \\ &= \sum_{j=0}^{n}\frac{1}{j!} + \frac{1}{n!}\left\lfloor \sum_{j=n+1}^{\infty}\frac{n!}{j!}\right\rfloor.\end{aligned} \qquad (7.23)$$

Now, we will prove that

$$0 < \sum_{j=n+1}^{\infty}\frac{n!}{j!} < 1. \qquad (7.24)$$

From (7.23) and (7.24), the result will follow. Using the infinite geometric series, we have

$$0 < \sum_{j=n+1}^{\infty} \frac{n!}{j!} = \sum_{j=1}^{\infty} \frac{1}{\prod_{k=1}^{j}(n+k)} < \sum_{j=1}^{\infty} \frac{1}{2^j} = 1.$$

[1.67] Using the infinite series for $1/e$, we get

$$\begin{aligned}
\frac{1}{n!}\left\lfloor \frac{n!}{e} \right\rfloor &= \frac{1}{n!}\left\lfloor n!\sum_{j=0}^{\infty} \frac{(-1)^j}{j!} \right\rfloor \\
&= \frac{1}{n!}\left\lfloor \sum_{j=0}^{n}(-1)^j \frac{n!}{j!} + \sum_{j=n+1}^{\infty}(-1)^j \frac{n!}{j!} \right\rfloor \\
&= \frac{1}{n!}\sum_{j=0}^{n}(-1)^j \frac{n!}{j!} + \frac{1}{n!}\left\lfloor \sum_{j=n+1}^{\infty}(-1)^j \frac{n!}{j!} \right\rfloor \\
&= \sum_{j=0}^{n} \frac{(-1)^j}{j!} + \frac{1}{n!}\left\lfloor \sum_{j=n+1}^{\infty}(-1)^j \frac{n!}{j!} \right\rfloor .
\end{aligned} \qquad (7.25)$$

Now, we will prove that

$$0 < \sum_{j=n+1}^{\infty} (-1)^j \frac{n!}{j!} < 1. \qquad (7.26)$$

From (7.25) and (7.26), the result will follow. From (7.24), we have

$$\sum_{j=n+1}^{\infty} (-1)^j \frac{n!}{j!} < \sum_{j=n+1}^{\infty} \frac{n!}{j!} < 1.$$

Let $n = 2m - 1$. We get

$$\begin{aligned}
\sum_{j=n+1}^{\infty}(-1)^j \frac{n!}{j!} &= \sum_{k=m}^{\infty}\left((-1)^{2k}\frac{(2m-1)!}{(2k)!} + (-1)^{2k+1}\frac{(2m-1)!}{(2k+1)!}\right) \\
&= \sum_{k=m}^{\infty}\left(\frac{(2m-1)!}{(2k)!} - \frac{(2m-1)!}{(2k+1)!}\right) > 0.
\end{aligned}$$

7 Solutions to Binomial Series

Once again, using the infinite series for $1/e$, we get

$$\frac{1}{(n+1)!}\left\lfloor\frac{(n+1)!}{e}\right\rfloor = \frac{1}{(n+1)!}\left\lfloor(n+1)!\sum_{j=0}^{\infty}\frac{(-1)^j}{j!}\right\rfloor$$

$$= \frac{1}{(n+1)!}\left\lfloor\sum_{j=0}^{n}(-1)^j\frac{(n+1)!}{j!} + \sum_{j=n+1}^{\infty}(-1)^j\frac{(n+1)!}{j!}\right\rfloor$$

$$= \frac{1}{(n+1)!}\sum_{j=0}^{n}(-1)^j\frac{(n+1)!}{j!} + \frac{1}{(n+1)!}\left\lfloor\sum_{j=n+1}^{\infty}(-1)^j\frac{(n+1)!}{j!}\right\rfloor$$

$$= \sum_{j=0}^{n}\frac{(-1)^j}{j!} + \frac{1}{(n+1)!}\left\lfloor\sum_{j=n+1}^{\infty}(-1)^j\frac{(n+1)!}{j!}\right\rfloor.$$

(7.27)

Now, we will prove that

$$0 < \sum_{j=n+1}^{\infty}\frac{(-1)^j(n+1)!}{j!} < 1. \tag{7.28}$$

From (7.27) and (7.28), the result will follow. Let $n = 2m - 1$. We get

$$\sum_{j=n+1}^{\infty}(-1)^j\frac{(n+1)!}{j!} = \sum_{k=m}^{\infty}\left((-1)^{2k}\frac{(2m)!}{(2k)!} + (-1)^{2k+1}\frac{(2m)!}{(2k+1)!}\right)$$

$$= \sum_{k=m}^{\infty}\left(\frac{(2m)!}{(2k)!} - \frac{(2m)!}{(2k+1)!}\right) > 0.$$

We also have

$$\sum_{j=n+1}^{\infty}(-1)^j\frac{(n+1)!}{j!} = 1 - \sum_{j=n+2}^{\infty}(-1)^{j-1}\frac{(n+1)!}{j!}$$

$$= 1 - \sum_{k=m}^{\infty}\left((-1)^{2k}\frac{(2m)!}{(2k+1)!} + (-1)^{2k+1}\frac{(2m)!}{(2k+2)!}\right)$$

$$= 1 - \sum_{k=m}^{\infty}\left(\frac{(2m)!}{(2k+1)!} - \frac{(2m)!}{(2k+2)!}\right) < 1.$$

[1.68] Put $x = 100$ in [1.44].

[1.69] Put $\theta = \pi^2$ in [2.46].

[1.70] We will evaluate the integral $\int_0^a (a^x - y^x)^n \, dy$ in two ways to get this equation. First, using the binomial theorem, we get

$$\int_0^a (a^x - y^x)^n \, dy = \sum_{j=0}^n (-1)^j \binom{n}{j} a^{(n-j)x} \int_0^a y^{jx} \, dy$$

$$= \sum_{j=0}^n (-1)^j \binom{n}{j} a^{(n-j)x} \left. \frac{y^{jx+1}}{jx+1} \right|_0^a \qquad (7.29)$$

$$= \sum_{j=0}^n (-1)^j \binom{n}{j} \frac{a^{nx+1}}{jx+1}.$$

Now we will evaluate the integral $\int_0^a (a^x - y^x)^n \, dy$ using integration by parts. Let

$$I_j = \int_0^a y^{jx} (a^x - y^x)^{n-j} \, dy, \quad \forall j \in [0..n].$$

We have to evaluate I_0. Integrating the integral I_j by parts, $\forall j \in [0..n-1]$, we get

$$I_j = \int_0^a y^{jx} (a^x - y^x)^{n-j} \, dy$$

$$= \left. \frac{y^{jx+1}}{jx+1} (a^x - y^x)^{n-j} \right|_0^a - \int_0^a \frac{y^{jx+1}}{jx+1} (n-j)(a^x - y^x)^{n-j-1} (-xy^{x-1}) \, dy$$

$$= \frac{(n-j)x}{jx+1} \int_0^a y^{(j+1)x} (a^x - y^x)^{n-j-1} \, dy$$

$$= \frac{(n-j)x}{jx+1} I_{j+1}.$$

(7.30)

Using (7.30), we get

$$I_0 = I_n \prod_{j=0}^{n-1} \frac{(n-j)x}{jx+1} = \frac{n! x^n I_n}{\prod_{j=0}^{n-1} (1+jx)}. \qquad (7.31)$$

Now we will evaluate I_n.

7 Solutions to Binomial Series   235

$$I_n = \int_0^a y^{nx} \, dy = \frac{y^{nx+1}}{nx+1}\bigg|_0^a = \frac{a^{nx+1}}{nx+1}. \qquad (7.32)$$

From (7.31) and (7.32), we get

$$I_0 = \frac{n! x^n a^{nx+1}}{\prod_{j=0}^{n}(1+jx)}. \qquad (7.33)$$

From (7.29) and (7.33), we get our result.

[1.71] For $m = n$, the series has only one term

$$\sum_{j=0}^{0} (-1)^j \binom{n+j}{n}\binom{n}{n+j} = \binom{n}{n}^2 = 1.$$

[1.72] In the case of $m < n$, the series sum is empty. Therefore, it has the sum of 0. Assuming $m > n$, using the binomial theorem, we have

$$(1+x)^m = \sum_{j=0}^{m} \binom{m}{m-j} x^j, \qquad (7.34)$$

and

$$(1+x)^{-(n+1)} = \sum_{j=0}^{\infty} (-1)^j \binom{n+j}{n} x^j. \qquad (7.35)$$

Multiplying (7.34) and (7.35), we get

$$(1+x)^{m-n-1} = \sum_{k=0}^{\infty} \left(\sum_{j=0}^{k} (-1)^j \binom{n+j}{n}\binom{m}{m-k+j} \right) x^k. \qquad (7.36)$$

Equating the coefficient of x^{m-n} in (7.36), we get

$$0 = \sum_{j=0}^{m-n} (-1)^j \binom{n+j}{n}\binom{m}{n+j}.$$

[1.73] Let

$$f(\theta, k) = \frac{1}{2^k} \cot \frac{1}{2^k \theta} + \sum_{j=k+1}^{\infty} \frac{1}{2^j} \tan \frac{1}{2^j \theta}.$$

We have

$$f(\theta, k) = \frac{1}{2^k} \cot \frac{1}{2^k \theta} + \sum_{j=k+1}^{\infty} \frac{1}{2^j} \tan \frac{1}{2^j \theta}$$

$$= \frac{1}{2^k} \cdot \frac{1 - \tan^2 \frac{1}{2^{k+1}\theta}}{2 \tan \frac{1}{2^{k+1}\theta}} + \frac{1}{2^{k+1}} \tan \frac{1}{2^{k+1}\theta} + \sum_{j=k+2}^{\infty} \frac{1}{2^j} \tan \frac{1}{2^j \theta} \quad (7.37)$$

$$= \frac{1}{2^{k+1}} \cot \frac{1}{2^{k+1}\theta} + \sum_{j=k+2}^{\infty} \frac{1}{2^j} \tan \frac{1}{2^j \theta}$$

$$= f(\theta, k+1).$$

Applying the recursion in (7.37) for $k \in [0..n-1]$, we get

$$f(\theta, 0) = f(\theta, n)$$

$$= \frac{1}{2^n} \cot \frac{1}{2^n \theta} + \sum_{j=n+1}^{\infty} \frac{1}{2^j} \tan \frac{1}{2^j \theta}. \quad (7.38)$$

Taking the limit as $n \to \infty$ in (7.38), we get

$$\cot \frac{1}{\theta} + \sum_{j=1}^{\infty} \frac{1}{2^j} \tan \frac{1}{2^j \theta} = \lim_{n \to \infty} \left(\frac{1}{2^n} \cot \frac{1}{2^n \theta} + \sum_{j=n+1}^{\infty} \frac{1}{2^j} \tan \frac{1}{2^j \theta} \right)$$

$$= \lim_{n \to \infty} \theta \frac{\frac{1}{2^n \theta}}{\tan \frac{1}{2^n \theta}} + \lim_{n \to \infty} \sum_{j=n+1}^{\infty} \frac{1}{2^j} \cdot \frac{1}{2^j \theta} \cdot \frac{\tan \frac{1}{2^j \theta}}{\frac{1}{2^j \theta}}$$

$$= \theta + \frac{1}{\theta} \lim_{n \to \infty} \sum_{j=n+1}^{\infty} \frac{1}{2^{2j}}$$

$$= \theta + \frac{1}{\theta} \lim_{n \to \infty} \frac{\frac{1}{2^{2n+2}}}{1 - \frac{1}{4}}$$

$$= \theta.$$

[1.74] Put $y = 1$ in [1.77].

7 Solutions to Binomial Series

[1.75] Let
$$f(x,n) = \sum_{j=0}^{n-1} \frac{2^{2j} x^{2^j}}{(x^{2^j}+1)^2} + \frac{2^{2n} x^{2^n}}{(x^{2^n}-1)^2}.$$

We have

$$\begin{aligned}
f(x,n) &= \sum_{j=0}^{n-1} \frac{2^{2j} x^{2^j}}{\left(x^{2^j}+1\right)^2} + \frac{2^{2n} x^{2^n}}{\left(x^{2^n}-1\right)^2} \\
&= \sum_{j=0}^{n-2} \frac{2^{2j} x^{2^j}}{\left(x^{2^j}+1\right)^2} + \frac{2^{2n-2} x^{2^{n-1}}}{\left(x^{2^{n-1}}+1\right)^2} + \frac{2^{2n} x^{2^n}}{\left(x^{2^{n-1}}-1\right)^2 \left(x^{2^{n-1}}+1\right)^2} \\
&= \sum_{j=0}^{n-2} \frac{2^{2j} x^{2^j}}{\left(x^{2^j}+1\right)^2} + \frac{2^{2n-2} x^{2^{n-1}}}{\left(x^{2^{n-1}}+1\right)^2} + \frac{2^{2n-2} x^{2^{n-1}}}{\left(x^{2^{n-1}}-1\right)^2} - \frac{2^{2n-2} x^{2^{n-1}}}{\left(x^{2^{n-1}}+1\right)^2} \\
&= \sum_{j=0}^{n-2} \frac{2^{2j} x^{2^j}}{\left(x^{2^j}+1\right)^2} + \frac{2^{2n-2} x^{2^{n-1}}}{\left(x^{2^{n-1}}-1\right)^2} \\
&= f(x, n-1). \quad\quad (7.39)
\end{aligned}$$

Applying the recursion in (7.39) for $n-1$ times, we get

$$\begin{aligned}
f(x,n) &= f(x,1) \\
&= \frac{x}{(x+1)^2} + \frac{4x^2}{\left(x^2-1\right)^2} \\
&= \frac{x}{(x+1)^2} + \frac{4x^2}{(x-1)^2(x+1)^2} \\
&= \frac{x}{(x+1)^2} + \frac{x}{(x-1)^2} - \frac{x}{(x+1)^2} \\
&= \frac{x}{(x-1)^2}.
\end{aligned}$$

[1.76] Let
$$f(x,y,n) = \sum_{j=0}^{n-1} \frac{2^j x^{2^j} \left(y^{2^{j+1}}-1\right)}{\left(x^{2^j}+y^{2^j}\right)\left(x^{2^j} y^{2^j}+1\right)} + \frac{2^n x^{2^n} \left(y^{2^{n+1}}-1\right)}{\left(x^{2^n}-y^{2^n}\right)\left(x^{2^n} y^{2^n}-1\right)}.$$

We have

$$f(x,y,n) = \sum_{j=0}^{n-1} \frac{2^j x^{2^j} \left(y^{2^{j+1}} - 1\right)}{\left(x^{2^j} + y^{2^j}\right)\left(x^{2^j} y^{2^j} + 1\right)} + \frac{2^n x^{2^n} \left(y^{2^{n+1}} - 1\right)}{\left(x^{2^n} - y^{2^n}\right)\left(x^{2^n} y^{2^n} - 1\right)}$$

$$= \sum_{j=0}^{n-2} \frac{2^j x^{2^j} \left(y^{2^{j+1}} - 1\right)}{\left(x^{2^j} + y^{2^j}\right)\left(x^{2^j} y^{2^j} + 1\right)} + \frac{2^{n-1} x^{2^{n-1}} \left(y^{2^n} - 1\right)}{\left(x^{2^{n-1}} + y^{2^{n-1}}\right)\left(x^{2^{n-1}} y^{2^{n-1}} + 1\right)}$$

$$+ \frac{2^n x^{2^n} \left(y^{2^{n+1}} - 1\right)}{\left(x^{2^{n-1}} - y^{2^{n-1}}\right)\left(x^{2^{n-1}} y^{2^{n-1}} - 1\right)\left(x^{2^{n-1}} + y^{2^{n-1}}\right)\left(x^{2^{n-1}} y^{2^{n-1}} + 1\right)}$$

$$= \sum_{j=0}^{n-2} \frac{2^j x^{2^j} \left(y^{2^{j+1}} - 1\right)}{\left(x^{2^j} + y^{2^j}\right)\left(x^{2^j} y^{2^j} + 1\right)} + \frac{2^{n-1} x^{2^{n-1}} \left(y^{2^n} - 1\right)}{\left(x^{2^{n-1}} + y^{2^{n-1}}\right)\left(x^{2^{n-1}} y^{2^{n-1}} + 1\right)}$$

$$+ \frac{2^{n-1} x^{2^{n-1}} \left(y^{2^n} - 1\right)}{\left(x^{2^{n-1}} - y^{2^{n-1}}\right)\left(x^{2^{n-1}} y^{2^{n-1}} - 1\right)} - \frac{2^{n-1} x^{2^{n-1}} \left(y^{2^n} - 1\right)}{\left(x^{2^{n-1}} + y^{2^{n-1}}\right)\left(x^{2^{n-1}} y^{2^{n-1}} + 1\right)}$$

$$= \sum_{j=0}^{n-2} \frac{2^j x^{2^j} \left(y^{2^{j+1}} - 1\right)}{\left(x^{2^j} + y^{2^j}\right)\left(x^{2^j} y^{2^j} + 1\right)} + \frac{2^{n-1} x^{2^{n-1}} \left(y^{2^n} - 1\right)}{\left(x^{2^{n-1}} - y^{2^{n-1}}\right)\left(x^{2^{n-1}} y^{2^{n-1}} - 1\right)}$$

$$= f(x, y, n - 1). \tag{7.40}$$

Applying the recursion in (7.40) for $n - 1$ times, we get

$$f(x, y, n) = f(x, y, 1)$$
$$= \frac{x\left(y^2 - 1\right)}{(x + y)(xy + 1)} + \frac{2x^2 \left(y^4 - 1\right)}{\left(x^2 - y^2\right)\left(x^2 y^2 - 1\right)}$$
$$= \frac{x\left(y^2 - 1\right)}{(x + y)(xy + 1)} + \frac{2x^2 \left(y^4 - 1\right)}{(x - y)(xy - 1)(x + y)(xy + 1)}$$
$$= \frac{x\left(y^2 - 1\right)}{(x + y)(xy + 1)} + \frac{x\left(y^2 - 1\right)}{(x - y)(xy - 1)} - \frac{x\left(y^2 - 1\right)}{(x + y)(xy + 1)}$$
$$= \frac{x\left(y^2 - 1\right)}{(x - y)(xy - 1)}.$$

[1.77] Let

$$f(x, y, n) = \sum_{j=0}^{n-1} \frac{2^j y^{2^j}}{x^{2^j} + y^{2^j}} + \frac{2^n y^{2^n}}{x^{2^n} - y^{2^n}}.$$

We have

7 Solutions to Binomial Series

$$f(x, y, n) = \sum_{j=0}^{n-1} \frac{2^j y^{2^j}}{x^{2^j} + y^{2^j}} + \frac{2^n y^{2^n}}{x^{2^n} - y^{2^n}}$$

$$= \sum_{j=0}^{n-2} \frac{2^j y^{2^j}}{x^{2^j} + y^{2^j}} + \frac{2^{n-1} y^{2^{n-1}}}{x^{2^{n-1}} + y^{2^{n-1}}} + \frac{2^n y^{2^n}}{\left(x^{2^{n-1}} - y^{2^{n-1}}\right)\left(x^{2^{n-1}} + y^{2^{n-1}}\right)}$$

$$= \sum_{j=0}^{n-2} \frac{2^j y^{2^j}}{x^{2^j} + y^{2^j}} + \frac{2^{n-1} y^{2^{n-1}}}{x^{2^{n-1}} + y^{2^{n-1}}} + \frac{2^{n-1} y^{2^{n-1}}}{x^{2^{n-1}} - y^{2^{n-1}}} - \frac{2^{n-1} y^{2^{n-1}}}{x^{2^{n-1}} + y^{2^{n-1}}}$$

$$= \sum_{j=0}^{n-2} \frac{2^j y^{2^j}}{x^{2^j} + y^{2^j}} + \frac{2^{n-1} y^{2^{n-1}}}{x^{2^{n-1}} - y^{2^{n-1}}}$$

$$= f(x, y, n - 1). \tag{7.41}$$

Applying the recursion in (7.41) for $n - 1$ times, we get

$$f(x, y, n) = f(x, y, 1)$$

$$= \frac{y}{x + y} + \frac{2y^2}{x^2 - y^2}$$

$$= \frac{y}{x + y} + \frac{2y^2}{(x - y)(x + y)}$$

$$= \frac{y}{x + y} + \frac{y}{x - y} - \frac{y}{x + y}$$

$$= \frac{y}{x - y}.$$

[1.78] Replacing x with xy in [1.74], we get

$$\frac{1}{xy - 1} = \sum_{j=0}^{n-1} \frac{2^j}{x^{2^j} y^{2^j} + 1} + \frac{2^n}{x^{2^n} y^{2^n} - 1}. \tag{7.42}$$

Adding [1.77] and (7.42), we get

$$\frac{y}{x-y} + \frac{1}{xy-1} = \sum_{j=0}^{n-1} 2^j \left(\frac{y^{2^j}}{x^{2^j} + y^{2^j}} + \frac{1}{x^{2^j} y^{2^j} + 1} \right)$$

$$+ 2^n \left(\frac{y^{2^n}}{x^{2^n} - y^{2^n}} + \frac{1}{x^{2^n} y^{2^n} - 1} \right)$$

$$\iff \frac{y(xy-1) + (x-y)}{(x-y)(xy-1)}$$

$$= \sum_{j=0}^{n-1} 2^j \frac{y^{2^j}(x^{2^j} y^{2^j} + 1) + (x^{2^j} + y^{2^j})}{(x^{2^j} + y^{2^j})(x^{2^j} y^{2^j} + 1)} + 2^n \frac{y^{2^n}(x^{2^n} y^{2^n} - 1) + (x^{2^n} - y^{2^n})}{(x^{2^n} - y^{2^n})(x^{2^n} y^{2^n} - 1)}.$$

[1.79] Let

$$f(x, \theta, n) = \sum_{j=0}^{n-1} \frac{2^j \left(x^{2^j} \cos 2^j \theta + 1 \right)}{x^{2^{j+1}} + 2x^{2^j} \cos 2^j \theta + 1} + \frac{2^n \left(x^{2^n} \cos 2^n \theta - 1 \right)}{x^{2^{n+1}} - 2x^{2^n} \cos 2^n \theta + 1}.$$

We have

$$f(x, \theta, n) = \sum_{j=0}^{n-1} \frac{2^j \left(x^{2^j} \cos 2^j \theta + 1 \right)}{x^{2^{j+1}} + 2x^{2^j} \cos 2^j \theta + 1} + \frac{2^n \left(x^{2^n} \cos 2^n \theta - 1 \right)}{x^{2^{n+1}} - 2x^{2^n} \cos 2^n \theta + 1}$$

$$= \sum_{j=0}^{n-2} \frac{2^j \left(x^{2^j} \cos 2^j \theta + 1 \right)}{x^{2^{j+1}} + 2x^{2^j} \cos 2^j \theta + 1} + \frac{2^{n-1} \left(x^{2^{n-1}} \cos 2^{n-1} \theta + 1 \right)}{x^{2^n} + 2x^{2^{n-1}} \cos 2^{n-1} \theta + 1}$$

$$+ \frac{2^n \left(x^{2^n} \cos 2^n \theta - 1 \right)}{x^{2^{n+1}} - 2x^{2^n} \cos 2^n \theta + 1}$$

$$= \sum_{j=0}^{n-2} \frac{2^j \left(x^{2^j} \cos 2^j \theta + 1 \right)}{x^{2^{j+1}} + 2x^{2^j} \cos 2^j \theta + 1} + \frac{2^{n-1} \left(x^{2^{n-1}} \cos 2^{n-1} \theta + 1 \right)}{x^{2^n} + 2x^{2^{n-1}} \cos 2^{n-1} \theta + 1}$$

$$+ \frac{2^n \left(2x^{2^n} \cos^2 2^{n-1} \theta - x^{2^n} - 1 \right)}{x^{2^{n+1}} - 4x^{2^n} \cos^2 2^{n-1} \theta + 2x^{2^n} + 1}$$

$$= \sum_{j=0}^{n-2} \frac{2^j \left(x^{2^j} \cos 2^j \theta + 1 \right)}{x^{2^{j+1}} + 2x^{2^j} \cos 2^j \theta + 1} + \frac{2^{n-1} \left(x^{2^{n-1}} \cos 2^{n-1} \theta + 1 \right)}{x^{2^n} + 2x^{2^{n-1}} \cos 2^{n-1} \theta + 1}$$

$$+ \frac{2^n \left(2x^{2^n} \cos^2 2^{n-1} \theta - x^{2^n} - 1 \right)}{\left(x^{2^n} - 2x^{2^{n-1}} \cos 2^{n-1} \theta + 1 \right)\left(x^{2^n} + 2x^{2^{n-1}} \cos 2^{n-1} \theta + 1 \right)}$$

$$= \sum_{j=0}^{n-2} \frac{2^j \left(x^{2^j} \cos 2^j \theta + 1 \right)}{x^{2^{j+1}} + 2x^{2^j} \cos 2^j \theta + 1} + \frac{2^{n-1} \left(x^{2^{n-1}} \cos 2^{n-1} \theta + 1 \right)}{x^{2^n} + 2x^{2^{n-1}} \cos 2^{n-1} \theta + 1}$$

7 Solutions to Binomial Series 241

$$+\frac{2^{n-1}\left(x^{2^{n-1}}\cos 2^{n-1}\theta - 1\right)}{x^{2^n} - 2x^{2^{n-1}}\cos 2^{n-1}\theta + 1} - \frac{2^{n-1}\left(x^{2^{n-1}}\cos 2^{n-1}\theta + 1\right)}{x^{2^n} + 2x^{2^{n-1}}\cos 2^{n-1}\theta + 1}$$

$$= \sum_{j=0}^{n-2} \frac{2^j\left(x^{2^j}\cos 2^j\theta + 1\right)}{x^{2^{j+1}} + 2x^{2^j}\cos 2^j\theta + 1} + \frac{2^{n-1}\left(x^{2^{n-1}}\cos 2^{n-1}\theta - 1\right)}{x^{2^n} - 2x^{2^{n-1}}\cos 2^{n-1}\theta + 1}$$

$$= f(x, \theta, n-1). \tag{7.43}$$

Applying the recursion in (7.43) for $n-1$ times, we get

$$f(x,\theta,n) = f(x,\theta,1) = \frac{x\cos\theta + 1}{x^2 + 2x\cos\theta + 1} + \frac{2\left(x^2\cos 2\theta - 1\right)}{x^4 - 2x^2\cos 2\theta + 1}$$

$$= \frac{x\cos\theta + 1}{x^2 + 2x\cos\theta + 1} + \frac{2\left(2x^2\cos^2\theta - x^2 - 1\right)}{x^4 - 4x^2\cos^2\theta + 2x^2 + 1}$$

$$= \frac{x\cos\theta + 1}{x^2 + 2x\cos\theta + 1} + \frac{2\left(2x^2\cos^2\theta - x^2 - 1\right)}{\left(x^2 - 2x\cos\theta + 1\right)\left(x^2 + 2x\cos\theta + 1\right)}$$

$$= \frac{x\cos\theta + 1}{x^2 + 2x\cos\theta + 1} + \frac{x\cos\theta - 1}{x^2 - 2x\cos\theta + 1} - \frac{x\cos\theta + 1}{x^2 + 2x\cos\theta + 1}$$

$$= \frac{x\cos\theta - 1}{x^2 - 2x\cos\theta + 1}.$$

[1.80] Let

$$f(x,\theta,n) = \sum_{j=0}^{n-1} \frac{2^j x^{2^j}\sin 2^j\theta}{x^{2^{j+1}} + 2x^{2^j}\cos 2^j\theta + 1} + \frac{2^n x^{2^n}\sin 2^n\theta}{x^{2^{n+1}} - 2x^{2^n}\cos 2^n\theta + 1}.$$

We have

$$f(x,\theta,n) = \sum_{j=0}^{n-1} \frac{2^j x^{2^j}\sin 2^j\theta}{x^{2^{j+1}} + 2x^{2^j}\cos 2^j\theta + 1} + \frac{2^n x^{2^n}\sin 2^n\theta}{x^{2^{n+1}} - 2x^{2^n}\cos 2^n\theta + 1}$$

$$= \sum_{j=0}^{n-2} \frac{2^j x^{2^j}\sin 2^j\theta}{x^{2^{j+1}} + 2x^{2^j}\cos 2^j\theta + 1} + \frac{2^{n-1} x^{2^{n-1}}\sin 2^{n-1}\theta}{x^{2^n} + 2x^{2^{n-1}}\cos 2^{n-1}\theta + 1}$$

$$+ \frac{2^n x^{2^n}\sin 2^n\theta}{x^{2^{n+1}} - 2x^{2^n}\cos 2^n\theta + 1}$$

$$= \sum_{j=0}^{n-2} \frac{2^j x^{2^j}\sin 2^j\theta}{x^{2^{j+1}} + 2x^{2^j}\cos 2^j\theta + 1} + \frac{2^{n-1} x^{2^{n-1}}\sin 2^{n-1}\theta}{x^{2^n} + 2x^{2^{n-1}}\cos 2^{n-1}\theta + 1}$$

$$+ \frac{2^{n+1}x^{2^n}\sin 2^{n-1}\theta \cos 2^{n-1}\theta}{x^{2^{n+1}} - 4x^{2^n}\cos^2 2^{n-1}\theta + 2x^{2^n} + 1}$$

$$= \sum_{j=0}^{n-2} \frac{2^j x^{2^j}\sin 2^j\theta}{x^{2^{j+1}} + 2x^{2^j}\cos 2^j\theta + 1} + \frac{2^{n-1}x^{2^{n-1}}\sin 2^{n-1}\theta}{x^{2^n} + 2x^{2^{n-1}}\cos 2^{n-1}\theta + 1}$$

$$+ \frac{2^{n+1}x^{2^n}\sin 2^{n-1}\theta \cos 2^{n-1}\theta}{\left(x^{2^n} - 2x^{2^{n-1}}\cos 2^{n-1}\theta + 1\right)\left(x^{2^n} + 2x^{2^{n-1}}\cos 2^{n-1}\theta + 1\right)}$$

$$= \sum_{j=0}^{n-2} \frac{2^j x^{2^j}\sin 2^j\theta}{x^{2^{j+1}} + 2x^{2^j}\cos 2^j\theta + 1} + \frac{2^{n-1}x^{2^{n-1}}\sin 2^{n-1}\theta}{x^{2^n} + 2x^{2^{n-1}}\cos 2^{n-1}\theta + 1}$$

$$+ \frac{2^{n-1}x^{2^{n-1}}\sin 2^{n-1}\theta}{x^{2^n} - 2x^{2^{n-1}}\cos 2^{n-1}\theta + 1} - \frac{2^{n-1}x^{2^{n-1}}\sin 2^{n-1}\theta}{x^{2^n} + 2x^{2^{n-1}}\cos 2^{n-1}\theta + 1}$$

$$= \sum_{j=0}^{n-2} \frac{2^j x^{2^j}\sin 2^j\theta}{x^{2^{j+1}} + 2x^{2^j}\cos 2^j\theta + 1} + \frac{2^{n-1}x^{2^{n-1}}\sin 2^{n-1}\theta}{x^{2^n} - 2x^{2^{n-1}}\cos 2^{n-1}\theta + 1}$$

$$= f(x, \theta, n-1). \tag{7.44}$$

Applying the recursion in (7.44) for $n-1$ times, we get

$$f(x, \theta, n) = f(x, \theta, 1) = \frac{x\sin\theta}{x^2 + 2x\cos\theta + 1} + \frac{2x^2\sin 2\theta}{x^4 - 2x^2\cos 2\theta + 1}$$

$$= \frac{x\sin\theta}{x^2 + 2x\cos\theta + 1} + \frac{4x^2\sin\theta\cos\theta}{x^4 - 4x^2\cos^2\theta + 2x^2 + 1}$$

$$= \frac{x\sin\theta}{x^2 + 2x\cos\theta + 1} + \frac{4x^2\sin\theta\cos\theta}{(x^2 - 2x\cos\theta + 1)(x^2 + 2x\cos\theta + 1)}$$

$$= \frac{x\sin\theta}{x^2 + 2x\cos\theta + 1} + \frac{x\sin\theta}{x^2 - 2x\cos\theta + 1} - \frac{x\sin\theta}{x^2 + 2x\cos\theta + 1}$$

$$= \frac{x\sin\theta}{x^2 - 2x\cos\theta + 1}.$$

[1.81] We will prove by induction. Let $P(n)$ be the statement that

$$\frac{x^{2^{n+2}} - 2x^{2^{n+1}}\cos 2^{n+1}\theta + 1}{x^2 - 2x\cos\theta + 1} = \prod_{j=0}^{n}\left(x^{2^{j+1}} + 2x^{2^j}\cos 2^j\theta + 1\right). \tag{7.45}$$

7 Solutions to Binomial Series

Basis Step: $P(0)$ states that

$$\frac{x^4 - 2x^2 \cos 2\theta + 1}{x^2 - 2x \cos \theta + 1} = x^2 + 2x \cos \theta + 1.$$

We have

$$\begin{aligned}
\frac{x^4 - 2x^2 \cos 2\theta + 1}{x^2 - 2x \cos \theta + 1} &= \frac{x^4 - 2x^2 \left(2\cos^2 \theta - 1\right) + 1}{x^2 - 2x \cos \theta + 1} \\
&= \frac{x^4 - 4x^2 \cos^2 \theta + 2x^2 + 1}{x^2 - 2x \cos \theta + 1} \\
&= \frac{\left(x^2 - 2x \cos \theta + 1\right)\left(x^2 + 2x \cos \theta + 1\right)}{x^2 - 2x \cos \theta + 1} \\
&= x^2 + 2x \cos \theta + 1,
\end{aligned}$$

proving $P(0)$ to be correct.

Induction Step: Assuming $P(n)$ to be true, using (7.45), we have

$$\begin{aligned}
&\prod_{j=0}^{n+1} \left(x^{2^{j+1}} + 2x^{2^j} \cos 2^j \theta + 1\right) \\
&= \left(\prod_{j=0}^{n} \left(x^{2^{j+1}} + 2x^{2^j} \cos 2^j \theta + 1\right)\right) \left(x^{2^{n+2}} + 2x^{2^{n+1}} \cos 2^{n+1} \theta + 1\right) \\
&= \frac{\left(x^{2^{n+2}} - 2x^{2^{n+1}} \cos 2^{n+1} \theta + 1\right)\left(x^{2^{n+2}} + 2x^{2^{n+1}} \cos 2^{n+1} \theta + 1\right)}{x^2 - 2x \cos \theta + 1} \\
&= \frac{x^{2^{n+3}} - 2x^{2^{n+2}} \left(2\cos^2 2^{n+1} \theta - 1\right) + 1}{x^2 - 2x \cos \theta + 1} \\
&= \frac{x^{2^{n+3}} - 2x^{2^{n+2}} \cos 2^{n+2} \theta + 1}{x^2 - 2x \cos \theta + 1},
\end{aligned}$$

proving $P(n+1)$ to be correct.

[1.82] We have

$$T(x,\theta,n) = \sum_{j=0}^{n-1} \tan^{-1} \frac{x^{2^j} + \cos 2^j\theta}{\sin 2^j\theta} + \tan^{-1} \frac{x^{2^n} - \cos 2^n\theta}{\sin 2^n\theta}$$

$$= \sum_{j=0}^{n-2} \tan^{-1} \frac{x^{2^j} + \cos 2^j\theta}{\sin 2^j\theta} + \tan^{-1} \frac{x^{2^{n-1}} + \cos 2^{n-1}\theta}{\sin 2^{n-1}\theta}$$

$$+ \tan^{-1} \frac{x^{2^n} - \cos 2^n\theta}{\sin 2^n\theta}$$

$$= \sum_{j=0}^{n-2} \tan^{-1} \frac{x^{2^j} + \cos 2^j\theta}{\sin 2^j\theta} + \{-1, 0, 1\}\pi$$

$$+ \tan^{-1} \frac{\left(x^{2^{n-1}} + \cos 2^{n-1}\theta\right)\sin 2^n\theta + \left(x^{2^n} - \cos 2^n\theta\right)\sin 2^{n-1}\theta}{\sin 2^{n-1}\theta \sin 2^n\theta - \left(x^{2^{n-1}} + \cos 2^{n-1}\theta\right)\left(x^{2^n} - \cos 2^n\theta\right)}$$

$$= \sum_{j=0}^{n-2} \tan^{-1} \frac{x^{2^j} + \cos 2^j\theta}{\sin 2^j\theta} + \{-1, 0, 1\}\pi$$

$$+ \tan^{-1} \left(-\frac{x^{2^n} \sin 2^{n-1}\theta + 2x^{2^{n-1}} \sin 2^{n-1}\theta \cos 2^{n-1}\theta + \sin 2^{n-1}\theta}{x^{2^n+2^{n-1}} + x^{2^n}\cos 2^{n-1}\theta - x^{2^{n-1}}\left(2\cos^2 2^{n-1}\theta - 1\right) - \cos 2^{n-1}\theta}\right)$$

$$= \sum_{j=0}^{n-2} \tan^{-1} \frac{x^{2^j} + \cos 2^j\theta}{\sin 2^j\theta} + \{-1, 0, 1\}\pi$$

$$+ \tan^{-1}\left(-\frac{\left(x^{2^n} + 2x^{2^{n-1}}\cos 2^{n-1}\theta + 1\right)\sin 2^{n-1}\theta}{\left(x^{2^{n-1}} - \cos 2^{n-1}\theta\right)\left(x^{2^n} + 2x^{2^{n-1}}\cos 2^{n-1}\theta + 1\right)}\right)$$

$$= \sum_{j=0}^{n-2} \tan^{-1} \frac{x^{2^j} + \cos 2^j\theta}{\sin 2^j\theta} + \tan^{-1}\left(-\frac{\sin 2^{n-1}\theta}{x^{2^{n-1}} - \cos 2^{n-1}\theta}\right) + \{-1, 0, 1\}\pi$$

$$= \sum_{j=0}^{n-2} \tan^{-1} \frac{x^{2^j} + \cos 2^j\theta}{\sin 2^j\theta} + \tan^{-1} \frac{x^{2^{n-1}} - \cos 2^{n-1}\theta}{\sin 2^{n-1}\theta} - \frac{\pi}{2} + \{-1, 0, 1\}\pi$$

$$= T(x, \theta, n-1) + \frac{\pi}{2}\{-3, -1, 1\}$$

$$\implies T(x, \theta, n) - T(x, \theta, n-1) \in \frac{\pi}{2}\mathbb{Z}.$$

(7.46)

Applying the recursion in (7.46) for $n-1$ times and adding all the equations, we get

7 Solutions to Binomial Series   245

$$T(x,\theta,n) - T(x,\theta,1) \in \frac{\pi}{2}\mathbb{Z}$$

$$\iff T(x,\theta,n) - \tan^{-1}\frac{x+\cos\theta}{\sin\theta} - \tan^{-1}\frac{x^2-\cos 2\theta}{\sin 2\theta} \in \frac{\pi}{2}\mathbb{Z}$$

$$\iff T(x,\theta,n) - \tan^{-1}\frac{(x+\cos\theta)\sin 2\theta + (x^2-\cos 2\theta)\sin\theta}{\sin\theta\sin 2\theta - (x+\cos\theta)(x^2-\cos 2\theta)}$$

$$+ \{-1,0,1\}\pi \in \frac{\pi}{2}\mathbb{Z}$$

$$\iff T(x,\theta,n) - \tan^{-1}\left(-\frac{x^2\sin\theta + 2x\sin\theta\cos\theta + \sin\theta}{x^3 + x^2\cos\theta - (2\cos^2\theta - 1)x + \cos\theta}\right) \in \frac{\pi}{2}\mathbb{Z}$$

$$\iff T(x,\theta,n) - \tan^{-1}\left(-\frac{(x^2+2x\cos\theta+1)\sin\theta}{(x-\cos\theta)(x^2+2x\cos\theta+1)}\right) \in \frac{\pi}{2}\mathbb{Z}$$

$$\iff T(x,\theta,n) - \tan^{-1}\frac{x-\cos\theta}{\sin\theta} + \frac{\pi}{2} \in \frac{\pi}{2}\mathbb{Z}$$

$$\iff T(x,\theta,n) - \tan^{-1}\frac{x-\cos\theta}{\sin\theta} \in \frac{\pi}{2}\mathbb{Z}.$$

[1.83] Let

$$f(x,k) = \frac{1}{2^k\left(x^{1/2^k}-1\right)} + \sum_{j=k+1}^{\infty}\frac{1}{2^j\left(1+x^{1/2^j}\right)}.$$

We have

$$f(x,k) = \frac{1}{2^k\left(x^{1/2^k}-1\right)} + \sum_{j=k+1}^{\infty}\frac{1}{2^j\left(1+x^{1/2^j}\right)}$$

$$= \frac{1}{2^k\left(x^{1/2^{k+1}}-1\right)\left(x^{1/2^{k+1}}+1\right)} + \frac{1}{2^{k+1}\left(1+x^{1/2^{k+1}}\right)}$$

$$+ \sum_{j=k+2}^{\infty}\frac{1}{2^j\left(1+x^{1/2^j}\right)}$$

$$= \frac{1}{2^{k+1}\left(x^{1/2^{k+1}}-1\right)} - \frac{1}{2^{k+1}\left(1+x^{1/2^{k+1}}\right)} + \frac{1}{2^{k+1}\left(1+x^{1/2^{k+1}}\right)}$$

$$+ \sum_{j=k+2}^{\infty}\frac{1}{2^j\left(1+x^{1/2^j}\right)}$$

$$= \frac{1}{2^{k+1}\left(x^{1/2^{k+1}}-1\right)} + \sum_{j=k+2}^{\infty}\frac{1}{2^j\left(1+x^{1/2^j}\right)}$$

$$= f(x,k+1).$$

(7.47)

Applying the recursion in (7.47) for $k \in [0..n-1]$, we get

$$f(x, 0) = \frac{1}{x-1} + \sum_{j=1}^{\infty} \frac{1}{2^j \left(1 + x^{1/2^j}\right)}$$
$$= f(x, n) = \frac{1}{2^n \left(x^{1/2^n} - 1\right)} + \sum_{j=n+1}^{\infty} \frac{1}{2^j \left(1 + x^{1/2^j}\right)}. \quad (7.48)$$

Taking the limit as $n \to \infty$ in (7.48) and using (1.19), we get

$$\frac{1}{x-1} + \sum_{j=1}^{\infty} \frac{1}{2^j \left(1 + x^{1/2^j}\right)} = \lim_{n \to \infty} \frac{1}{2^n \left(x^{1/2^n} - 1\right)} + \lim_{n \to \infty} \sum_{j=n+1}^{\infty} \frac{1}{2^j \left(1 + x^{1/2^j}\right)}$$
$$= \frac{1}{\log x} + \lim_{n \to \infty} \sum_{j=n+1}^{\infty} \frac{1}{2^{j+1}}$$
$$= \frac{1}{\log x} + \lim_{n \to \infty} \frac{1}{2^{n+2}\left(1 - \frac{1}{2}\right)}$$
$$= \frac{1}{\log x}. \quad (7.49)$$

[1.84] Differentiating (7.49) with respect to x, we get

$$\frac{d}{dx}\left(\frac{1}{\log x}\right) = \frac{d}{dx}\left(\frac{1}{x-1}\right) + \sum_{j=1}^{\infty} \frac{d}{dx}\left(\frac{1}{2^j\left(1+x^{1/2^j}\right)}\right)$$
$$\iff -\frac{1}{x \log^2 x} = -\frac{1}{(x-1)^2} - \sum_{j=1}^{\infty} \frac{x^{1/2^j}}{2^{2j} x \left(1 + x^{1/2^j}\right)^2}$$
$$\iff \frac{1}{\log^2 x} = \frac{x}{(x-1)^2} + \sum_{j=1}^{\infty} \frac{x^{1/2^j}}{2^{2j} \left(1 + x^{1/2^j}\right)^2}.$$

[1.85] We will prove by induction. Let $P(n)$ be the statement that $\forall n \in \mathbb{N}$, $\exists (a_k)_{k=1}^n \in \mathbb{R}^n$ and $\exists (b_k)_{k=1}^n \in \mathbb{R}^n$ such that

$$\frac{1}{\log^n x} = \sum_{k=1}^{n} \frac{a_k}{(x-1)^k} + \sum_{k=1}^{n} \sum_{j=1}^{\infty} \frac{b_k}{2^{jn}\left(1 + x^{1/2^j}\right)^k}. \quad (7.50)$$

7 Solutions to Binomial Series 247

Basis Step: $P(1)$ is proved in [1.83].
Induction Step: Assuming $P(n)$ to be true, differentiating (7.50) with respect to x, we get

$$\frac{d}{dx}\left(\frac{1}{\log^n x}\right) = \sum_{k=1}^{n}\frac{d}{dx}\left(\frac{a_k}{(x-1)^k}\right) + \sum_{k=1}^{n}\sum_{j=1}^{\infty}\frac{d}{dx}\left(\frac{b_k}{2^{jn}\left(1+x^{1/2^j}\right)^k}\right)$$

$$\iff -\frac{n}{x\log^{n+1} x} = -\sum_{k=1}^{n}\frac{ka_k}{(x-1)^{k+1}} - \sum_{k=1}^{n}\sum_{j=1}^{\infty}\frac{kb_k x^{1/2^j}}{2^{j(n+1)}x\left(1+x^{1/2^j}\right)^{k+1}}$$

$$\iff \frac{1}{\log^{n+1} x} = \sum_{k=1}^{n}\frac{ka_k x}{n(x-1)^{k+1}} + \sum_{k=1}^{n}\sum_{j=1}^{\infty}\frac{kb_k x^{1/2^j}}{n2^{j(n+1)}\left(1+x^{1/2^j}\right)^{k+1}}$$

$$= \sum_{k=1}^{n}\left(\frac{ka_k}{n(x-1)^k} + \frac{ka_k}{n(x-1)^{k+1}}\right)$$

$$+ \sum_{k=1}^{n}\sum_{j=1}^{\infty}\left(\frac{kb_k}{n2^{j(n+1)}\left(1+x^{1/2^j}\right)^k} - \frac{kb_k}{n2^{j(n+1)}\left(1+x^{1/2^j}\right)^{k+1}}\right)$$

$$= \frac{a_1}{n(x-1)} + \sum_{k=2}^{n}\frac{(k-1)a_{k-1}+ka_k}{n(x-1)^k} + \frac{a_n}{(x-1)^{n+1}}$$

$$+ \sum_{j=1}^{\infty}\frac{b_1}{n2^{j(n+1)}\left(1+x^{1/2^j}\right)} + \sum_{k=2}^{n}\sum_{j=1}^{\infty}\frac{kb_k - (k-1)b_{k-1}}{n2^{j(n+1)}\left(1+x^{1/2^j}\right)^k}$$

$$- \sum_{j=1}^{\infty}\frac{b_n}{2^{j(n+1)}\left(1+x^{1/2^j}\right)^{n+1}},$$

proving $P(n+1)$ to be correct.

[1.86] Let

$$f(x,\theta,k) = \frac{x^{1/2^k}\cos\dfrac{\theta}{2^k} - 1}{2^k\left(x^{1/2^{k-1}} - 2x^{1/2^k}\cos\dfrac{\theta}{2^k} + 1\right)} + \sum_{j=k+1}^{\infty}\frac{x^{1/2^j}\cos\dfrac{\theta}{2^j} + 1}{2^j\left(x^{1/2^{j-1}} + 2x^{1/2^j}\cos\dfrac{\theta}{2^j} + 1\right)}.$$

We have

$f(x,\theta,k)$

$$= \frac{x^{1/2^k}\cos\dfrac{\theta}{2^k} - 1}{2^k\left(x^{1/2^{k-1}} - 2x^{1/2^k}\cos\dfrac{\theta}{2^k} + 1\right)} + \sum_{j=k+1}^{\infty} \frac{x^{1/2^j}\cos\dfrac{\theta}{2^j} + 1}{2^j\left(x^{1/2^{j-1}} + 2x^{1/2^j}\cos\dfrac{\theta}{2^j} + 1\right)}$$

$$= \frac{2x^{1/2^k}\cos^2\dfrac{\theta}{2^k} - x^{1/2^k} - 1}{2^k\left(x^{1/2^{k-1}} - 4x^{1/2^k}\cos^2\dfrac{\theta}{2^k} + 2x^{1/2^k} + 1\right)}$$

$$+ \frac{x^{1/2^{k+1}}\cos\dfrac{\theta}{2^{k+1}} + 1}{2^{k+1}\left(x^{1/2^k} + 2x^{1/2^{k+1}}\cos\dfrac{\theta}{2^{k+1}} + 1\right)}$$

$$+ \sum_{j=k+2}^{\infty} \frac{x^{1/2^j}\cos\dfrac{\theta}{2^j} + 1}{2^j\left(x^{1/2^{j-1}} + 2x^{1/2^j}\cos\dfrac{\theta}{2^j} + 1\right)}$$

$$= \frac{2x^{1/2^k}\cos^2\dfrac{\theta}{2^k} - x^{1/2^k} - 1}{2^k\left(x^{1/2^k} - 2x^{1/2^{k+1}}\cos^2\dfrac{\theta}{2^{k+1}} + 1\right)\left(x^{1/2^k} + 2x^{1/2^{k+1}}\cos^2\dfrac{\theta}{2^{k+1}} + 1\right)}$$

$$+ \frac{x^{1/2^{k+1}}\cos\dfrac{\theta}{2^{k+1}} + 1}{2^{k+1}\left(x^{1/2^k} + 2x^{1/2^{k+1}}\cos\dfrac{\theta}{2^{k+1}} + 1\right)}$$

$$+ \sum_{j=k+2}^{\infty} \frac{x^{1/2^j}\cos\dfrac{\theta}{2^j} + 1}{2^j\left(x^{1/2^{j-1}} + 2x^{1/2^j}\cos\dfrac{\theta}{2^j} + 1\right)}$$

$$= \frac{x^{1/2^{k+1}}\cos\dfrac{\theta}{2^{k+1}} - 1}{2^{k+1}\left(x^{1/2^k} - 2x^{1/2^{k+1}}\cos\dfrac{\theta}{2^{k+1}} + 1\right)}$$

$$- \frac{x^{1/2^{k+1}}\cos\dfrac{\theta}{2^{k+1}} + 1}{2^{k+1}\left(x^{1/2^k} + 2x^{1/2^{k+1}}\cos\dfrac{\theta}{2^{k+1}} + 1\right)}$$

$$+ \frac{x^{1/2^{k+1}}\cos\dfrac{\theta}{2^{k+1}} + 1}{2^{k+1}\left(x^{1/2^k} + 2x^{1/2^{k+1}}\cos\dfrac{\theta}{2^{k+1}} + 1\right)}$$

$$+ \sum_{j=k+2}^{\infty} \frac{x^{1/2^j} \cos \frac{\theta}{2^j} + 1}{2^j \left(x^{1/2^{j-1}} + 2x^{1/2^j} \cos \frac{\theta}{2^j} + 1 \right)}$$

$$= \frac{x^{1/2^{k+1}} \cos \frac{\theta}{2^{k+1}} - 1}{2^{k+1} \left(x^{1/2^k} - 2x^{1/2^{k+1}} \cos \frac{\theta}{2^{k+1}} + 1 \right)} \quad (7.51)$$

$$+ \sum_{j=k+2}^{\infty} \frac{x^{1/2^j} \cos \frac{\theta}{2^j} + 1}{2^j \left(x^{1/2^{j-1}} + 2x^{1/2^j} \cos \frac{\theta}{2^j} + 1 \right)}$$

$$= f(x, \theta, k+1).$$

Applying the recursion in (7.51) for $k \in [0..n-1]$, we get

$$f(x, \theta, 0) = \frac{x \cos \theta - 1}{x^2 - 2x \cos \theta + 1} + \sum_{j=1}^{\infty} \frac{x^{1/2^j} \cos \frac{\theta}{2^j} + 1}{2^j \left(x^{1/2^{j-1}} + 2x^{1/2^j} \cos \frac{\theta}{2^j} + 1 \right)}$$

$$= f(x, \theta, n)$$

$$= \frac{x^{1/2^n} \cos \frac{\theta}{2^n} - 1}{2^n \left(x^{1/2^{n-1}} - 2x^{1/2^n} \cos \frac{\theta}{2^n} + 1 \right)}$$

$$+ \sum_{j=n+1}^{\infty} \frac{x^{1/2^j} \cos \frac{\theta}{2^j} + 1}{2^j \left(x^{1/2^{j-1}} + 2x^{1/2^j} \cos \frac{\theta}{2^j} + 1 \right)}.$$

(7.52)

Taking the limit as $n \to \infty$ in (7.52) and using (1.19), we get

$$\frac{x\cos\theta - 1}{x^2 - 2x\cos\theta + 1} + \sum_{j=1}^{\infty} \frac{x^{1/2^j}\cos\frac{\theta}{2^j} + 1}{2^j\left(x^{1/2^{j-1}} + 2x^{1/2^j}\cos\frac{\theta}{2^j} + 1\right)}$$

$$= \lim_{n\to\infty} \frac{x^{1/2^n}\cos\frac{\theta}{2^n} - 1}{2^n\left(x^{1/2^{n-1}} - 2x^{1/2^n}\cos\frac{\theta}{2^n} + 1\right)}$$

$$+ \lim_{n\to\infty} \sum_{j=n+1}^{\infty} \frac{x^{1/2^j}\cos\frac{\theta}{2^j} + 1}{2^j\left(x^{1/2^{j-1}} + 2x^{1/2^j}\cos\frac{\theta}{2^j} + 1\right)}$$

$$= \lim_{n\to\infty} \frac{x^{1/2^n}\left(1 - \frac{\theta^2}{2^{2n+1}}\right) - 1}{2^n\left(x^{1/2^{n-1}} - 2x^{1/2^n}\left(1 - \frac{\theta^2}{2^{2n+1}}\right) + 1\right)}$$

$$+ \lim_{n\to\infty} \sum_{j=n+1}^{\infty} \frac{x^{1/2^j}\left(1 - \frac{\theta^2}{2^{2j+1}}\right) + 1}{2^j\left(x^{1/2^{j-1}} + 2x^{1/2^j}\left(1 - \frac{\theta^2}{2^{2j+1}}\right) + 1\right)}$$

$$= \lim_{n\to\infty} \frac{2^n\left(x^{1/2^n} - 1\right) - \frac{\theta^2 x^{1/2^n}}{2^{n+1}}}{\theta^2 x^{1/2^n} + 2^{2n}\left(x^{1/2^n} - 1\right)^2} + \lim_{n\to\infty} \sum_{j=n+1}^{\infty} \frac{x^{1/2^j} + 1 - \frac{\theta^2 x^{1/2^j}}{2^{2j+1}}}{2^j\left(x^{1/2^j} + 1\right)^2 - \frac{\theta^2 x^{1/2^j}}{2^j}}$$

$$= \frac{\log x}{\theta^2 + \log^2 x} + \lim_{n\to\infty} \sum_{j=n+1}^{\infty} \frac{1}{2^{j+1}}$$

$$= \frac{\log x}{\theta^2 + \log^2 x} + \lim_{n\to\infty} \frac{1}{2^{n+2}\left(1 - \frac{1}{2}\right)}$$

$$= \frac{\log x}{\theta^2 + \log^2 x}.$$

[1.87] Let

$$f(x, \theta, k)$$

$$= \frac{x^{1/2^k}\sin\frac{\theta}{2^k}}{2^k\left(x^{1/2^{k-1}} - 2x^{1/2^k}\cos\frac{\theta}{2^k} + 1\right)} + \sum_{j=k+1}^{\infty} \frac{x^{1/2^j}\sin\frac{\theta}{2^j}}{2^j\left(x^{1/2^{j-1}} + 2x^{1/2^j}\cos\frac{\theta}{2^j} + 1\right)}.$$

7 Solutions to Binomial Series

We have

$$f(x,\theta,k) = \frac{x^{1/2^k}\sin\frac{\theta}{2^k}}{2^k\left(x^{1/2^{k-1}} - 2x^{1/2^k}\cos\frac{\theta}{2^k} + 1\right)}$$

$$+ \sum_{j=k+1}^{\infty} \frac{x^{1/2^j}\sin\frac{\theta}{2^j}}{2^j\left(x^{1/2^{j-1}} + 2x^{1/2^j}\cos\frac{\theta}{2^j} + 1\right)}$$

$$= \frac{2x^{1/2^k}\sin\frac{\theta}{2^{k+1}}\cos\frac{\theta}{2^{k+1}}}{2^k\left(x^{1/2^{k-1}} - 4x^{1/2^k}\cos^2\frac{\theta}{2^k} + 2x^{1/2^k} + 1\right)}$$

$$+ \frac{x^{1/2^{k+1}}\sin\frac{\theta}{2^{k+1}}}{2^{k+1}\left(x^{1/2^k} + 2x^{1/2^{k+1}}\cos\frac{\theta}{2^{k+1}} + 1\right)}$$

$$+ \sum_{j=k+2}^{\infty} \frac{x^{1/2^j}\sin\frac{\theta}{2^j}}{2^j\left(x^{1/2^{j-1}} + 2x^{1/2^j}\cos\frac{\theta}{2^j} + 1\right)}$$

$$= \frac{2x^{1/2^k}\sin\frac{\theta}{2^{k+1}}\cos\frac{\theta}{2^{k+1}}}{2^k\left(x^{1/2^k} - 2x^{1/2^{k+1}}\cos^2\frac{\theta}{2^{k+1}} + 1\right)\left(x^{1/2^k} + 2x^{1/2^{k+1}}\cos^2\frac{\theta}{2^{k+1}} + 1\right)}$$

$$+ \frac{x^{1/2^{k+1}}\sin\frac{\theta}{2^{k+1}}}{2^{k+1}\left(x^{1/2^k} + 2x^{1/2^{k+1}}\cos\frac{\theta}{2^{k+1}} + 1\right)}$$

$$+ \sum_{j=k+2}^{\infty} \frac{x^{1/2^j}\sin\frac{\theta}{2^j}}{2^j\left(x^{1/2^{j-1}} + 2x^{1/2^j}\cos\frac{\theta}{2^j} + 1\right)}$$

$$= \frac{x^{1/2^{k+1}}\sin\frac{\theta}{2^{k+1}}}{2^{k+1}\left(x^{1/2^k} - 2x^{1/2^{k+1}}\cos\frac{\theta}{2^{k+1}} + 1\right)}$$

$$-\frac{x^{1/2^{k+1}}\sin\frac{\theta}{2^{k+1}}}{2^{k+1}\left(x^{1/2^k}+2x^{1/2^{k+1}}\cos\frac{\theta}{2^{k+1}}+1\right)}$$

$$+\frac{x^{1/2^{k+1}}\sin\frac{\theta}{2^{k+1}}}{2^{k+1}\left(x^{1/2^k}+2x^{1/2^{k+1}}\cos\frac{\theta}{2^{k+1}}+1\right)}$$

$$+\sum_{j=k+2}^{\infty}\frac{x^{1/2^j}\sin\frac{\theta}{2^j}}{2^j\left(x^{1/2^{j-1}}+2x^{1/2^j}\cos\frac{\theta}{2^j}+1\right)}$$

$$=\frac{x^{1/2^{k+1}}\sin\frac{\theta}{2^{k+1}}}{2^{k+1}\left(x^{1/2^k}-2x^{1/2^{k+1}}\cos\frac{\theta}{2^{k+1}}+1\right)}$$

$$+\sum_{j=k+2}^{\infty}\frac{x^{1/2^j}\sin\frac{\theta}{2^j}}{2^j\left(x^{1/2^{j-1}}+2x^{1/2^j}\cos\frac{\theta}{2^j}+1\right)}$$

$$= f(x,\theta,k+1). \tag{7.53}$$

Applying the recursion in (7.53) for $k \in [0..n-1]$, we get

$$f(x,\theta,0) = \frac{x\sin\theta}{x^2 - 2x\cos\theta + 1} + \sum_{j=1}^{\infty}\frac{x^{1/2^j}\sin\frac{\theta}{2^j}}{2^j\left(x^{1/2^{j-1}}+2x^{1/2^j}\cos\frac{\theta}{2^j}+1\right)}$$

$$= f(x,\theta,n)$$

$$= \frac{x^{1/2^n}\sin\frac{\theta}{2^n}}{2^n\left(x^{1/2^{n-1}}-2x^{1/2^n}\cos\frac{\theta}{2^n}+1\right)} + \sum_{j=n+1}^{\infty}\frac{x^{1/2^j}\sin\frac{\theta}{2^j}}{2^j\left(x^{1/2^{j-1}}+2x^{1/2^j}\cos\frac{\theta}{2^j}+1\right)}.$$
$$\tag{7.54}$$

Taking the limit as $n \to \infty$ in (7.54) and using (1.19), we get

$$\frac{x\sin\theta}{x^2 - 2x\cos\theta + 1} + \sum_{j=1}^{\infty}\frac{x^{1/2^j}\sin\frac{\theta}{2^j}}{2^j\left(x^{1/2^{j-1}}+2x^{1/2^j}\cos\frac{\theta}{2^j}+1\right)}$$

$$= \lim_{n\to\infty} \frac{x^{1/2^n} \sin\frac{\theta}{2^n}}{2^n \left(x^{1/2^{n-1}} - 2x^{1/2^n}\cos\frac{\theta}{2^n} + 1\right)}$$

$$+ \lim_{n\to\infty} \sum_{j=n+1}^{\infty} \frac{x^{1/2^j} \sin\frac{\theta}{2^j}}{2^j \left(x^{1/2^{j-1}} + 2x^{1/2^j}\cos\frac{\theta}{2^j} + 1\right)}$$

$$= \lim_{n\to\infty} \frac{\frac{\theta x^{1/2^n}}{2^n}}{2^n \left(x^{1/2^{n-1}} - 2x^{1/2^n}\left(1 - \frac{\theta^2}{2^{2n+1}}\right) + 1\right)}$$

$$+ \lim_{n\to\infty} \sum_{j=n+1}^{\infty} \frac{\frac{\theta x^{1/2^j}}{2^j}}{2^j \left(x^{1/2^{j-1}} + 2x^{1/2^j}\left(1 - \frac{\theta^2}{2^{2j+1}}\right) + 1\right)}$$

$$= \lim_{n\to\infty} \frac{\theta x^{1/2^n}}{\theta^2 x^{1/2^n} + 2^{2n}\left(x^{1/2^n} - 1\right)^2} + \lim_{n\to\infty} \sum_{j=n+1}^{\infty} \frac{\frac{\theta x^{1/2^j}}{2^j}}{2^j \left(x^{1/2^j} + 1\right)^2 - \frac{\theta^2 x^{1/2^j}}{2^j}}$$

$$= \frac{\theta}{\theta^2 + \log^2 x} + \theta \lim_{n\to\infty} \sum_{j=n+1}^{\infty} \frac{1}{2^{2j+2}}$$

$$= \frac{\theta}{\theta^2 + \log^2 x} + \theta \lim_{n\to\infty} \frac{1}{2^{2n+4}\left(1 - \frac{1}{4}\right)}$$

$$= \frac{\theta}{\theta^2 + \log^2 x}.$$

[1.88] Replace θ with $i\theta$ in [1.86].

[1.89] Replace θ with $i\theta$ in [1.87].

[1.90] Putting $\theta = \log y$ in [1.88], we get

$$\frac{\log x}{\log^2 x - \log^2 y} = \frac{x\left(\frac{y+y^{-1}}{2}\right) - 1}{x^2 - x\left(y + y^{-1}\right) + 1}$$

$$+ \sum_{j=1}^{\infty} \frac{x^{1/2^j}\left(\frac{y^{1/2^j} + y^{-1/2^j}}{2}\right) + 1}{2^j \left(x^{1/2^{j-1}} + x^{1/2^j}\left(y^{1/2^j} + y^{-1/2^j}\right) + 1\right)}$$

$$\iff \frac{\log x^2}{\log^2 x - \log^2 y} = \frac{x\left(y^2 + 1\right) - 2y}{x^2 y - x\left(y^2 + 1\right) + y}$$

$$+ \sum_{j=1}^{\infty} \frac{x^{1/2^j}\left(y^{1/2^{j-1}} + 1\right) + 2y^{1/2^j}}{2^j \left(x^{1/2^{j-1}} y^{1/2^j} + x^{1/2^j}\left(y^{1/2^{j-1}} + 1\right) + y^{1/2^j}\right)}$$

$$= \frac{y(xy - 1) + x - y}{(x - y)(xy - 1)} + \sum_{j=1}^{\infty} \frac{y^{1/2^j}\left(x^{1/2^j} y^{1/2^j} + 1\right) + x^{1/2^j} + y^{1/2^j}}{2^j \left(x^{1/2^j} + y^{1/2^j}\right)\left(x^{1/2^j} y^{1/2^j} + 1\right)}.$$

[1.91] Putting $\theta = \log y$ in [1.89], we get

$$\frac{\log y}{\log^2 x - \log^2 y} = \frac{x\left(\frac{y - y^{-1}}{2}\right)}{x^2 - x\left(y + y^{-1}\right) + 1}$$

$$+ \sum_{j=1}^{\infty} \frac{x^{1/2^j}\left(\frac{y^{1/2^j} - y^{-1/2^j}}{2}\right)}{2^j \left(x^{1/2^{j-1}} + x^{1/2^j}\left(y^{1/2^j} + y^{-1/2^j}\right) + 1\right)}$$

$$\iff \frac{\log y^2}{\log^2 x - \log^2 y} = \frac{x\left(y^2 - 1\right)}{x^2 y - x\left(y^2 + 1\right) + y}$$

$$+ \sum_{j=1}^{\infty} \frac{x^{1/2^j}\left(y^{1/2^{j-1}} - 1\right)}{2^j \left(x^{1/2^{j-1}} y^{1/2^j} + x^{1/2^j}\left(y^{1/2^{j-1}} + 1\right) + y^{1/2^j}\right)}$$

$$= \frac{x(y^2 - 1)}{(x - y)(xy - 1)} + \sum_{j=1}^{\infty} \frac{x^{1/2^j}\left(y^{1/2^{j-1}} - 1\right)}{2^j \left(x^{1/2^j} + y^{1/2^j}\right)\left(x^{1/2^j} y^{1/2^j} + 1\right)}.$$

7 Solutions to Binomial Series

[1.92] From [1.74], we have

$$\frac{1}{x-1} = \sum_{j=0}^{n-1} \frac{2^j}{x^{2^j}+1} + \frac{2^n}{x^{2^n}-1}$$

$$\iff -1 - \frac{2}{x-1} = -1 - \sum_{j=0}^{n-1} \frac{2^{j+1}}{x^{2^j}+1} - \frac{2^{n+1}}{x^{2^n}-1}$$

$$\iff \frac{1+x}{1-x} = -1 - \sum_{j=0}^{n-1} 2^j \frac{\left(x^{2^j}+1\right) - \left(x^{2^j}-1\right)}{x^{2^j}+1}$$

$$+ 2^n \frac{\left(1-x^{2^n}\right) + \left(1+x^{2^n}\right)}{1-x^{2^n}}$$

$$\iff \frac{1+x}{1-x} = -1 - \sum_{j=0}^{n-1} 2^j + \sum_{j=0}^{n-1} 2^j \frac{x^{2^j}-1}{x^{2^j}+1} + 2^n + 2^n \frac{1+x^{2^n}}{1-x^{2^n}}$$

$$= -1 - (2^n - 1) + 2^n + \sum_{j=0}^{n-1} 2^j \frac{x^{2^j}-1}{x^{2^j}+1} + 2^n \frac{1+x^{2^j}}{1-x^{2^n}}$$

$$= \sum_{j=0}^{n-1} 2^j \frac{x^{2^j}-1}{x^{2^j}+1} + 2^n \frac{1+x^{2^n}}{1-x^{2^n}}.$$

Chapter 8
Solutions to Trigonometrical Series

[2.1] Differentiating (2.12) with respect to θ, we get

$$\frac{n\cos n\theta \sin\theta - \sin n\theta \cos\theta}{\sin^2\theta}$$

$$= (-1)^{n/2} n \sum_{j=0}^{n/2-1} (-1)^j \frac{\prod_{k=1}^{j}(n^2 - (2k)^2)}{(2j)!} \cos^{2j}\theta \sin\theta$$

$$\iff (-1)^{n/2} \frac{n\cos n\theta \sin\theta - \sin n\theta \cos\theta}{n\sin^3\theta}$$

$$= \sum_{j=0}^{n/2-1} (-1)^j \frac{\prod_{k=1}^{j}(n^2 - (2k)^2)}{(2j)!} \cos^{2j}\theta.$$

[2.2] Differentiating (2.15) with respect to θ, we get

$$\frac{n\cos n\theta \sin\theta - \sin n\theta \cos\theta}{\sin^2\theta}$$

$$= (-1)^{(n-1)/2} \sum_{j=1}^{(n-1)/2} (-1)^{j-1} \frac{\prod_{k=0}^{j-1}(n^2 - (2k+1)^2)}{(2j-1)!} \cos^{2j-1}\theta \sin\theta$$

© The Author(s), under exclusive license to Springer Nature Singapore Pte Ltd. 2025
A. Mishra, *Special Integrals*, University Texts in the Mathematical Sciences,
https://doi.org/10.1007/978-981-97-7514-9_8

$$\Longrightarrow (-1)^{(n-1)/2} \frac{n \cos n\theta \sin \theta - \sin n\theta \cos \theta}{\sin^3 \theta}$$

$$= \sum_{j=1}^{(n-1)/2} (-1)^{j-1} \frac{\prod_{k=0}^{j-1} (n^2 - (2k+1)^2)}{(2j-1)!} \cos^{2j-1} \theta.$$

[2.3] Replace θ with $\pi/2 - \theta$ in [2.2].

[2.4] Replace θ with $\pi/2 - \theta$ in [2.1].

[2.5] Replacing θ with $\theta/2$ in [2.21], we get

$$\frac{\tan \frac{\theta}{2}}{4\theta} = \sum_{j=0}^{\infty} \frac{1}{(2j+1)^2 \pi^2 - \theta^2} \tag{8.1}$$

$$\Longleftrightarrow \frac{\theta}{2} \tan \frac{\theta}{2} = 2\theta^2 \sum_{j=0}^{\infty} \frac{1}{(2j+1)^2 \pi^2 - \theta^2}.$$

Replacing θ with $\theta/2$ in [2.20], we get

$$\frac{2 - \theta \cot \frac{\theta}{2}}{\theta^2} = 4 \sum_{j=0}^{\infty} \frac{1}{(2j)^2 \pi^2 - \theta^2} \tag{8.2}$$

$$\Longleftrightarrow \frac{\theta}{2} \cot \frac{\theta}{2} = 1 - 2\theta^2 \sum_{j=0}^{\infty} \frac{1}{(2j)^2 \pi^2 - \theta^2}.$$

Adding (8.1) and (8.2), we get

$$\frac{\theta}{\sin \theta} = 1 + 2\theta^2 \sum_{j=1}^{\infty} \frac{(-1)^{j-1}}{j^2 \pi^2 - \theta^2}. \tag{8.3}$$

Putting $\theta = \pi/x$ in (8.3), we get

$$\frac{\pi}{x \sin \frac{\pi}{x}} = 1 + 2 \sum_{j=1}^{\infty} \frac{(-1)^{j-1}}{j^2 x^2 - 1}.$$

[2.6] Using the binomial theorem, we have

$$\left(1 - \sin^2 \theta\right)^{-3/2} = \sum_{j=0}^{\infty} (-1)^{2j} \frac{\sin^{2j} \theta}{j!} \prod_{k=1}^{j} \left(\frac{3}{2} + k - 1\right)$$

$$\iff \sec^3 \theta = \sum_{j=0}^{\infty} \frac{\sin^{2j} \theta}{2^j j!} \prod_{k=1}^{j} (2k+1)$$

$$= \sum_{j=0}^{\infty} \frac{\sin^{2j} \theta}{2^{2j} (j!)^2} \prod_{k=1}^{j} (2k)(2k+1)$$

$$= \sum_{j=0}^{\infty} \frac{2j+1}{2^{2j}} \binom{2j}{j} \sin^{2j} \theta$$

$$= \sum_{j=1}^{\infty} \frac{2j-1}{2^{2j-2}} \binom{2j-2}{j-1} \sin^{2j-2} \theta$$

$$= \sum_{j=1}^{\infty} \frac{2j-1}{2^{2j-2}} \cdot \frac{j^2}{(2j-1)(2j)} \binom{2j}{j} \sin^{2j-2} \theta$$

$$= \sum_{j=1}^{\infty} \frac{j}{2^{2j-1}} \binom{2j}{j} \sin^{2j-2} \theta.$$

[2.7] Using the binomial theorem, we have

$$\left(1 - \sin^2 \theta\right)^{-1/2} = \sum_{j=0}^{\infty} (-1)^{2j} \frac{\sin^{2j} \theta}{j!} \prod_{k=1}^{j} \left(\frac{1}{2} + k - 1\right)$$

$$\iff \sec \theta = \sum_{j=0}^{\infty} \frac{\sin^{2j} \theta}{2^j j!} \prod_{k=1}^{j} (2k - 1)$$

$$= \sum_{j=0}^{\infty} \frac{\sin^{2j} \theta}{2^{2j} (j!)^2} \prod_{k=1}^{j} (2k-1)(2k)$$

$$= \sum_{j=0}^{\infty} \binom{2j}{j} \frac{\sin^{2j} \theta}{2^{2j}}.$$

[2.8] Multiplying [2.7] by $\cos\theta$, we get

$$1 = \sum_{j=0}^{\infty} \binom{2j}{j} \frac{\sin^{2j}\theta \cos\theta}{2^{2j}}. \tag{8.4}$$

Integrating (8.4) with respect to θ from 0 to θ, we get

$$\theta\Big|_0^\theta = \sum_{j=0}^{\infty} \binom{2j}{j} \frac{\sin^{2j+1}\theta}{2^{2j}(2j+1)} \Big|_0^\theta$$

$$\iff \theta = \sum_{j=0}^{\infty} \binom{2j}{j} \frac{\sin^{2j+1}\theta}{2^{2j}(2j+1)}.$$

[2.9] Putting $x = \tan\theta$ in (1.13), multiplying it by $2\tan\theta$, and using [1.37], we get

$$2\theta \tan\theta = \sum_{l=0}^{\infty} (-1)^l \frac{2}{2l+1} \tan^{2l+2}\theta$$

$$= \sum_{l=0}^{\infty} (-1)^l \frac{2}{2l+1} \sin^{2l+2}\theta \left(1 - \sin^2\theta\right)^{-(l+1)}$$

$$= \sum_{l=0}^{\infty} (-1)^l \frac{2}{2l+1} \sin^{2l+2}\theta \sum_{k=0}^{\infty} (-1)^{2k} \frac{(l+k)!}{l!k!} \sin^{2k}\theta$$

$$= \sum_{l=0}^{\infty} (-1)^l \frac{2}{2l+1} \sin^{2l+2}\theta \sum_{k=0}^{\infty} \binom{l+k}{l} \sin^{2k}\theta$$

$$= \sum_{j=1}^{\infty} \left(\sum_{l=0}^{j-1} (-1)^l \frac{2}{2l+1} \binom{j-1}{l} \right) \sin^{2j}\theta$$

$$= \sum_{j=1}^{\infty} 2 \frac{2^{2j-2}}{2j-1} \binom{2j-2}{j-1}^{-1} \sin^{2j}\theta$$

$$= \sum_{j=1}^{\infty} \frac{2^{2j-1} \cdot 2j}{2j(2j-1)} \frac{((j-1)!)^2}{(2j-2)!} \sin^{2j}\theta$$

$$= \sum_{j=1}^{\infty} \frac{2^{2j}(j!)^2}{j(2j)!} \sin^{2j}\theta$$

$$= \sum_{j=1}^{\infty} \frac{2^{2j}}{j} \binom{2j}{j}^{-1} \sin^{2j}\theta.$$

8 Solutions to Trigonometrical Series

[2.10] The identity is true for $\theta = 0$. Assuming $\theta \neq 0$, multiplying [2.9] by $2\cot\theta$, we get

$$4\theta = \sum_{j=1}^{\infty} \frac{2^{2j+1}}{j}\binom{2j}{j}^{-1} \sin^{2j-1}\theta \cos\theta. \tag{8.5}$$

Integrating (8.5) with respect to θ from 0 to θ, we get

$$2\theta^2\Big|_0^\theta = \sum_{j=1}^{\infty} \frac{2^{2j+1}}{2j^2}\binom{2j}{j}^{-1} \sin^{2j}\theta\Big|_0^\theta$$

$$\iff 2\theta^2 = \sum_{j=1}^{\infty} \frac{2^{2j}}{j^2}\binom{2j}{j}^{-1} \sin^{2j}\theta.$$

[2.11] Multiplying [2.8] by $\cos\theta$, we get

$$\theta \cos\theta = \sum_{j=0}^{\infty} \binom{2j}{j} \frac{\sin^{2j+1}\theta \cos\theta}{2^{2j}(2j+1)}. \tag{8.6}$$

Integrating (8.6) with respect to θ from 0 to θ, we get

$$\theta \sin\theta + \cos\theta\Big|_0^\theta = \sum_{j=0}^{\infty} \binom{2j}{j} \frac{\sin^{2j+2}\theta}{2^{2j}(2j+1)(2j+2)}\Big|_0^\theta$$

$$\iff \theta\sin\theta + \cos\theta = 1 + \sum_{j=0}^{\infty} \binom{2j}{j}\frac{\sin^{2j+2}\theta}{2^{2j}(2j+1)(2j+2)}.$$

[2.12] Multiplying [2.9] by $(1/4)\cos\theta$, we get

$$\frac{\theta}{2}\sin\theta = \sum_{j=1}^{\infty} \frac{2^{2j-2}}{j}\binom{2j}{j}^{-1} \sin^{2j}\theta \cos\theta. \tag{8.7}$$

Integrating (8.7) with respect to θ from 0 to θ, we get

$$\frac{\sin\theta - \theta\cos\theta}{2}\Big|_0^\theta = \sum_{j=1}^{\infty} \frac{2^{2j-2}}{j(2j+1)}\binom{2j}{j}^{-1} \sin^{2j+1}\theta\Big|_0^\theta$$

$$\iff \frac{\sin\theta - \theta\cos\theta}{2} = \sum_{j=1}^{\infty} \frac{2^{2j-2}}{j(2j+1)}\binom{2j}{j}^{-1} \sin^{2j+1}\theta.$$

[2.13] Multiplying [2.12] by $\cos\theta$, we get

$$\sum_{j=1}^{\infty} \frac{2^{2j-2}}{j(2j+1)} \binom{2j}{j}^{-1} \sin^{2j+1}\theta \cos\theta = \frac{\sin\theta\cos\theta - \theta\cos^2\theta}{2} \qquad (8.8)$$
$$= \frac{\sin 2\theta - \theta - \theta\cos 2\theta}{4}.$$

Integrating (8.8) with respect to θ from 0 to θ, we get

$$\left.\frac{-\frac{3}{2}\cos 2\theta - \theta^2 - \theta\sin 2\theta}{8}\right|_0^\theta = \sum_{j=1}^{\infty} \frac{2^{2j-2}}{j(2j+1)(2j+2)} \binom{2j}{j}^{-1} \sin^{2j+2}\theta \bigg|_0^\theta$$

$$\iff \frac{3\sin^2\theta - \theta^2 - \theta\sin 2\theta}{8} = \sum_{j=1}^{\infty} \frac{2^{2j-2}}{j(2j+1)(2j+2)} \binom{2j}{j}^{-1} \sin^{2j+2}\theta.$$

[2.14] Differentiating [2.9] with respect to θ, we get

$$2\theta\sec^2\theta + 2\tan\theta = \sum_{j=1}^{\infty} 2^{2j+1} \binom{2j}{j}^{-1} \sin^{2j-1}\theta \cos\theta$$

$$\iff \theta\tan\theta\sec^2\theta + \tan^2\theta = \sum_{j=1}^{\infty} 2^{2j} \binom{2j}{j}^{-1} \sin^{2j}\theta$$

$$\iff \theta\tan\theta\sec^2\theta + \sec^2\theta - 1 = \sum_{j=1}^{\infty} 2^{2j} \binom{2j}{j}^{-1} \sin^{2j}\theta$$

$$\iff \sec^2\theta(1 + \theta\tan\theta) = \sum_{j=0}^{\infty} 2^{2j} \binom{2j}{j}^{-1} \sin^{2j}\theta.$$

[2.15] Multiplying the second part of [2.6] by $\sin\theta$, we get

$$\frac{\sin\theta}{\cos^3\theta} = \sum_{j=0}^{\infty} \frac{2j+1}{2^{2j}} \binom{2j}{j} \sin^{2j+1}\theta. \qquad (8.9)$$

Differentiating (8.9) with respect to θ, we get

8 Solutions to Trigonometrical Series

$$\frac{\cos^4\theta + 3\cos^2\theta \sin^2\theta}{\cos^6\theta} = \sum_{j=0}^{\infty} \frac{(2j+1)^2}{2^{2j}} \binom{2j}{j} \sin^{2j}\theta \cos\theta$$

$$\iff \sum_{j=0}^{\infty} \frac{(2j+1)^2}{2^{2j}} \binom{2j}{j} \sin^{2j}\theta = \frac{\cos^2\theta + 3\sin^2\theta}{\cos^5\theta}$$

$$= \frac{1 + 2\sin^2\theta}{\cos^5\theta}.$$

[2.16] Multiplying [2.15] by $\sin\theta$, we get

$$\frac{\sin\theta + 2\sin^3\theta}{\cos^5\theta} = \sum_{j=0}^{\infty} \frac{(2j+1)^2}{2^{2j}} \binom{2j}{j} \sin^{2j+1}\theta. \qquad (8.10)$$

Differentiating (8.10) with respect to θ, we get

$$\frac{\left(\cos\theta + 6\sin^2\theta \cos\theta\right)\cos^5\theta + 5\cos^4\theta \sin\theta \left(\sin\theta + 2\sin^3\theta\right)}{\cos^{10}\theta}$$

$$= \sum_{j=0}^{\infty} \frac{(2j+1)^3}{2^{2j}} \binom{2j}{j} \sin^{2j}\theta \cos\theta$$

$$\iff \sum_{j=0}^{\infty} \frac{(2j+1)^3}{2^{2j}} \binom{2j}{j} \sin^{2j}\theta$$

$$= \frac{\cos^2\theta + 6\sin^2\theta \cos^2\theta + 5\sin^2\theta + 10\sin^4\theta}{\cos^7\theta}$$

$$= \frac{1 + 2\sin^2\theta \left(5 + 2\sin^2\theta\right)}{\cos^7\theta}.$$

[2.17] Multiplying [2.16] by $\sin\theta$, we get

$$\frac{\sin\theta + 10\sin^3\theta + 4\sin^5\theta}{\cos^7\theta} = \sum_{j=0}^{\infty} \frac{(2j+1)^3}{2^{2j}} \binom{2j}{j} \sin^{2j+1}\theta. \qquad (8.11)$$

Differentiating (8.11) with respect to θ, we get

$$\sum_{j=0}^{\infty} \frac{(2j+1)^4}{2^{2j}} \binom{2j}{j} \sin^{2j}\theta \cos\theta$$
$$= \left((\cos\theta + 30\sin^2\theta\cos\theta + 20\sin^4\theta\cos\theta)\cos^7\theta \right.$$
$$\left. + 7\cos^6\theta\sin\theta (\sin\theta + 10\sin^3\theta + 4\sin^5\theta) \right) / \cos^{14}\theta$$
$$\iff \sum_{j=0}^{\infty} \frac{(2j+1)^4}{2^{2j}} \binom{2j}{j} \sin^{2j}\theta$$
$$= \frac{\cos^2\theta (1 + 30\sin^2\theta + 20\sin^4\theta) + 7\sin^2\theta (1 + 10\sin^2\theta + 4\sin^4\theta)}{\cos^9\theta}$$
$$= \frac{1 + 4\sin^2\theta (9 + 15\sin^2\theta + 2\sin^4\theta)}{\cos^9\theta}.$$

[2.18] Multiplying [2.10] by $\cos\theta$, we get

$$2\theta^2 \cos\theta = \sum_{j=1}^{\infty} \frac{2^{2j}}{j^2} \binom{2j}{j}^{-1} \sin^{2j}\theta \cos\theta. \tag{8.12}$$

Integrating (8.12) with respect to θ from 0 to θ, we get

$$2\theta^2 \sin\theta + 4\theta\cos\theta - 4\sin\theta \Big|_0^\theta = \sum_{j=1}^{\infty} \frac{2^{2j}}{j^2(2j+1)} \binom{2j}{j}^{-1} \sin^{2j+1}\theta \Big|_0^\theta$$

$$\iff \frac{\theta^2}{2}\sin\theta + \theta\cos\theta - \sin\theta = \sum_{j=1}^{\infty} \frac{2^{2j-2}}{j^2(2j+1)} \binom{2j}{j}^{-1} \sin^{2j+1}\theta.$$

[2.19] Multiplying [2.11] by $\cot\theta$, we get

$$\theta\cos\theta + \frac{\cos^2\theta}{\sin\theta} = \frac{\cos\theta}{\sin\theta} + \sum_{j=0}^{\infty} \binom{2j}{j} \frac{\sin^{2j+1}\theta \cos\theta}{2^{2j}(2j+1)(2j+2)}$$

$$\iff \sum_{j=0}^{\infty} \binom{2j}{j} \frac{\sin^{2j+1}\theta \cos\theta}{2^{2j}(2j+1)(2j+2)} = \theta\cos\theta - \sin\theta + \frac{1-\cos\theta}{\sin\theta} \tag{8.13}$$

$$= \theta\cos\theta - \sin\theta + \tan\frac{\theta}{2}.$$

Integrating (8.13) with respect to θ from 0 to θ, we get

8 Solutions to Trigonometrical Series

$$\theta \sin\theta + 2\cos\theta - 2\log\cos\frac{\theta}{2}\bigg|_0^\theta = \sum_{j=0}^\infty \binom{2j}{j}\frac{\sin^{2j+2}\theta}{2^{2j}(2j+1)(2j+2)^2}\bigg|_0^\theta$$

$$\iff \sum_{j=0}^\infty \binom{2j}{j}\frac{\sin^{2j+2}\theta}{2^{2j}(2j+1)(2j+2)^2} = \theta\sin\theta + 2\cos\theta - 2 - 2\log\cos\frac{\theta}{2}$$

$$= \theta\sin\theta + 2\cos\theta - \log\left(\frac{e^2}{2}(1+\cos\theta)\right).$$

[2.20] Taking log of (2.22), we get

$$\log\sin\theta = \log\theta + \sum_{j=1}^\infty \log\left(1 - \frac{\theta^2}{j^2\pi^2}\right). \tag{8.14}$$

Differentiating (8.14) with respect to θ, we get

$$\cot\theta = \frac{1}{\theta} - \sum_{j=1}^\infty \frac{2\theta}{j^2\pi^2 - \theta^2}$$

$$\iff \frac{1-\theta\cot\theta}{2\theta^2} = \sum_{j=1}^\infty \frac{1}{j^2\pi^2 - \theta^2}.$$

[2.21] Taking log of (2.29), we get

$$\log\cos\theta = \sum_{j=0}^\infty \log\left(1 - \frac{4\theta^2}{(2j+1)^2\pi^2}\right). \tag{8.15}$$

Differentiating (8.15) with respect to θ, we get

$$-\tan\theta = -\sum_{j=0}^\infty \frac{8\theta}{(2j+1)^2\pi^2 - 4\theta^2}$$

$$\iff \frac{\tan\theta}{8\theta} = \sum_{j=0}^\infty \frac{1}{(2j+1)^2\pi^2 - 4\theta^2}.$$

[2.22] Differentiating [2.20] with respect to θ, we get

$$\frac{\left(\theta \csc^2 \theta - \cot \theta\right) \theta^2 - 2\theta(1 - \theta \cot \theta)}{2\theta^4} = \sum_{j=1}^{\infty} \frac{2\theta}{\left(j^2\pi^2 - \theta^2\right)^2}$$

$$\iff \sum_{j=1}^{\infty} \frac{1}{\left(j^2\pi^2 - \theta^2\right)^2} = \frac{\theta^2 - \theta \cos\theta \sin\theta - 2\sin^2\theta + 2\theta \cos\theta \sin\theta}{4\theta^4 \sin^2\theta}$$

$$= \frac{2\theta^2 + \theta \sin 2\theta - 4\sin^2\theta}{8\theta^4 \sin^2\theta}.$$

[2.23] Differentiating [2.21] with respect to θ, we get

$$\frac{\theta \sec^2\theta - \tan\theta}{8\theta^2} = \sum_{j=0}^{\infty} \frac{8\theta}{\left((2j+1)^2\pi^2 - 4\theta^2\right)^2}$$

$$\iff \sum_{j=0}^{\infty} \frac{1}{\left((2j+1)^2\pi^2 - 4\theta^2\right)^2} = \frac{\theta - \sin\theta \cos\theta}{64\theta^3 \cos^2\theta}$$

$$= \frac{2\theta - \sin 2\theta}{128\theta^3 \cos^2\theta}.$$

[2.24] The sum of the GP $\left((-1)^{j-1} \tan^{2j-1} \theta\right)_{j=1}^{\infty}$ is given by

$$\sum_{j=1}^{\infty} (-1)^{j-1} \tan^{2j-1} \theta = \frac{\tan\theta}{1 + \tan^2\theta} = \sin\theta \cos\theta. \qquad (8.16)$$

Multiplying (8.16) by $\tan\theta$, we get

$$\sin^2\theta = \sum_{j=1}^{\infty} (-1)^{j-1} \tan^{2j} \theta. \qquad (8.17)$$

Differentiating (8.17) with respect to θ, we get

$$2\sin\theta \cos\theta = \sum_{j=1}^{\infty} (-1)^{j-1} 2j \tan^{2j-1}\theta \sec^2\theta$$

$$\iff \sin\theta \cos^3\theta = \sum_{j=1}^{\infty} (-1)^{j-1} j \tan^{2j-1}\theta. \qquad (8.18)$$

8 Solutions to Trigonometrical Series

Multiplying (8.18) by $\tan\theta$, we get

$$\sin^2\theta\cos^2\theta = \sum_{j=1}^{\infty}(-1)^{j-1}j\tan^{2j}\theta. \qquad (8.19)$$

Differentiating (8.19) with respect to θ, we get

$$2\left(\sin\theta\cos^3\theta - \sin^3\theta\cos\theta\right) = \sum_{j=1}^{\infty}(-1)^{j-1}2j^2\tan^{2j-1}\theta\sec^2\theta$$

$$\iff \sum_{j=1}^{\infty}(-1)^{j-1}j^2\tan^{2j-1}\theta = \sin\theta\cos^3\theta\cos 2\theta$$

$$= \frac{1}{2}\cos^2\theta\sin 2\theta\cos 2\theta$$

$$= \frac{1}{4}\sin 4\theta\cos^2\theta.$$

[2.25] Multiplying [2.24] by $\tan\theta$, we get

$$\sum_{j=1}^{\infty}(-1)^{j-1}j^2\tan^{2j}\theta = \frac{1}{4}\sin 4\theta\sin\theta\cos\theta$$

$$= \frac{1}{8}\sin 4\theta\sin 2\theta. \qquad (8.20)$$

Differentiating (8.20) with respect to θ, we get

$$\sum_{j=1}^{\infty}(-1)^{j-1}2j^3\tan^{2j-1}\theta\sec^2\theta = \frac{1}{4}\left(2\cos 4\theta\sin 2\theta + \sin 4\theta\cos 2\theta\right)$$

$$\iff \sum_{j=1}^{\infty}(-1)^{j-1}j^3\tan^{2j-1}\theta = \frac{1}{8}\cos^2\theta\left(\sin 6\theta + \sin 2\theta\cos 4\theta\right).$$

[2.26] Multiplying [2.25] by $\tan\theta$, we get

$$\sum_{j=1}^{\infty}(-1)^{j-1}j^3\tan^{2j}\theta = \frac{1}{8}\sin\theta\cos\theta\,(\sin 6\theta+\sin 2\theta\cos 4\theta)$$
$$= \frac{1}{16}\sin 2\theta\,(\sin 6\theta+\sin 2\theta\cos 4\theta)$$
$$= \frac{1}{32}(\cos 4\theta-\cos 8\theta+(1-\cos 4\theta)\cos 4\theta) \quad (8.21)$$
$$= \frac{1}{32}\left(2\cos 4\theta-\cos 8\theta-\cos^2 4\theta\right)$$
$$= \frac{1}{32}\left(2\cos 4\theta-\frac{3}{2}\cos 8\theta-\frac{1}{2}\right).$$

Differentiating (8.21) with respect to θ, we get

$$\sum_{j=1}^{\infty}(-1)^{j-1}2j^4\tan^{2j-1}\theta\sec^2\theta = \frac{1}{8}(3\sin 8\theta-2\sin 4\theta)$$
$$\iff \sum_{j=1}^{\infty}(-1)^{j-1}j^4\tan^{2j-1}\theta = \frac{1}{16}\cos^2\theta\,(3\sin 8\theta-2\sin 4\theta).$$

[2.27] Replace θ with $i\theta$ in [2.1].

[2.28] Replace θ with $i\theta$ in [2.2].

[2.29] Replace θ with $i\theta$ in [2.3].

[2.30] Replace θ with $i\theta$ in [2.4].

[2.31] Replace x with ix in [2.5].

[2.32] Replace θ with $i\theta$ in [2.6].

[2.33] Replace θ with $i\theta$ in [2.7].

[2.34] Replace θ with $i\theta$ in [2.8].

[2.35] Replace θ with $i\theta$ in [2.9].

[2.36] Replace θ with $i\theta$ in [2.10].

8 Solutions to Trigonometrical Series

[2.37] Replace θ with $i\theta$ in [2.11].

[2.38] Replace θ with $i\theta$ in [2.12].

[2.39] Replace θ with $i\theta$ in [2.13].

[2.40] Replace θ with $i\theta$ in [2.14].

[2.41] Replace θ with $i\theta$ in [2.15].

[2.42] Replace θ with $i\theta$ in [2.16].

[2.43] Replace θ with $i\theta$ in [2.17].

[2.44] Replace θ with $i\theta$ in [2.18].

[2.45] Replace θ with $i\theta$ in [2.19].

[2.46] Replace θ with $i\theta$ in [2.20].

[2.47] Replace θ with $i\theta$ in [2.21].

[2.48] Replace θ with $i\theta$ in [2.22].

[2.49] Replace θ with $i\theta$ in [2.23].

[2.50] Replace θ with $i\theta$ in [2.24].

[2.51] Replace θ with $i\theta$ in [2.25].

[2.52] Replace θ with $i\theta$ in [2.26].

Chapter 9
Solutions to Bessel Functions

[3.1] Using the binomial theorem and (1.11), we have

$$\int_0^1 \frac{1-(1-t)^n}{t} dt = \int_0^1 \left(1 - \sum_{j=0}^n (-1)^j t^j \binom{n}{j}\right) \frac{dt}{t}$$

$$= \int_0^1 \left(\sum_{j=1}^n (-1)^{j-1} t^{j-1} \binom{n}{j}\right) dt$$

$$= \sum_{j=1}^n \frac{(-1)^{j-1} t^j}{j} \binom{n}{j} \bigg|_0^1$$

$$= \sum_{j=1}^n \frac{(-1)^{j-1}}{j} \binom{n}{j}$$

$$= \sum_{j=1}^n \frac{1}{j}$$

$$= H(n).$$

272   9 Solutions to Bessel Functions

[3.2] From [3.1], we have

$$\begin{aligned}
H(n) &= \int_0^1 \frac{1-(1-t)^n}{t}\,dt \\
&= \int_0^n \frac{1-(1-t/n)^n}{t/n}\,d\left(\frac{t}{n}\right) \\
&= \int_0^n \frac{1-(1-t/n)^n}{t}\,dt \\
&= \int_0^1 \frac{1-(1-t/n)^n}{t}\,dt + \int_1^n \frac{1-(1-t/n)^n}{t}\,dt \qquad (9.1)\\
&= \int_0^1 \frac{1-(1-t/n)^n}{t}\,dt + \log x\big|_1^n - \int_1^n \frac{(1-t/n)^n}{t}\,dt \\
&= \int_0^1 \frac{1-(1-t/n)^n}{t}\,dt + \log n - \int_1^n \frac{(1-t/n)^n}{t}\,dt \\
\Longrightarrow H(n) - \log n &= \int_0^1 \frac{1-(1-t/n)^n}{t}\,dt - \int_1^n \frac{(1-t/n)^n}{t}\,dt.
\end{aligned}$$

Taking the limit as $n \to \infty$ in (9.1) and using (3.3), we get the result.

[3.3] Take the limit as $n \to \infty$ in [3.2].

[3.4] From (3.3) and (3.4), we have

$$\begin{aligned}
\Gamma(1) &= \lim_{n\to\infty} \frac{e^{\sum_{j=1}^n 1/j}}{e^{\sum_{j=1}^n 1/j - \log n} \prod_{j=1}^n \left(1 + \frac{1}{j}\right)} \\
&= \lim_{n\to\infty} \frac{n}{\prod_{j=1}^n \frac{j+1}{j}} \\
&= \lim_{n\to\infty} \frac{n}{n+1} \\
&= 1.
\end{aligned}$$

[3.5] Taking log of (3.4), we get

$$\log \Gamma(z) = -\log z - \gamma z + \sum_{j=1}^\infty \left(\frac{z}{j} - \log\left(1 + \frac{z}{j}\right)\right). \qquad (9.2)$$

Differentiating (9.2) with respect to z, we get

9 Solutions to Bessel Functions 273

$$\frac{\Gamma'(z)}{\Gamma(z)} = -\frac{1}{z} - \gamma + \sum_{j=1}^{\infty}\left(\frac{1}{j} - \frac{1/j}{1+z/j}\right)$$
$$= -\frac{1}{z} - \gamma + \sum_{j=1}^{\infty}\left(\frac{1}{j} - \frac{1}{j+z}\right). \tag{9.3}$$

Putting $z = 1$ in (9.3) and using [3.4], we get

$$\Gamma'(1) = -1 - \gamma + \sum_{j=1}^{\infty}\left(\frac{1}{j} - \frac{1}{j+1}\right)$$
$$= -1 - \gamma + 1$$
$$= -\gamma.$$

[3.6] The result follows from [3.4] and (3.6).

[3.7] From (3.4), (3.6), and (2.22), we have

$$\Gamma^2\left(\frac{1}{2}\right) = \Gamma\left(1 - \frac{1}{2}\right)\Gamma\left(\frac{1}{2}\right)$$
$$= -\frac{1}{2}\Gamma\left(\frac{1}{2}\right)\Gamma\left(-\frac{1}{2}\right)$$
$$= -\frac{1}{2}\left(2e^{-\gamma/2}\prod_{j=1}^{\infty}e^{1/(2j)}\left(1+\frac{1}{2j}\right)^{-1}\right)\left(-2e^{\gamma/2}\prod_{j=1}^{\infty}e^{-1/(2j)}\left(1-\frac{1}{2j}\right)^{-1}\right)$$
$$= 2\prod_{j=1}^{\infty}\left(1 - \frac{1}{4j^2}\right)^{-1}$$
$$= 2\prod_{j=1}^{\infty}\left(1 - \frac{(\pi/2)^2}{j^2\pi^2}\right)^{-1}$$
$$= 2\frac{\pi/2}{\sin(\pi/2)}$$
$$= \pi$$
$$\implies \Gamma\left(\frac{1}{2}\right) = \sqrt{\pi}.$$

[3.8] From (3.5), we have

$$\Gamma(z) = \frac{1}{z} \lim_{n\to\infty} \prod_{j=1}^{n-1} \left(1 + \frac{1}{j}\right)^z \left(1 + \frac{z}{j}\right)^{-1}$$

$$= \frac{1}{z} \lim_{n\to\infty} \prod_{j=1}^{n-1} \left(\frac{j+1}{j}\right)^z \left(\frac{j}{z+j}\right)$$

$$= \frac{1}{z} \lim_{n\to\infty} n^z \prod_{j=1}^{n-1} \frac{j}{z+j}.$$

[3.9] Replacing t with $-1/t$ in (3.7), we get

$$e^{z(-1/t+t)/2} = \sum_{n=-\infty}^{\infty} (-1)^n J_n(z) t^{-n}$$

$$= \sum_{n=-\infty}^{\infty} (-1)^{-n} J_{-n}(z) t^n \qquad (9.4)$$

$$= \sum_{n=-\infty}^{\infty} (-1)^n J_{-n}(z) t^n.$$

We get [3.9] by equating the coefficients of t^n in (3.7) and (9.4).

[3.10] Put $n = 0$ in (3.12) and use [3.9].

[3.11] The case of $n = 0$ is [3.10]. Assuming $n \neq 0$, using (3.10) and (3.12), we have

$$\frac{d}{dz}\left(z^n J_n(z)\right) = n z^{n-1} J_n(z) + z^n J_n'(z)$$

$$= n z^{n-1} \frac{z}{2n} (J_{n-1}(z) + J_{n+1}(z)) + \frac{z^n}{2} (J_{n-1}(z) - J_{n+1}(z))$$

$$= \frac{z^n}{2} (2 J_{n-1}(z))$$

$$= z^n J_{n-1}(z).$$

9 Solutions to Bessel Functions   275

[3.12] The case of $n = 0$ is [3.10]. Assuming $n \neq 0$, using (3.10) and (3.12), we have

$$\frac{d}{dz}\left(z^{-n} J_n(z)\right) = -n z^{-n-1} J_n(z) + z^{-n} J_n'(z)$$

$$= -n z^{-n-1} \frac{z}{2n} \left(J_{n-1}(z) + J_{n+1}(z)\right) + \frac{z^{-n}}{2} \left(J_{n-1}(z) - J_{n+1}(z)\right)$$

$$= \frac{z^{-n}}{2} \left(2 J_{n+1}(z)\right)$$

$$= z^{-n} J_{n+1}(z).$$

[3.13] Putting $t = e^{i\theta}$ in (3.7), we get

$$e^{z(e^{i\theta} - e^{-i\theta})/2} = \sum_{j=-\infty}^{\infty} J_j(z) e^{ij\theta}$$

$$\iff e^{iz \sin \theta} = \sum_{j=-\infty}^{\infty} J_j(z) e^{ij\theta}.$$

(9.5)

Putting $t = e^{-i\theta}$ in (3.7), we get

$$e^{z(e^{-i\theta} - e^{i\theta})/2} = \sum_{j=-\infty}^{\infty} J_j(z) e^{-ij\theta}$$

$$\iff e^{-iz \sin \theta} = \sum_{j=-\infty}^{\infty} J_j(z) e^{-ij\theta}.$$

(9.6)

Taking the average of (9.5) and (9.6), and using [3.9], we get

$$\frac{e^{iz \sin \theta} + e^{-iz \sin \theta}}{2} = \sum_{j=-\infty}^{\infty} J_j(z) \frac{e^{ij\theta} + e^{ij\theta}}{2}$$

$$\iff \cos(z \sin \theta) = \sum_{j=-\infty}^{\infty} J_j(z) \cos j\theta$$

$$= J_0(z) + \sum_{j=1}^{\infty} \left(J_j(z) + J_{-j}(z)\right) \cos j\theta$$

$$= J_0(z) + \sum_{j=1}^{\infty} \left(1 + (-1)^j\right) J_j(z) \cos j\theta$$

$$= J_0(z) + 2 \sum_{j=1}^{\infty} J_{2j}(z) \cos 2j\theta.$$

[3.14] Subtracting (9.6) from (9.5), dividing the result by $2i$, and using [3.9], we get

$$\frac{e^{iz\sin\theta} - e^{-iz\sin\theta}}{2i} = \sum_{j=-\infty}^{\infty} J_j(z) \frac{e^{ij\theta} - e^{ij\theta}}{2i}$$

$$\iff \sin(z\sin\theta) = \sum_{j=-\infty}^{\infty} J_j(z) \sin j\theta$$

$$= \sum_{j=1}^{\infty} \left(J_j(z) - J_{-j}(z)\right) \sin j\theta$$

$$= \sum_{j=1}^{\infty} \left(1 - (-1)^j\right) J_j(z) \sin j\theta$$

$$= 2\sum_{j=0}^{\infty} J_{2j+1}(z) \sin(2j+1)\theta.$$

[3.15] Replace θ with $\pi/2 - \theta$ in [3.13].

[3.16] Replace θ with $\pi/2 - \theta$ in [3.14].

[3.17] Differentiating [3.15] with respect to θ, we get

$$(-z\sin\theta)(-\sin(z\cos\theta)) = 4\sum_{j=1}^{\infty} (-1)^{j-1} j J_{2j}(z) \sin 2j\theta$$

$$\iff z\sin\theta \sin(z\cos\theta) = 4\sum_{j=1}^{\infty} (-1)^{j-1} j J_{2j}(z) \sin 2j\theta.$$

(9.7)

Differentiating (9.7) with respect to θ, we get

$$z\cos\theta \sin(z\cos\theta) + (z\sin\theta)(-z\sin\theta)\cos(z\cos\theta)$$

$$= 8\sum_{j=1}^{\infty} (-1)^{j-1} j^2 J_{2j}(z) \cos 2j\theta$$

$$\iff z\cos\theta \sin(z\cos\theta) - z^2 \sin^2\theta \cos(z\cos\theta)$$

$$= 2\sum_{j=1}^{\infty} (-1)^{j-1} (2j)^2 J_{2j}(z) \cos 2j\theta.$$

(9.8)

We get the result by putting $\theta = 0$ in (9.8).

9 Solutions to Bessel Functions

[3.18] Differentiating [3.16] with respect to θ, we get

$$(-z\sin\theta)\cos(z\cos\theta) = 2\sum_{j=0}^{\infty}(-1)^{j-1}(2j+1)J_{2j+1}(z)\sin(2j+1)\theta$$

$$\iff z\sin\theta\cos(z\cos\theta) = 2\sum_{j=0}^{\infty}(-1)^{j}(2j+1)J_{2j+1}(z)\sin(2j+1)\theta. \tag{9.9}$$

Differentiating (9.9) with respect to θ, we get

$$z\cos\theta\cos(z\cos\theta) + (z\sin\theta)(-z\sin\theta)(-\sin(z\cos\theta))$$

$$= 2\sum_{j=0}^{\infty}(-1)^{j}(2j+1)^{2}J_{2j+1}(z)\cos(2j+1)\theta$$

$$\iff z\cos\theta\cos(z\cos\theta) + z^{2}\sin^{2}\theta\sin(z\cos\theta)$$

$$= 2\sum_{j=0}^{\infty}(-1)^{j}(2j+1)^{2}J_{2j+1}(z)\cos(2j+1)\theta. \tag{9.10}$$

We get the result by putting $\theta = 0$ in (9.10).

[3.19] Using (3.17) and [3.9], we have

$$\begin{aligned}
I_{-n}(z) &= i^{n}J_{-n}(iz) \\
&= i^{n}\left((-1)^{n}J_{n}(iz)\right) \\
&= (-i)^{n}\left(i^{n}I_{n}(z)\right) \\
&= I_{n}(z).
\end{aligned}$$

[3.20] Using (3.17) and [3.10], we have

$$\begin{aligned}
I_{0}'(z) &= \frac{dJ_{0}(iz)}{dz} \\
&= iJ_{0}'(iz) \\
&= i(-J_{1}(iz)) \\
&= i(-iI_{1}(z)) \\
&= I_{1}(z).
\end{aligned}$$

[3.21] Replace z with iz in [3.11] and use (3.17).

[3.22] Replace z with iz in [3.12] and use (3.17).

[3.23] Replace z with iz in [3.13] and use (3.17).

[3.24] Replace z with iz in [3.14] and use (3.17).

[3.25] Replace z with iz in [3.15] and use (3.17).

[3.26] Replace z with iz in [3.16] and use (3.17).

[3.27] Replace z with iz in [3.17] and use (3.17).

[3.28] Replace z with iz in [3.18] and use (3.17).

[3.29] Using (3.27) and (3.6), we have

$$\frac{d}{dz}(z^v J_v(z)) = \frac{d}{dz}\left(\sum_{j=0}^{\infty}(-1)^j \frac{z^{2v+2j}}{j!\Gamma(v+j+1)2^{v+2j}}\right)$$

$$= \sum_{j=0}^{\infty}(-1)^j \frac{(2v+2j)z^{2v+2j-1}}{j!(v+j)\Gamma(v+j)2^{v+2j}}$$

$$= z^v \sum_{j=0}^{\infty}(-1)^j \frac{z^{v-1+2j}}{j!\Gamma(v-1+j+1)2^{v-1+2j}}$$

$$= z^v J_{v-1}(z).$$

[3.30] Using (3.27), we have

$$\frac{d}{dz}(z^{-v} J_v(z)) = \frac{d}{dz}\left(\sum_{j=0}^{\infty}(-1)^j \frac{z^{2j}}{j!\Gamma(v+j+1)2^{v+2j}}\right)$$

$$= \sum_{j=1}^{\infty}(-1)^j \frac{(2j)z^{2j-1}}{j!\Gamma(v+j+1)2^{v+2j}}$$

$$= \sum_{j=1}^{\infty}(-1)^j \frac{z^{2j-1}}{(j-1)!\Gamma(v+j+1)2^{v+2j-1}}$$

$$= z^{-v}\sum_{j=0}^{\infty}(-1)^{j+1} \frac{z^{v+1+2j}}{j!\Gamma(v+1+j+1)2^{v+1+2j}}$$

$$= -z^{-v} J_{v+1}(z).$$

9 Solutions to Bessel Functions

[3.31] Using (3.32) and (3.6), we have

$$\frac{d}{dz}(z^v I_v(z)) = \frac{d}{dz}\left(\sum_{j=0}^{\infty}\frac{z^{2v+2j}}{j!\Gamma(v+j+1)2^{v+2j}}\right)$$

$$= \sum_{j=0}^{\infty}\frac{(2v+2j)z^{2v+2j-1}}{j!(v+j)\Gamma(v+j)2^{v+2j}}$$

$$= z^v\sum_{j=0}^{\infty}\frac{z^{v-1+2j}}{j!\Gamma(v-1+j+1)2^{v-1+2j}}$$

$$= z^v I_{v-1}(z).$$

[3.32] Using (3.32), we have

$$\frac{d}{dz}\left(z^{-v} I_v(z)\right) = \frac{d}{dz}\left(\sum_{j=0}^{\infty}\frac{z^{2j}}{j!\Gamma(v+j+1)2^{v+2j}}\right)$$

$$= \sum_{j=1}^{\infty}\frac{(2j)z^{2j-1}}{j!\Gamma(v+j+1)2^{v+2j}}$$

$$= \sum_{j=1}^{\infty}\frac{z^{2j-1}}{(j-1)!\Gamma(v+j+1)2^{v+2j-1}}$$

$$= z^{-v}\sum_{j=0}^{\infty}\frac{z^{v+1+2j}}{j!\Gamma(v+1+j+1)2^{v+1+2j}}$$

$$= z^{-v} J_{v+1}(z).$$

Chapter 10
Solutions to Special Integrals

[4.1] Putting $n = 0$ in [4.2], we get

$$\int_0^\pi e^{\cos\theta} \cos(-\sin\theta)\, d\theta = \int_0^\pi e^{\cos\theta} \cos(\sin\theta)\, d\theta = \pi. \qquad (10.1)$$

Putting $n = 0$ in [4.4], we get

$$\int_0^\pi e^{-\cos\theta} \cos(\sin\theta)\, d\theta = \pi. \qquad (10.2)$$

Taking the average of (10.1) and (10.2), we get

$$\int_0^\pi \frac{e^{\cos\theta} + e^{-\cos\theta}}{2} \cos(\sin\theta)\, d\theta = \int_0^\pi \cosh(\cos\theta) \cos(\sin\theta)\, d\theta = \pi.$$

[4.2] Put $x = 1$ in [4.6].

[4.3] Put $x = 1$ in [4.7].

[4.4] Put $x = -1$ in [4.6].

[4.5] Put $x = -1$ in [4.7].

[4.6] Applying Theorem 4.2 to $f(z) = e^{xz}$ for $(x, n) \in \mathbb{R} \times (\mathbb{N} - 1)$, we get

$$\pi a_n = \frac{\pi}{n!} x^n$$
$$= \int_0^\pi \frac{e^{-in\theta} e^{xe^{i\theta}} + e^{in\theta} e^{xe^{-i\theta}}}{2} d\theta$$
$$= \int_0^\pi \frac{e^{-in\theta} e^{x(\cos\theta + i\sin\theta)} + e^{in\theta} e^{x(\cos\theta - i\sin\theta)}}{2} d\theta \qquad (10.3)$$
$$= \int_0^\pi e^{x\cos\theta} \frac{e^{-i(n\theta - x\sin\theta)} + e^{i(n\theta - x\sin\theta)}}{2} d\theta$$
$$= \int_0^\pi e^{x\cos\theta} \cos(n\theta - x\sin\theta) d\theta.$$

[4.7] Applying Theorem 4.2 to $f(z) = e^{xz}$ for $(x, n) \in \mathbb{R} \times (-\mathbb{N})$ and using (10.3), we get

$$\pi a_n = 0 = \int_0^\pi e^{x\cos\theta} \cos(n\theta - x\sin\theta) d\theta.$$

[4.8] Applying Theorem 4.2 to $f(z) = \sin(xz)$ for $(x, n) \in \mathbb{R} \times (2\mathbb{N} - 1)$, we get

$$\pi a_n = (-1)^{(n-1)/2} \frac{\pi}{n!} x^n$$
$$= \int_0^\pi \frac{e^{-in\theta} \sin\left(xe^{i\theta}\right) + e^{in\theta} \sin\left(xe^{-i\theta}\right)}{2} d\theta$$
$$= \int_0^\pi \frac{e^{-in\theta} \sin(x(\cos\theta + i\sin\theta)) + e^{in\theta} \sin(x(\cos\theta - i\sin\theta))}{2} d\theta$$
$$= \frac{1}{2} \int_0^\pi \left(e^{-in\theta} \left(\sin(x\cos\theta)\cos(ix\sin\theta) + \cos(x\cos\theta)\sin(ix\sin\theta)\right)\right.$$
$$\left. + e^{in\theta} \left(\sin(x\cos\theta)\cos(ix\sin\theta) - \cos(x\cos\theta)\sin(ix\sin\theta)\right)\right) d\theta$$
$$= \int_0^\pi \left(\frac{e^{-in\theta} + e^{in\theta}}{2} \sin(x\cos\theta)\cosh(x\sin\theta)\right.$$
$$\left. - \frac{e^{-in\theta} - e^{in\theta}}{2i} \cos(x\cos\theta)\sinh(x\sin\theta)\right) d\theta$$
$$= \int_0^\pi (\cos(n\theta) \sin(x\cos\theta) \cosh(x\sin\theta)$$
$$+ \sin(n\theta) \cos(x\cos\theta) \sinh(x\sin\theta)) d\theta.$$
$$(10.4)$$

[4.9] Applying Theorem 4.2 to $f(z) = \sin(xz)$ for $(x, n) \in \mathbb{R} \times (-\mathbb{N} \cup (2\mathbb{N} - 2))$ and using (10.4), we get

$$\pi a_n = 0$$
$$= \int_0^\pi (\cos(n\theta) \sin(x \cos \theta) \cosh(x \sin \theta) + \sin(n\theta) \cos(x \cos \theta)$$
$$\sinh(x \sin \theta))\, d\theta.$$

[4.10] Applying Theorem 4.2 to $f(z) = \cos(xz)$ for $(x, n) \in \mathbb{R} \times (2\mathbb{N} - 2)$, we get

$$\pi a_n = (-1)^{n/2} \frac{\pi}{n!} x^n$$
$$= \int_0^\pi \frac{e^{-in\theta} \cos\left(xe^{i\theta}\right) + e^{in\theta} \cos\left(xe^{-i\theta}\right)}{2}\, d\theta$$
$$= \int_0^\pi \frac{e^{-in\theta} \cos(x(\cos \theta + i \sin \theta)) + e^{in\theta} \cos(x(\cos \theta - i \sin \theta))}{2}\, d\theta$$
$$= \frac{1}{2} \int_0^\pi \left(e^{-in\theta} (\cos(x \cos \theta) \cos(ix \sin \theta) - \sin(x \cos \theta) \sin(ix \sin \theta))\right.$$
$$\left. + e^{in\theta} (\cos(x \cos \theta) \cos(ix \sin \theta) + \sin(x \cos \theta) \sin(ix \sin \theta))\right) d\theta$$
$$= \int_0^\pi \left(\frac{e^{-in\theta} + e^{in\theta}}{2} \cos(x \cos \theta) \cosh(x \sin \theta)\right.$$
$$\left. - \frac{e^{in\theta} - e^{-in\theta}}{2i} \sin(x \cos \theta) \sinh(x \sin \theta)\right) d\theta$$
$$= \int_0^\pi (\cos(n\theta) \cos(x \cos \theta) \cosh(x \sin \theta) - \sin(n\theta) \sin(x \cos \theta)$$
$$\sinh(x \sin \theta))\, d\theta.$$
(10.5)

[4.11] Applying Theorem 4.2 to $f(z) = \cos(xz)$ for $(x, n) \in \mathbb{R} \times (-\mathbb{N} \cup (2\mathbb{N} - 1))$, and using (10.5), we get

$$\pi a_n = 0$$
$$= \int_0^\pi (\cos(n\theta) \cos(x \cos \theta) \cosh(x \sin \theta) - \sin(n\theta) \sin(x \cos \theta)$$
$$\sinh(x \sin \theta))\, d\theta.$$

[4.12] Applying Theorem 4.2 to $f(z) = \sinh(xz)$ for $(x, n) \in \mathbb{R} \times (2\mathbb{N} - 1)$, we get

$$\pi a_n = \frac{\pi}{n!} x^n$$

$$= \int_0^\pi \frac{e^{-in\theta} \sinh\left(xe^{i\theta}\right) + e^{in\theta} \sinh\left(xe^{-i\theta}\right)}{2} d\theta$$

$$= \int_0^\pi \frac{e^{-in\theta} \sinh(x(\cos\theta + i\sin\theta)) + e^{in\theta} \sinh(x(\cos\theta - i\sin\theta))}{2} d\theta$$

$$= \frac{1}{2} \int_0^\pi \left(e^{-in\theta} (\sinh(x\cos\theta)\cosh(ix\sin\theta) + \cosh(x\cos\theta)\sinh(ix\sin\theta)) \right.$$

$$\left. + e^{in\theta} (\sinh(x\cos\theta)\cosh(ix\sin\theta) - \cosh(x\cos\theta)\sinh(ix\sin\theta)) \right) d\theta$$

$$= \int_0^\pi \left(\frac{e^{-in\theta} + e^{in\theta}}{2} \sinh(x\cos\theta)\cos(x\sin\theta) \right.$$

$$\left. + \frac{e^{in\theta} - e^{-in\theta}}{2i} \cosh(x\cos\theta)\sin(x\sin\theta) \right) d\theta$$

$$= \int_0^\pi (\cos(n\theta)\sinh(x\cos\theta)\cos(x\sin\theta) + \sin(n\theta)\cosh(x\cos\theta)$$

$$\sin(x\sin\theta)) d\theta.$$

(10.6)

[4.13] Applying Theorem 4.2 to $f(z) = \sinh(xz)$ for $(x, n) \in \mathbb{R} \times (-\mathbb{N} \cup (2\mathbb{N} - 2))$ and using (10.6), we get

$$\pi a_n = 0$$

$$= \int_0^\pi (\cos(n\theta)\sinh(x\cos\theta)\cos(x\sin\theta) + \sin(n\theta)\cosh(x\cos\theta)$$

$$\sin(x\sin\theta)) d\theta.$$

[4.14] Applying Theorem 4.2 to $f(z) = \cosh(xz)$ for $(x, n) \in \mathbb{R} \times (2\mathbb{N} - 2)$, we get

$$\pi a_n = \frac{\pi}{n!} x^n$$

$$= \int_0^\pi \frac{e^{-in\theta} \cosh\left(xe^{i\theta}\right) + e^{in\theta} \cosh\left(xe^{-i\theta}\right)}{2} d\theta$$

$$= \int_0^\pi \frac{e^{-in\theta} \cosh(x(\cos\theta + i\sin\theta)) + e^{in\theta} \cosh(x(\cos\theta - i\sin\theta))}{2} d\theta$$

10 Solutions to Special Integrals   285

$$
\begin{aligned}
&= \frac{1}{2}\int_0^\pi \left(e^{-in\theta}\left(\cosh(x\cos\theta)\cosh(ix\sin\theta)+\sinh(x\cos\theta)\sinh(ix\sin\theta)\right)\right.\\
&\quad \left.+e^{in\theta}\left(\cosh(x\cos\theta)\cosh(ix\sin\theta)-\sinh(x\cos\theta)\sinh(ix\sin\theta)\right)\right)d\theta\\
&= \int_0^\pi \left(\frac{e^{-in\theta}+e^{in\theta}}{2}\cosh(x\cos\theta)\cos(x\sin\theta)\right.\\
&\quad \left.+\frac{e^{in\theta}-e^{-in\theta}}{2i}\sinh(x\cos\theta)\sin(x\sin\theta)\right)d\theta\\
&= \int_0^\pi (\cos(n\theta)\cosh(x\cos\theta)\cos(x\sin\theta)+\sin(n\theta)\sinh(x\cos\theta)\\
&\quad \sin(x\sin\theta))\, d\theta.
\end{aligned}
$$
(10.7)

[4.15] Applying Theorem 4.2 to $f(z)=\cosh(xz)$ for $(x,n)\in\mathbb{R}\times(-\mathbb{N}\cup(2\mathbb{N}-1))$ and using (10.7), we get

$$
\begin{aligned}
\pi a_n &= 0\\
&= \int_0^\pi (\cos(n\theta)\cosh(x\cos\theta)\cos(x\sin\theta)+\sin(n\theta)\sinh(x\cos\theta)\\
&\quad \sin(x\sin\theta))\, d\theta.
\end{aligned}
$$

[4.16] Applying Theorem 4.2 to $f(z)=(1+z)^m$ for $(m,n)\in(\mathbb{N}-1)\times[0..m]$, we get

$$
\begin{aligned}
\frac{\pi}{2^m}a_n &= \frac{\pi}{2^m}\binom{m}{n}\\
&= \frac{1}{2^m}\int_0^\pi \frac{e^{-in\theta}(1+e^{i\theta})^m + e^{in\theta}(1+e^{-i\theta})^m}{2}\, d\theta\\
&= \frac{1}{2^m}\int_0^\pi \left(\frac{e^{-in\theta}((1+\cos\theta)+i\sin\theta)^m}{2}\right.\\
&\quad \left.+\frac{e^{in\theta}((1+\cos\theta)-i\sin\theta)^m}{2}\right)d\theta\\
&= \frac{1}{2^m}\int_0^\pi \left(\frac{e^{-in\theta}\left(2\cos^2\frac{\theta}{2}+2i\cos\frac{\theta}{2}\sin\frac{\theta}{2}\right)^m}{2}\right.
\end{aligned}
$$

$$+\frac{e^{in\theta}\left(2\cos^2\frac{\theta}{2} - 2i\cos\frac{\theta}{2}\sin\frac{\theta}{2}\right)^m}{2}\Bigg)d\theta$$

$$= \frac{1}{2^m}\int_0^\pi \Bigg(\frac{2^m e^{-in\theta}\cos^m\frac{\theta}{2}\left(\cos\frac{\theta}{2} + i\sin\frac{\theta}{2}\right)^m}{2}$$

$$+\frac{2^m e^{in\theta}\cos^m\frac{\theta}{2}\left(\cos\frac{\theta}{2} - i\sin\frac{\theta}{2}\right)^m}{2}\Bigg)d\theta$$

$$= \int_0^\pi \frac{e^{-in\theta}e^{im\theta/2}\cos^m\frac{\theta}{2} + e^{in\theta}e^{-im\theta/2}\cos^m\frac{\theta}{2}}{2}d\theta$$

$$= \int_0^\pi \frac{e^{-i(2n-m)\theta/2} + e^{i(2n-m)\theta/2}}{2}\cos^m\frac{\theta}{2}d\theta$$

$$= \int_0^\pi \cos\frac{(2n-m)\theta}{2}\cos^m\frac{\theta}{2}d\theta. \tag{10.8}$$

[4.17] Applying Theorem 4.2 to $f(z) = (1+z)^m$ for $(m,n) \in (\mathbb{N}-1) \times (\mathbb{Z} - [0..m])$ and using (10.8), we get

$$\frac{\pi}{2^m}a_n = 0 = \int_0^\pi \cos\frac{(2n-m)\theta}{2}\cos^m\frac{\theta}{2}d\theta.$$

[4.18] Applying Theorem 4.2 to $f(z) = (1-z)^m$ for $(m,n) \in (\mathbb{N}-1) \times [0..m]$, we get

$$\frac{\pi}{2^m}a_n = (-1)^n \frac{\pi}{2^m}\binom{m}{n}$$

$$= \frac{1}{2^m}\int_0^\pi \frac{e^{-in\theta}\left(1-e^{i\theta}\right)^m + e^{in\theta}\left(1-e^{-i\theta}\right)^m}{2}d\theta$$

$$= \frac{1}{2^m}\int_0^\pi \Bigg(\frac{e^{-in\theta}((1-\cos\theta) - i\sin\theta)^m}{2}$$

$$+\frac{e^{in\theta}((1-\cos\theta) + i\sin\theta)^m}{2}\Bigg)d\theta$$

$$= \frac{1}{2^m} \int_0^\pi \left(\frac{e^{-in\theta} \left(2\sin^2\frac{\theta}{2} - 2i\sin\frac{\theta}{2}\cos\frac{\theta}{2}\right)^m}{2} \right.$$

$$\left. + \frac{e^{in\theta} \left(2\sin^2\frac{\theta}{2} + 2i\sin\frac{\theta}{2}\cos\frac{\theta}{2}\right)^m}{2} \right) d\theta$$

$$= \frac{1}{2^m} \int_0^\pi \left(\frac{2^m e^{-in\theta} \sin^m\frac{\theta}{2} \left(\sin\frac{\theta}{2} - i\cos\frac{\theta}{2}\right)^m}{2} \right.$$

$$\left. + \frac{2^m e^{in\theta} \sin^m\frac{\theta}{2} \left(\sin\frac{\theta}{2} + i\cos\frac{\theta}{2}\right)^m}{2} \right) d\theta$$

$$= \int_0^\pi \frac{e^{-in\theta} e^{-i(\pi-\theta)m/2} \sin^m\frac{\theta}{2} + e^{in\theta} e^{i(\pi-\theta)m/2} \sin^m\frac{\theta}{2}}{2} d\theta$$

$$= \int_0^\pi \frac{e^{-i((2n-m)\theta + m\pi)/2} + e^{i((2n-m)\theta + m\pi)/2}}{2} \sin^m\frac{\theta}{2} d\theta$$

$$= \int_0^\pi \cos\frac{(2n-m)\theta + m\pi}{2} \sin^m\frac{\theta}{2} d\theta. \tag{10.9}$$

[4.19] Applying Theorem 4.2 to $f(z) = (1-z)^m$ for $(m,n) \in (\mathbb{N}-1) \times (\mathbb{Z} - [0..m])$ and using (10.9), we get

$$\frac{\pi}{2^m} a_n = 0 = \int_0^\pi \cos\frac{(2n-m)\theta + m\pi}{2} \sin^m\frac{\theta}{2} d\theta.$$

[4.20] Applying Theorem 4.2 to $f(z) = (1+xz)^m$ for $(x, m, n) \in [-1, 1] \times (\mathbb{N} - 1) \times [0..m]$, we get

$$\pi a_n = \pi x^n \binom{m}{n}$$

$$= \int_0^\pi \frac{e^{-in\theta}\left(1 + xe^{i\theta}\right)^m + e^{in\theta}\left(1 + xe^{-i\theta}\right)^m}{2} d\theta$$

$$= \int_0^\pi \left(\frac{e^{-in\theta}((1 + x\cos\theta) + ix\sin\theta)^m}{2} + \frac{e^{in\theta}((1 + x\cos\theta) - ix\sin\theta)^m}{2} \right) d\theta$$

$$= \int_0^\pi \frac{1}{2} \left(e^{-in\theta} \left(\frac{1 + x\cos\theta}{\sqrt{1 + x^2 + 2x\cos\theta}} + i\frac{x\sin\theta}{\sqrt{1 + x^2 + 2x\cos\theta}} \right)^m \right.$$

$$\left. + e^{in\theta} \left(\frac{1 + x\cos\theta}{\sqrt{1 + x^2 + 2x\cos\theta}} - i\frac{x\sin\theta}{\sqrt{1 + x^2 + 2x\cos\theta}} \right)^m \right)$$

$$\left(1 + x^2 + 2x\cos\theta \right)^{m/2} d\theta$$

$$= \int_0^\pi \frac{1}{2} \left(e^{-in\theta} e^{im \tan^{-1}(x\sin\theta/(1+x\cos\theta))} \right.$$

$$\left. + e^{in\theta} e^{-im \tan^{-1}(x\sin\theta/(1+x\cos\theta))} \right) \left(1 + x^2 + 2x\cos\theta\right)^{m/2} d\theta$$

$$= \int_0^\pi \left(1 + x^2 + 2x\cos\theta\right)^{m/2} \cos\left(n\theta - m\tan^{-1}\left(\frac{x\sin\theta}{1 + x\cos\theta}\right)\right) d\theta. \tag{10.10}$$

[4.21] Applying Theorem 4.2 to $f(z) = (1+xz)^m$ for $(x, m, n) \in [-1, 1] \times (\mathbb{N} - 1) \times (\mathbb{Z} - [0..m])$ and using (10.10), we get

$$\pi a_n = 0$$

$$= \int_0^\pi \left(1 + x^2 + 2x\cos\theta\right)^{m/2} \cos\left(n\theta - m\tan^{-1}\left(\frac{x\sin\theta}{1 + x\cos\theta}\right)\right) d\theta.$$

[4.22] Put $x = 1$ in [4.24].

[4.23] Put $x = 1$ in [4.25].

[4.24] Applying Theorem 4.5 to $f(z) = e^{xz}$, and using (3.19), we get

$$\pi \sum_{j=0}^{\infty} a_j^2 = \pi \sum_{j=0}^{\infty} \frac{x^{2j}}{(j!)^2}$$
$$= \pi I_0(2x)$$
$$= \int_0^{\pi} e^{xe^{i\theta}} e^{xe^{-i\theta}} d\theta$$
$$= \int_0^{\pi} e^{x(e^{i\theta}+e^{-i\theta})} d\theta$$
$$= \int_0^{\pi} e^{2x\cos\theta} d\theta.$$

[4.25] Applying Theorem 4.6 to $f(z) = e^{xz}$, and using (3.19), we get

$$\pi \sum_{j=0}^{\infty} a_j a_{n+j} = \pi \sum_{j=0}^{\infty} \frac{x^{2j}}{j!(n+j)!}$$
$$= \pi I_n(2x)$$
$$= \int_0^{\pi} e^{xe^{i\theta}} e^{xe^{-i\theta}} \cos n\theta \, d\theta$$
$$= \int_0^{\pi} e^{x(e^{i\theta}+e^{-i\theta})} \cos n\theta \, d\theta$$
$$= \int_0^{\pi} e^{2x\cos\theta} \cos n\theta \, d\theta.$$

[4.26] Applying Theorem 4.5 to $f(z) = \sum_{j=0}^{n} a_j z^j$, we get

$$\pi \sum_{j=0}^{n} a_j^2 = \int_0^{\pi} \left(\sum_{j=0}^{n} a_j e^{ji\theta} \right) \left(\sum_{j=0}^{n} a_j e^{-ji\theta} \right) d\theta$$
$$= \int_0^{\pi} \left(\sum_{j=0}^{n} a_j \cos j\theta + i \sum_{j=1}^{n} a_j \sin j\theta \right) \left(\sum_{j=0}^{n} a_j \cos j\theta - i \sum_{j=1}^{n} a_j \sin j\theta \right) d\theta$$
$$= \int_0^{\pi} \left(\left(\sum_{j=0}^{n} a_j \cos j\theta \right)^2 + \left(\sum_{j=1}^{n} a_j \sin j\theta \right)^2 \right) d\theta$$
$$= \int_0^{\pi} \left(C^2(\theta) + S^2(\theta) \right) d\theta, \tag{10.11}$$

where $C(\theta)$ and $S(\theta)$ are as defined in (4.46) and (4.47), respectively.

[4.27] Applying Theorem 4.6 to $f(z) = \sum_{j=0}^{n} a_j z^j$ and using (10.11), we get

$$\pi \sum_{j=0}^{\infty} a_j a_{m+j} = \pi \sum_{j=0}^{n-m} a_j a_{m+j}$$

$$= \int_0^{\pi} \left(\sum_{j=0}^{n} a_j e^{ji\theta} \right) \left(\sum_{j=0}^{n} a_j e^{-ji\theta} \right) \cos m\theta \, d\theta$$

$$= \int_0^{\pi} \left(C^2(\theta) + S^2(\theta) \right) \cos m\theta \, d\theta,$$

where $C(\theta)$ and $S(\theta)$ are as defined in (4.46) and (4.47), respectively.

[4.28] Applying Theorem 4.5 to $f(z) = (1-z)^n$, we get

$$\frac{\pi}{2^{2n}} \sum_{j=0}^{\infty} a_j^2 = \frac{\pi}{2^{2n}} \sum_{j=0}^{n} \binom{n}{j}^2$$

$$= \frac{1}{2^{2n}} \int_0^{\pi} \left(1 - e^{i\theta} \right)^n \left(1 - e^{-i\theta} \right)^n d\theta$$

$$= \frac{1}{2^{2n}} \int_0^{\pi} \left((1 - \cos\theta) - i\sin\theta \right)^n \left((1 - \cos\theta) + i\sin\theta \right)^n d\theta$$

$$= \frac{1}{2^{2n}} \int_0^{\pi} \left(2\sin^2 \frac{\theta}{2} - 2i \cos\frac{\theta}{2} \sin\frac{\theta}{2} \right)^n \left(2\sin^2 \frac{\theta}{2} + 2i \cos\frac{\theta}{2} \sin\frac{\theta}{2} \right)^n d\theta$$

$$= \frac{1}{2^{2n}} \int_0^{\pi} 2^{2n} \sin^{2n} \frac{\theta}{2} e^{-in(\pi-\theta)/2} e^{in(\pi-\theta)/2} d\theta$$

$$= \int_0^{\pi} \sin^{2n} \frac{\theta}{2} d\theta. \tag{10.12}$$

[4.29] Applying Theorem 4.6 to $f(z) = (1-z)^n$ for $(n, m) \in \mathbb{N} \times [0..n]$ and using (10.12), we get

$$\frac{\pi}{2^{2n}} \sum_{j=0}^{\infty} a_j a_{m+j} = \frac{\pi}{2^{2n}} \sum_{j=0}^{n-m} a_j a_{m+j}$$

$$= (-1)^m \frac{\pi}{2^{2n}} \sum_{j=0}^{n-m} \binom{n}{j} \binom{n}{m+j}$$

$$= \frac{1}{2^{2n}} \int_0^\pi \left(1 - e^{i\theta}\right)^n \left(1 - e^{-i\theta}\right)^n \cos m\theta \, d\theta$$

$$= \int_0^\pi \sin^{2n} \frac{\theta}{2} \cos m\theta \, d\theta.$$

[4.30] Applying Theorem 4.6 to $f(z) = (1-z)^n$ for $(n, m) \in \mathbb{N} \times ((\mathbb{N} - 1) - [0..n])$ and using (10.12), we get

$$\frac{\pi}{2^{2n}} \sum_{j=0}^{\infty} a_j a_{m+j} = \frac{\pi}{2^{2n}} \sum_{j=0}^{n-m} a_j a_{m+j}$$

$$= 0$$

$$= \frac{1}{2^{2n}} \int_0^\pi \left(1 - e^{i\theta}\right)^n \left(1 - e^{-i\theta}\right)^n \cos m\theta \, d\theta$$

$$= \int_0^\pi \sin^{2n} \frac{\theta}{2} \cos m\theta \, d\theta.$$

[4.31] Applying Theorem 4.5 to $f(z) = \sin xz$, we get

$$\pi \sum_{j=0}^{\infty} a_j^2 = \pi \sum_{j=0}^{\infty} \frac{x^{4j+2}}{((2j+1)!)^2}$$

$$= \int_0^\pi \sin\left(xe^{i\theta}\right) \sin\left(xe^{-i\theta}\right) d\theta$$

$$= \int_0^\pi \sin(x\cos\theta + ix\sin\theta) \sin(x\cos\theta - ix\sin\theta) d\theta$$

$$= \int_0^\pi (\sin(x\cos\theta)\cos(ix\sin\theta) + \cos(x\cos\theta)\sin(ix\sin\theta))$$
$$(\sin(x\cos\theta)\cos(ix\sin\theta) - \cos(x\cos\theta)\sin(ix\sin\theta)) d\theta$$

$$= \int_0^\pi (\sin(x\cos\theta)\cosh(x\sin\theta) + i\cos(x\cos\theta)\sinh(x\sin\theta))$$
$$(\sin(x\cos\theta)\cosh(x\sin\theta) - i\cos(x\cos\theta)\sinh(x\sin\theta)) d\theta$$

$$= \int_0^\pi \left(\sin^2(x\cos\theta)\cosh^2(x\sin\theta) + \cos^2(x\cos\theta)\sinh^2(x\sin\theta)\right) d\theta.$$
(10.13)

[4.32] Applying Theorem 4.5 to $f(z) = \sinh xz$, we get

$$\pi \sum_{j=0}^\infty a_j^2 = \pi \sum_{j=0}^\infty \frac{x^{4j+2}}{((2j+1)!)^2}$$

$$= \int_0^\pi \sinh\left(xe^{i\theta}\right) \sinh\left(xe^{-i\theta}\right) d\theta$$

$$= \int_0^\pi \sinh(x\cos\theta + ix\sin\theta) \sinh(x\cos\theta - ix\sin\theta) d\theta$$

$$= \int_0^\pi (\sinh(x\cos\theta)\cosh(ix\sin\theta) + \cosh(x\cos\theta)\sinh(ix\sin\theta))$$
$$(\sinh(x\cos\theta)\cosh(ix\sin\theta) - \cosh(x\cos\theta)\sinh(ix\sin\theta)) d\theta$$

$$= \int_0^\pi (\sinh(x\cos\theta)\cos(x\sin\theta) + i\cosh(x\cos\theta)\sin(x\sin\theta))$$
$$(\sinh(x\cos\theta)\cos(x\sin\theta) - i\cosh(x\cos\theta)\sin(x\sin\theta)) d\theta$$

$$= \int_0^\pi \left(\sinh^2(x\cos\theta)\cos^2(x\sin\theta) + \cosh^2(x\cos\theta)\sin^2(x\sin\theta)\right) d\theta.$$
(10.14)

[4.33] Applying Theorem 4.6 to $f(z) = \sin xz$ for $(x, n) \in \mathbb{R} \times (2\mathbb{N} - 2)$ and using (10.13), we get

$$\pi \sum_{j=0}^\infty a_j a_{n+j} = (-1)^{n/2} \pi x^n \sum_{j=0}^\infty \frac{x^{4j+2}}{(2j+1)!(n+2j+1)!}$$

$$= \int_0^\pi \sin\left(xe^{i\theta}\right) \sin\left(xe^{-i\theta}\right) \cos n\theta \, d\theta$$

$$= \int_0^\pi \left(\sin^2(x\cos\theta)\cosh^2(x\sin\theta) + \cos^2(x\cos\theta)\sinh^2(x\sin\theta)\right) \cos n\theta \, d\theta.$$

10 Solutions to Special Integrals

[4.34] Applying Theorem 4.6 to $f(z) = \sin xz$ for $(x, n) \in \mathbb{R} \times (2\mathbb{N} - 1)$ and using (10.13), we get

$$\pi \sum_{j=0}^{\infty} a_j a_{n+j} = 0$$

$$= \int_0^{\pi} \sin\left(xe^{i\theta}\right) \sin\left(xe^{-i\theta}\right) \cos n\theta \, d\theta$$

$$= \int_0^{\pi} \left(\sin^2(x\cos\theta) \cosh^2(x\sin\theta) + \cos^2(x\cos\theta) \sinh^2(x\sin\theta)\right) \cos n\theta \, d\theta.$$

The above identity is also true for $n \in 2\mathbb{Z} + 1$ because $\cos n\theta$ is an even function.

[4.35] Applying Theorem 4.6 to $f(z) = \sinh xz$ for $(x, n) \in \mathbb{R} \times (2\mathbb{N} - 2)$ and using (10.14), we get

$$\pi \sum_{j=0}^{\infty} a_j a_{n+j} = \pi x^n \sum_{j=0}^{\infty} \frac{x^{4j+2}}{(2j+1)!(n+2j+1)!}$$

$$= \int_0^{\pi} \sinh\left(xe^{i\theta}\right) \sinh\left(xe^{-i\theta}\right) \cos n\theta \, d\theta$$

$$= \int_0^{\pi} \left(\sinh^2(x\cos\theta) \cos^2(x\sin\theta) + \cosh^2(x\cos\theta) \sin^2(x\sin\theta)\right) \cos n\theta \, d\theta.$$

[4.36] Applying Theorem 4.6 to $f(z) = \sinh xz$ for $(x, n) \in \mathbb{R} \times (2\mathbb{N} - 1)$ and using (10.14), we get

$$\pi \sum_{j=0}^{\infty} a_j a_{n+j} = 0$$

$$= \int_0^{\pi} \sinh\left(xe^{i\theta}\right) \sinh\left(xe^{-i\theta}\right) \cos n\theta \, d\theta$$

$$= \int_0^{\pi} \left(\sinh^2(x\cos\theta) \cos^2(x\sin\theta) + \cosh^2(x\cos\theta) \sin^2(x\sin\theta)\right) \cos n\theta \, d\theta.$$

The above identity is also true for $n \in 2\mathbb{Z} + 1$ because $\cos n\theta$ is an even function.

[4.37] Applying Theorem 4.5 to $f(z) = \cos xz$, we get

$$\pi \sum_{j=0}^{\infty} a_j^2 = \pi \sum_{j=0}^{\infty} \frac{x^{4j}}{((2j)!)^2}$$

$$= \int_0^{\pi} \cos\left(xe^{i\theta}\right) \cos\left(xe^{-i\theta}\right) d\theta$$

$$= \int_0^{\pi} \cos(x\cos\theta + ix\sin\theta)\cos(x\cos\theta - ix\sin\theta) d\theta$$

$$= \int_0^{\pi} (\cos(x\cos\theta)\cos(ix\sin\theta) - \sin(x\cos\theta)\sin(ix\sin\theta))$$
$$(\cos(x\cos\theta)\cos(ix\sin\theta) + \sin(x\cos\theta)\sin(ix\sin\theta)) d\theta$$

$$= \int_0^{\pi} (\cos(x\cos\theta)\cosh(x\sin\theta) - i\sin(x\cos\theta)\sinh(x\sin\theta))$$
$$(\cos(x\cos\theta)\cosh(x\sin\theta) + i\sin(x\cos\theta)\sinh(x\sin\theta)) d\theta$$

$$= \int_0^{\pi} \left(\cos^2(x\cos\theta)\cosh^2(x\sin\theta) + \sin^2(x\cos\theta)\sinh^2(x\sin\theta)\right) d\theta.$$
(10.15)

[4.38] Applying Theorem 4.5 to $f(z) = \cosh xz$, we get

$$\pi \sum_{j=0}^{\infty} a_j^2 = \pi \sum_{j=0}^{\infty} \frac{x^{4j}}{((2j)!)^2}$$

$$= \int_0^{\pi} \cosh\left(xe^{i\theta}\right) \cosh\left(xe^{-i\theta}\right) d\theta$$

$$= \int_0^{\pi} \cosh(x\cos\theta + ix\sin\theta)\cosh(x\cos\theta - ix\sin\theta) d\theta$$

$$= \int_0^{\pi} (\cosh(x\cos\theta)\cosh(ix\sin\theta) + \sinh(x\cos\theta)\sinh(ix\sin\theta))$$
$$(\cosh(x\cos\theta)\cosh(ix\sin\theta) - \sinh(x\cos\theta)\sinh(ix\sin\theta)) d\theta$$

$$= \int_0^{\pi} (\cosh(x\cos\theta)\cos(x\sin\theta) + i\sinh(x\cos\theta)\sin(x\sin\theta))$$
$$(\cosh(x\cos\theta)\cos(x\sin\theta) - i\sinh(x\cos\theta)\sin(x\sin\theta)) d\theta$$

$$= \int_0^{\pi} \left(\cosh^2(x\cos\theta)\cos^2(x\sin\theta) + \sinh^2(x\cos\theta)\sin^2(x\sin\theta)\right) d\theta.$$
(10.16)

10 Solutions to Special Integrals   295

[4.39] Applying Theorem 4.6 to $f(z) = \cos xz$ for $(x, n) \in \mathbb{R} \times (2\mathbb{N} - 2)$ and using (10.15), we get

$$\pi \sum_{j=0}^{\infty} a_j a_{n+j} = (-1)^{n/2} \pi x^n \sum_{j=0}^{\infty} \frac{x^{4j}}{(2j)!(n+2j)!}$$

$$= \int_0^{\pi} \cos\left(xe^{i\theta}\right) \cos\left(xe^{-i\theta}\right) \cos n\theta \, d\theta$$

$$= \int_0^{\pi} \left(\cos^2(x \cos \theta) \cosh^2(x \sin \theta) + \sin^2(x \cos \theta) \sinh^2(x \sin \theta)\right) \cos n\theta \, d\theta.$$

[4.40] Applying Theorem 4.6 to $f(z) = \cos xz$ for $(x, n) \in \mathbb{R} \times (2\mathbb{N} - 1)$ and using (10.15), we get

$$\pi \sum_{j=0}^{\infty} a_j a_{n+j} = 0$$

$$= \int_0^{\pi} \cos\left(xe^{i\theta}\right) \cos\left(xe^{-i\theta}\right) \cos n\theta \, d\theta$$

$$= \int_0^{\pi} \left(\cos^2(x \cos \theta) \cosh^2(x \sin \theta) + \sin^2(x \cos \theta) \sinh^2(x \sin \theta)\right) \cos n\theta \, d\theta.$$

The above identity is also true for $n \in 2\mathbb{Z} + 1$ because $\cos n\theta$ is an even function.

[4.41] Applying Theorem 4.6 to $f(z) = \cosh xz$ for $(x, n) \in \mathbb{R} \times (2\mathbb{N} - 2)$ and using (10.16), we get

$$\pi \sum_{j=0}^{\infty} a_j a_{n+j} = \pi x^n \sum_{j=0}^{\infty} \frac{x^{4j}}{(2j)!(n+2j)!}$$

$$= \int_0^{\pi} \cosh\left(xe^{i\theta}\right) \cosh\left(xe^{-i\theta}\right) \cos n\theta \, d\theta$$

$$= \int_0^{\pi} \left(\cosh^2(x \cos \theta) \cos^2(x \sin \theta) + \sinh^2(x \cos \theta) \sin^2(x \sin \theta)\right) \cos n\theta \, d\theta.$$

[4.42] Applying Theorem 4.6 to $f(z) = \cosh xz$ for $(x, n) \in \mathbb{R} \times (2\mathbb{N} - 1)$ and using (10.16), we get

$$\pi \sum_{j=0}^{\infty} a_j a_{n+j} = 0$$
$$= \int_0^{\pi} \cosh\left(xe^{i\theta}\right) \cosh\left(xe^{-i\theta}\right) \cos n\theta \, d\theta$$
$$= \int_0^{\pi} \left(\cosh^2(x\cos\theta)\cos^2(x\sin\theta) + \sinh^2(x\cos\theta)\sin^2(x\sin\theta)\right) \cos n\theta \, d\theta.$$

The above identity is also true for $n \in 2\mathbb{Z} + 1$ because $\cos n\theta$ is an even function.

[4.43] Applying Theorem 4.2 to $f(z) = \log(1+z)$ for $n \in \mathbb{N}$, we get

$$\pi a_n = (-1)^{n+1} \frac{\pi}{n}$$
$$= \frac{1}{2} \int_0^{\pi} \left(e^{-in\theta} \log\left(1 + e^{i\theta}\right) + e^{in\theta} \log\left(1 + e^{-i\theta}\right)\right) d\theta$$
$$= \frac{1}{2} \int_0^{\pi} \left(e^{-in\theta} \log\left((1 + \cos\theta) + i\sin\theta\right)\right.$$
$$\left. + e^{in\theta} \log\left((1 + \cos\theta) - i\sin\theta\right)\right) d\theta \qquad (10.17)$$
$$= \frac{1}{2} \int_0^{\pi} \left(e^{-in\theta} \log\left(2\cos^2\frac{\theta}{2} + 2i\cos\frac{\theta}{2}\sin\frac{\theta}{2}\right)\right.$$
$$\left. + e^{in\theta} \log\left(2\cos^2\frac{\theta}{2} - 2i\cos\frac{\theta}{2}\sin\frac{\theta}{2}\right)\right) d\theta$$

$$= \frac{1}{2}\int_0^\pi \left(e^{-in\theta}\left(\log\left(2\cos\frac{\theta}{2}\right) + \log\left(\cos\frac{\theta}{2} + i\sin\frac{\theta}{2}\right)\right)\right.$$
$$\left. + e^{in\theta}\left(\log\left(2\cos\frac{\theta}{2}\right) + \log\left(\cos\frac{\theta}{2} - i\sin\frac{\theta}{2}\right)\right)\right)d\theta$$
$$= \int_0^\pi \left(\frac{e^{-in\theta} + e^{in\theta}}{2}\log\left(2\cos\frac{\theta}{2}\right) + \frac{e^{-in\theta}\log e^{i\theta/2} + e^{in\theta}\log e^{-i\theta/2}}{2}\right)d\theta$$
$$= \int_0^\pi \left(\cos n\theta \log\left(2\cos\frac{\theta}{2}\right)\right.$$
$$\left. + \frac{i\frac{\theta}{2}(\cos n\theta - i\sin n\theta) - i\frac{\theta}{2}(\cos n\theta + i\sin n\theta)}{2}\right)d\theta$$
$$= \int_0^\pi \left(\cos n\theta \log\left(2\cos\frac{\theta}{2}\right) + \frac{\theta}{2}\sin n\theta\right)d\theta.$$

[4.44] Applying Theorem 4.2 to $f(z) = \log(1+z)$ for $n \in -\mathbb{N}\cup\{0\}$ and using (10.17), we get

$$\pi a_n = 0 = \int_0^\pi \left(\cos n\theta \log\left(2\cos\frac{\theta}{2}\right) + \frac{\theta}{2}\sin n\theta\right)d\theta.$$

[4.45] Applying Theorem 4.2 to $f(z) = \log(1 + z/x)$ for $(x, n) \in (\mathbb{R} - (-1, 1)) \times \mathbb{N}$, we get

$$\pi a_n = (-1)^{n+1}\frac{\pi}{nx^n}$$
$$= \frac{1}{2}\int_0^\pi \left(e^{-in\theta}\log\left(1 + \frac{e^{i\theta}}{x}\right) + e^{in\theta}\log\left(1 + \frac{e^{-i\theta}}{x}\right)\right)d\theta \quad (10.18)$$
$$= \frac{1}{2}\int_0^\pi \left(e^{-in\theta}\log\left(1 + \frac{\cos\theta}{x} + \frac{i\sin\theta}{x}\right)\right.$$
$$\left. + e^{in\theta}\log\left(1 + \frac{\cos\theta}{x} - \frac{i\sin\theta}{x}\right)\right)d\theta$$

$$= \frac{1}{2} \int_0^\pi \left(e^{-in\theta} \left(\log \sqrt{\left(1 + \frac{\cos\theta}{x}\right)^2 + \left(\frac{\sin\theta}{x}\right)^2} + \right.\right.$$

$$\left. \log \left(\frac{1 + \frac{\cos\theta}{x} + \frac{i\sin\theta}{x}}{\sqrt{\left(1 + \frac{\cos\theta}{x}\right)^2 + \left(\frac{\sin\theta}{x}\right)^2}} \right) \right)$$

$$+ e^{in\theta} \left(\log \sqrt{\left(1 + \frac{\cos\theta}{x}\right)^2 + \left(\frac{\sin\theta}{x}\right)^2} \right.$$

$$\left.\left. + \log \left(\frac{1 + \frac{\cos\theta}{x} - \frac{i\sin\theta}{x}}{\sqrt{\left(1 + \frac{\cos\theta}{x}\right)^2 + \left(\frac{\sin\theta}{x}\right)^2}} \right) \right) \right) d\theta$$

$$= \int_0^\pi \left(\left(\frac{e^{-in\theta} + e^{in\theta}}{2} \right) \log \sqrt{1 + \frac{1}{x^2} + \frac{2}{x}\cos\theta} \right.$$

$$\left. + \frac{1}{2} \left(e^{-in\theta} \log e^{i\tan^{-1}(\sin\theta/(x+\cos\theta))} + e^{in\theta} \log e^{-i\tan^{-1}(\sin\theta/(x+\cos\theta))} \right) \right) d\theta$$

$$= \int_0^\pi \left(\cos n\theta \log \sqrt{1 + \frac{1}{x^2} + \frac{2}{x}\cos\theta} + \left(\frac{e^{in\theta} - e^{-in\theta}}{2i} \right) \right.$$

$$\left. \tan^{-1}\left(\frac{\sin\theta}{x + \cos\theta} \right) \right) d\theta$$

$$= \int_0^\pi \left(\cos n\theta \log \sqrt{1 + \frac{1}{x^2} + \frac{2}{x}\cos\theta} + \sin n\theta \tan^{-1}\left(\frac{\sin\theta}{x + \cos\theta} \right) \right) d\theta.$$

[4.46] Applying Theorem 4.2 to $f(z) = \log(1 + z/x)$ for $(x, n) \in (\mathbb{R} - (-1, 1)) \times (-\mathbb{N} \cup \{0\})$ and using (10.18), we get

$$\pi a_n = 0$$

$$= \frac{1}{2} \int_0^\pi \left(e^{-in\theta} \log\left(1 + \frac{e^{i\theta}}{x}\right) + e^{in\theta} \log\left(1 + \frac{e^{-i\theta}}{x}\right) \right) d\theta$$

$$= \int_0^\pi \left(\cos n\theta \log \sqrt{1 + \frac{1}{x^2} + \frac{2}{x}\cos\theta} + \sin n\theta \tan^{-1}\left(\frac{\sin\theta}{x + \cos\theta} \right) \right) d\theta.$$

10 Solutions to Special Integrals

[4.47] Applying Theorem 4.2 to $f(z) = (1+z)^y$ for $(y, n) \in \mathbb{R} \times (\mathbb{N} - 1)$, we get

$$\frac{\pi}{2^y} a_n = \frac{\pi}{2^y n!} \prod_{j=0}^{n-1} (y-j)$$

$$= \frac{1}{2^{y+1}} \int_0^\pi \left(e^{-in\theta} \left(1 + e^{i\theta}\right)^y + e^{in\theta} \left(1 + e^{-i\theta}\right)^y \right) d\theta$$

$$= \frac{1}{2^{y+1}} \int_0^\pi \left(e^{-in\theta} ((1+\cos\theta) + i\sin\theta)^y + e^{in\theta} ((1+\cos\theta) - i\sin\theta)^y \right) d\theta$$

$$= \frac{1}{2^{y+1}} \int_0^\pi \left(e^{-in\theta} \left(2\cos^2\frac{\theta}{2} + 2i\cos\frac{\theta}{2}\sin\frac{\theta}{2}\right)^y \right.$$
$$\left. + e^{in\theta} \left(2\cos^2\frac{\theta}{2} - 2i\cos\frac{\theta}{2}\sin\frac{\theta}{2}\right)^y \right) d\theta$$

$$= \frac{1}{2^{y+1}} \int_0^\pi 2^y \cos^y\frac{\theta}{2} \left(e^{-in\theta} \left(\cos\frac{\theta}{2} + i\sin\frac{\theta}{2}\right)^y \right.$$
$$\left. + e^{-in\theta} \left(\cos\frac{\theta}{2} - i\sin\frac{\theta}{2}\right)^y \right) d\theta$$

$$= \frac{1}{2} \int_0^\pi \left(e^{-in\theta} e^{iy\theta/2} + e^{in\theta} e^{-iy\theta/2} \right) \cos^y\frac{\theta}{2} d\theta$$

$$= \frac{1}{2} \int_0^\pi \left(e^{-i(2n-y)\theta/2} + e^{i(2n-y)\theta/2} \right) \cos^y\frac{\theta}{2} d\theta$$

$$= \int_0^\pi \cos\frac{(2n-y)\theta}{2} \cos^y\frac{\theta}{2} d\theta.$$

(10.19)

[4.48] Applying Theorem 4.2 to $f(z) = (1+z)^y$ for $(y, n) \in \mathbb{R} \times (-\mathbb{N})$ and using (10.19), we get

$$\frac{\pi}{2^y} a_n = 0$$

$$= \frac{1}{2^{y+1}} \int_0^\pi \left(e^{-in\theta} \left(1 + e^{i\theta}\right)^y + e^{in\theta} \left(1 + e^{-i\theta}\right)^y \right) d\theta$$

$$= \int_0^\pi \cos\frac{(2n-y)\theta}{2} \cos^y\frac{\theta}{2} d\theta.$$

[4.49] Applying Theorem 4.2 to $f(z) = (1+z/x)^y$ for $(x, y, n) \in (\mathbb{R} - [-1, 1]) \times \mathbb{R} \times \mathbb{N}$, we get

$$\pi a_n = \frac{\pi}{n! x^n} \prod_{j=0}^{n-1} (y-j)$$

$$= \frac{1}{2} \int_0^\pi \left(e^{-in\theta} \left(1 + \frac{e^{i\theta}}{x}\right)^y + e^{in\theta} \left(1 + \frac{e^{-i\theta}}{x}\right)^y \right) d\theta$$

$$= \frac{1}{2} \int_0^\pi \left(e^{-in\theta} \left(1 + \frac{\cos\theta}{x} + \frac{i\sin\theta}{x}\right)^y + e^{in\theta} \left(1 + \frac{\cos\theta}{x} - \frac{i\sin\theta}{x}\right)^y \right) d\theta$$

$$= \frac{1}{2} \int_0^\pi \left(1 + \frac{1}{x^2} + \frac{2}{x}\cos\theta\right)^{y/2} \left(e^{-in\theta} \left(\frac{1 + \frac{\cos\theta}{x} + \frac{i\sin\theta}{x}}{\sqrt{1 + \frac{1}{x^2} + \frac{2}{x}\cos\theta}} \right)^y \right.$$

$$\left. + e^{in\theta} \left(\frac{1 + \frac{\cos\theta}{x} - \frac{i\sin\theta}{x}}{\sqrt{1 + \frac{1}{x^2} + \frac{2}{x}\cos\theta}} \right)^y \right) d\theta$$

$$= \frac{1}{2} \int_0^\pi \left(1 + \frac{1}{x^2} + \frac{2}{x}\cos\theta\right)^{y/2} \left(e^{-in\theta} e^{iy \tan^{-1}(\sin\theta/(x+\cos\theta))} \right.$$

$$\left. + e^{in\theta} e^{-iy \tan^{-1}(\sin\theta/(x+\cos\theta))} \right) d\theta$$

$$= \frac{1}{2} \int_0^\pi \left(1 + \frac{1}{x^2} + \frac{2}{x}\cos\theta\right)^{y/2} \left(e^{-i(n\theta - y \tan^{-1}(\sin\theta/(x+\cos\theta)))} \right.$$

$$\left. + e^{i(n\theta - y \tan^{-1}(\sin\theta/(x+\cos\theta)))} \right) d\theta$$

$$= \int_0^\pi \left(1 + \frac{1}{x^2} + \frac{2}{x}\cos\theta\right)^{y/2} \cos\left(n\theta - \tan^{-1}\left(\frac{\sin\theta}{x+\cos\theta}\right)\right) d\theta. \tag{10.20}$$

[4.50] Applying Theorem 4.2 to $f(z) = (1+z/x)^y$ for $(x, y, n) \in (\mathbb{R} - [-1, 1]) \times \mathbb{R} \times (-\mathbb{N} \cup \{0\})$ and using (10.20), we get

$$\pi a_n = 0$$

$$= \frac{1}{2} \int_0^\pi \left(e^{-in\theta} \left(1 + \frac{e^{i\theta}}{x}\right)^y + e^{in\theta} \left(1 + \frac{e^{-i\theta}}{x}\right)^y \right) d\theta$$

$$= \int_0^\pi \left(1 + \frac{1}{x^2} + \frac{2}{x}\cos\theta\right)^{y/2} \cos\left(n\theta - \tan^{-1}\left(\frac{\sin\theta}{x+\cos\theta}\right)\right) d\theta.$$

[4.51] Putting $n = 0$ in [4.52] and using (2.30), we get

$$\int_0^\pi \left(\log^2 \left(2 \cos \frac{\theta}{2} \right) + \frac{\theta^2}{4} \right) d\theta = \pi \sum_{j=1}^\infty \frac{1}{j^2} = \frac{\pi^3}{6}$$

$$\iff \int_0^\pi \log^2 \left(2 \cos \frac{\theta}{2} \right) d\theta = \frac{\pi^3}{6} - \int_0^\pi \frac{\theta^2}{4} d\theta$$

$$= \frac{\pi^3}{6} - \frac{\theta^3}{12} \Big|_0^\pi$$

$$= \frac{\pi^3}{6} - \frac{\pi^3}{12} = \frac{\pi^3}{12}.$$

(10.21)

Using (10.21), we also have

$$\int_0^\pi \log^2 \left(2 \cos \frac{\theta}{2} \right) d\theta = \int_0^\pi \log^2 \left(2 \cos \left(\frac{\pi - \theta}{2} \right) \right) d\theta$$

$$= \int_0^\pi \log^2 \left(2 \sin \frac{\theta}{2} \right) d\theta$$

$$= \frac{\pi^3}{12}.$$

[4.52] Applying Theorem 4.6 to $f(z) = \log(1+z)$ for $n \in 2\mathbb{N} - 2$, we get

$$\pi \sum_{j=0}^{\infty} a_j a_{n+j} = \pi \sum_{j=1}^{\infty} \frac{1}{j(n+j)}$$

$$= \int_0^{\pi} \log\left(1+e^{i\theta}\right) \log\left(1+e^{-i\theta}\right) \cos n\theta \, d\theta$$

$$= \int_0^{\pi} \log((1+\cos\theta) + i\sin\theta) \log((1+\cos\theta) - i\sin\theta) \cos n\theta \, d\theta$$

$$= \int_0^{\pi} \log\left(2\cos^2\frac{\theta}{2} + 2i\cos\frac{\theta}{2}\sin\frac{\theta}{2}\right)$$

$$\log\left(2\cos^2\frac{\theta}{2} - 2i\cos\frac{\theta}{2}\sin\frac{\theta}{2}\right) \cos n\theta \, d\theta$$

$$= \int_0^{\pi} \left(\log\left(2\cos\frac{\theta}{2}\right) + \log\left(\cos\frac{\theta}{2} + i\sin\frac{\theta}{2}\right)\right)$$

$$\left(\log\left(2\cos\frac{\theta}{2}\right) + \log\left(\cos\frac{\theta}{2} - i\sin\frac{\theta}{2}\right)\right) \cos n\theta \, d\theta$$

$$= \int_0^{\pi} \left(\log\left(2\cos\frac{\theta}{2}\right) + \log e^{i\theta/2}\right) \left(\log\left(2\cos\frac{\theta}{2}\right) + \log e^{-i\theta/2}\right) \cos n\theta \, d\theta$$

$$= \int_0^{\pi} \left(\log\left(2\cos\frac{\theta}{2}\right) + i\frac{\theta}{2}\right) \left(\log\left(2\cos\frac{\theta}{2}\right) - i\frac{\theta}{2}\right) \cos n\theta \, d\theta$$

$$= \int_0^{\pi} \left(\log^2\left(2\cos\frac{\theta}{2}\right) + \frac{\theta^2}{4}\right) \cos n\theta \, d\theta.$$

(10.22)

[4.53] Applying Theorem 4.6 to $f(z) = \log(1+z)$ for $n \in 2\mathbb{N} - 1$ and using (10.22), we get

$$\pi \sum_{j=0}^{\infty} a_j a_{n+j} = -\pi \sum_{j=1}^{\infty} \frac{1}{j(n+j)}$$

$$= \int_0^{\pi} \log\left(1+e^{i\theta}\right) \log\left(1+e^{-i\theta}\right) \cos n\theta \, d\theta$$

$$= \int_0^{\pi} \left(\log^2\left(2\cos\frac{\theta}{2}\right) + \frac{\theta^2}{4}\right) \cos n\theta \, d\theta.$$

[4.54] Put $n = 0$ in [4.55].

[4.55] Applying Theorem 4.6 to $f(z) = \log(1 + z/x)$ for $(x, n) \in [1, \infty) \times (2\mathbb{N} - 2)$, we get

$$\pi \sum_{j=0}^{\infty} a_j a_{n+j} = \frac{\pi}{x^n} \sum_{j=1}^{\infty} \frac{1}{j(n+j)x^{2j}}$$

$$= \int_0^{\pi} \log\left(1 + \frac{e^{i\theta}}{x}\right) \log\left(1 + \frac{e^{-i\theta}}{x}\right) \cos n\theta \, d\theta$$

$$= \int_0^{\pi} \log\left(1 + \frac{\cos\theta}{x} + \frac{i\sin\theta}{x}\right) \log\left(1 + \frac{\cos\theta}{x} - \frac{i\sin\theta}{x}\right) \cos n\theta \, d\theta$$

$$= \int_0^{\pi} \left(\log\sqrt{1 + \frac{1}{x^2} + \frac{2\cos\theta}{x}} + \log\left(\frac{1 + \frac{\cos\theta}{x} + \frac{i\sin\theta}{x}}{\sqrt{1 + \frac{1}{x^2} + \frac{2\cos\theta}{x}}}\right)\right)$$

$$\left(\log\sqrt{1 + \frac{1}{x^2} + \frac{2\cos\theta}{x}} + \log\left(\frac{1 + \frac{\cos\theta}{x} - \frac{i\sin\theta}{x}}{\sqrt{1 + \frac{1}{x^2} + \frac{2\cos\theta}{x}}}\right)\right) \cos n\theta \, d\theta$$

$$= \int_0^{\pi} \left(\log\sqrt{1 + \frac{1}{x^2} + \frac{2\cos\theta}{x}} + \log e^{i\tan^{-1}(\sin\theta/(x+\cos\theta))}\right)$$

$$\left(\log\sqrt{1 + \frac{1}{x^2} + \frac{2\cos\theta}{x}} + \log e^{-i\tan^{-1}(\sin\theta/(x+\cos\theta))}\right) \cos n\theta \, d\theta \qquad (10.23)$$

$$= \int_0^{\pi} \left(\log\sqrt{1 + \frac{1}{x^2} + \frac{2\cos\theta}{x}} + i\tan^{-1}\left(\frac{\sin\theta}{x+\cos\theta}\right)\right)$$

$$\left(\log\sqrt{1 + \frac{1}{x^2} + \frac{2\cos\theta}{x}} - i\tan^{-1}\left(\frac{\sin\theta}{x+\cos\theta}\right)\right) \cos n\theta \, d\theta$$

$$= \int_0^{\pi} \left(\log^2\sqrt{1 + \frac{1}{x^2} + \frac{2}{x}\cos\theta} + \left(\tan^{-1}\left(\frac{\sin\theta}{x+\cos\theta}\right)\right)^2\right) \cos n\theta \, d\theta.$$

[4.56] Applying Theorem 4.6 to $f(z) = \log(1 + z/x)$ for $(x, n) \in [1, \infty) \times (2\mathbb{N} - 1)$ and using (10.23), we get

$$\pi \sum_{j=0}^{\infty} a_j a_{n+j} = -\frac{\pi}{x^n} \sum_{j=1}^{\infty} \frac{1}{j(n+j)x^{2j}}$$

$$= \int_0^{\pi} \log\left(1 + \frac{e^{i\theta}}{x}\right) \log\left(1 + \frac{e^{-i\theta}}{x}\right) \cos n\theta \, d\theta$$

$$= \int_0^{\pi} \left(\log^2 \sqrt{1 + \frac{1}{x^2} + \frac{2}{x}\cos\theta} + \left(\tan^{-1}\left(\frac{\sin\theta}{x + \cos\theta}\right)\right)^2\right) \cos n\theta \, d\theta.$$

[4.57] Put $n = 0$ in [4.58].

[4.58] Applying Theorem 4.6 to $f(z) = (1 + z/x)^y$ for $(x, y, n) \in (1, \infty) \times \mathbb{R} \times (\mathbb{N} - 1)$, we get

$$\pi \sum_{j=0}^{\infty} a_j a_{n+j} = \pi \sum_{j=0}^{\infty} \frac{\prod_{k=0}^{j-1}(y-k)}{j! x^j} \cdot \frac{\prod_{k=0}^{n+j-1}(y-k)}{(n+j)! x^{n+j}}$$

$$= \int_0^{\pi} \left(1 + \frac{e^{i\theta}}{x}\right)^y \left(1 + \frac{e^{-i\theta}}{x}\right)^y \cos n\theta \, d\theta$$

$$= \int_0^{\pi} \left(1 + \frac{\cos\theta}{x} + \frac{i\sin\theta}{x}\right)^y \left(1 + \frac{\cos\theta}{x} - \frac{i\sin\theta}{x}\right)^y \cos n\theta \, d\theta$$

$$= \int_0^{\pi} \left(1 + \frac{1}{x^2} + \frac{2\cos\theta}{x}\right)^y \cos n\theta \, d\theta.$$

[4.59] Replacing n with $(m - n)/2$ in [4.16], we get

$$\int_0^{\pi} \cos\frac{n\theta}{2} \cos^m\frac{\theta}{2} \, d\theta = \frac{\pi}{2^m} \binom{m}{(m-n)/2}. \qquad (10.24)$$

Substituting $\theta = 2x\phi$ in (10.24), we get

$$\int_0^{\pi/2x} 2x \cos(nx\phi) \cos^m(x\phi) \, d\phi = \frac{\pi}{2^m} \binom{m}{(m-n)/2}$$

$$\iff \int_0^{\pi/2x} \cos(nx\phi) \cos^m(x\phi) \, d\phi = \frac{\pi}{2^{m+1}x} \binom{m}{(m-n)/2}. \qquad (10.25)$$

Consider the integral

10 Solutions to Special Integrals

$$\int_{\pi/2x}^{\pi/x} \cos(nx\phi) \cos^m(x\phi)\, d\phi.$$

Substituting $\theta = \phi - \pi/2x$ in this integral, and using (10.25), we get

$$\int_{\pi/2x}^{\pi/x} \cos(nx\phi) \cos^m(x\phi)\, d\phi$$

$$= \int_0^{\pi/2x} \cos\left(nx\left(\theta + \frac{\pi}{2x}\right)\right) \cos^m\left(x\left(\theta + \frac{\pi}{2x}\right)\right) d\theta$$

$$= \int_0^{\pi/2x} \cos\left(nx\theta + \frac{n\pi}{2}\right) \cos^m\left(x\theta + \frac{\pi}{2}\right) d\theta$$

$$= \int_0^{\pi/2x} \cos\left(nx\left(\frac{\pi}{2x} - \theta\right) + \frac{n\pi}{2}\right) \cos^m\left(x\left(\frac{\pi}{2x} - \theta\right) + \frac{\pi}{2}\right) d\theta \quad (10.26)$$

$$= \int_0^{\pi/2x} \cos(n\pi - nx\theta) \cos^m(\pi - x\theta)\, d\theta$$

$$= (-1)^{m+n} \int_0^{\pi/2x} \cos(nx\theta) \cos^m(x\theta)\, d\theta$$

$$= (-1)^{m+n} \frac{\pi}{2^{m+1}x} \binom{m}{(m-n)/2}.$$

Adding (10.25) and (10.26), we get

$$\int_0^{\pi/x} \cos(nx\theta) \cos^m(x\theta)\, d\theta = \left(1 + (-1)^{m+n}\right) \int_0^{\pi/2x} \cos(nx\theta) \cos^m(x\theta)\, d\theta$$

$$= \left(1 + (-1)^{m+n}\right) \frac{\pi}{2^{m+1}x} \binom{m}{(m-n)/2}$$

$$= \frac{\pi}{2^m x} \binom{m}{(m-n)/2},$$

(10.27)

for $(x, m, n) \in (\mathbb{R} - \{0\}) \times (2\mathbb{N} - 2) \times ((2\mathbb{N} - 2) \cap [0..m]) \cup (\mathbb{R} - \{0\}) \times (2\mathbb{N} - 1) \times ((2\mathbb{N} - 1) \cap [0..m])$.

[4.60] Using (10.27) for $(x, m, n) \in (\mathbb{R} - \{0\}) \times (2\mathbb{N} - 2) \times (2\mathbb{N} - 1) \cup (\mathbb{R} - \{0\}) \times (2\mathbb{N} - 1) \times (2\mathbb{N} - 2)$, we get

$$\int_0^{\pi/x} \cos(nx\theta) \cos^m(x\theta)\, d\theta = 0.$$

[4.61] Applying Theorem 4.8 to $f(z) = \sin^{-1} z$ and using [2.8] for $(x, n) \in \left[-\frac{1}{2}, \frac{1}{2}\right] \times (2\mathbb{N} - 1)$, we get

$$\int_0^\pi \sin^{-1}(2x\cos\theta) \cos n\theta \, d\theta = \pi A_n(x)$$

$$= \pi \sum_{j=0}^\infty \binom{n+2j}{j} a_{n+2j} x^{n+2j}$$

$$= \pi \sum_{j=0}^\infty \binom{n+2j}{j} \binom{n+2j-1}{(n+2j-1)/2} \frac{x^{n+2j}}{(n+2j)2^{n+2j-1}}.$$

[4.62] Applying Theorem 4.8 to $f(z) = \sin^{-1} z$ for $(x, n) \in \left[-\frac{1}{2}, \frac{1}{2}\right] \times (2\mathbb{N} - 2)$, we get

$$\int_0^\pi \sin^{-1}(2x\cos\theta) \cos n\theta \, d\theta = \pi A_n(x)$$

$$= \pi \sum_{j=0}^\infty \binom{n+2j}{j} a_{n+2j} x^{n+2j}$$

$$= 0.$$

The above identity is also true for $n \in 2\mathbb{Z}$ because $\cos n\theta$ is an even function.

[4.63] Applying Theorem 4.13 to $f(z) = \sin^{-1} z$ for $(x, n) \in \left[-\frac{1}{2}, \frac{1}{2}\right] \times (2\mathbb{N} - 1)$, and using [2.8], we get

$$\int_0^\pi -i \sin^{-1}(2ix\sin\theta) \sin n\theta \, d\theta = \int_0^\pi \sinh^{-1}(2x\sin\theta) \sin n\theta \, d\theta$$

$$= \pi B_n(x)$$

$$= \pi \sum_{j=0}^\infty (-1)^j \binom{n+2j}{j} a_{n+2j} x^{n+2j}$$

$$= \pi \sum_{j=0}^\infty (-1)^j \binom{n+2j}{j} \binom{n+2j-1}{(n+2j-1)/2} \frac{x^{n+2j}}{(n+2j)2^{n+2j-1}}. \quad (10.28)$$

10 Solutions to Special Integrals 307

[4.64] Applying Theorem 4.13 to $f(z) = \sin^{-1} z$ for $(x, n) \in \mathbb{R} \times (2\mathbb{N} - 2)$, and using [2.8] and (10.28), we get

$$\int_0^\pi \sinh^{-1}(2x \sin \theta) \sin n\theta \, d\theta = \pi B_n(x)$$

$$= \pi \sum_{j=0}^\infty (-1)^j \binom{n+2j}{j} a_{n+2j} x^{n+2j}$$

$$= 0.$$

The above identity is also true for $n \in 2\mathbb{Z}$ because $\sin n\theta$ is an odd function.

[4.65] Applying Theorem 4.8 to $f(z) = \sinh^{-1} z$ for $(x, n) \in \left[-\frac{1}{2}, \frac{1}{2}\right] \times (2\mathbb{N} - 1)$, and using [2.34], we get

$$\int_0^\pi \sinh^{-1}(2x \cos \theta) \cos n\theta \, d\theta = \pi A_n(x)$$

$$= \pi \sum_{j=0}^\infty \binom{n+2j}{j} a_{n+2j} x^{n+2j}$$

$$= \pi \sum_{j=0}^\infty (-1)^{(n+2j-1)/2} \binom{n+2j}{j} \binom{n+2j-1}{(n+2j-1)/2} \frac{x^{n+2j}}{(n+2j)2^{n+2j-1}}$$

$$= (-1)^{(n-1)/2} \pi \sum_{j=0}^\infty (-1)^j \binom{n+2j}{j} \binom{n+2j-1}{(n+2j-1)/2} \frac{x^{n+2j}}{(n+2j)2^{n+2j-1}}.$$

[4.66] Applying Theorem 4.8 to $f(z) = \sinh^{-1} z$ for $(x, n) \in \mathbb{R} \times (2\mathbb{N} - 2)$, and using [2.34], we get

$$\int_0^\pi \sinh^{-1}(2x \cos \theta) \cos n\theta \, d\theta = \pi A_n(x)$$

$$= \pi \sum_{j=0}^\infty \binom{n+2j}{j} a_{n+2j} x^{n+2j}$$

$$= 0.$$

The above identity is also true for $n \in 2\mathbb{Z}$ because $\cos n\theta$ is an even function.

[4.67] Applying Theorem 4.13 to $f(z) = \sinh^{-1} z$ for $(x, n) \in \left[-\frac{1}{2}, \frac{1}{2}\right] \times (2\mathbb{N} - 1)$, and using [2.34], we get

$$\int_0^\pi -i \sinh^{-1}(2ix \sin \theta) \sin n\theta \, d\theta = \int_0^\pi \sin^{-1}(2x \sin \theta) \sin n\theta \, d\theta$$
$$= \pi B_n(x)$$
$$= \pi \sum_{j=0}^\infty (-1)^j \binom{n+2j}{j} a_{n+2j} x^{n+2j}$$
$$= \pi \sum_{j=0}^\infty (-1)^j (-1)^{(n+2j-1)/2} \binom{n+2j}{j} \binom{n+2j-1}{(n+2j-1)/2} \frac{x^{n+2j}}{(n+2j)2^{n+2j-1}}$$
$$= (-1)^{(n-1)/2} \pi \sum_{j=0}^\infty \binom{n+2j}{j} \binom{n+2j-1}{(n+2j-1)/2} \frac{x^{n+2j}}{(n+2j)2^{n+2j-1}}.$$
(10.29)

[4.68] Applying Theorem 4.13 to $f(z) = \sinh^{-1} z$ for $(x, n) \in \left[-\frac{1}{2}, \frac{1}{2}\right] \times (2\mathbb{N} - 2)$ and using [2.34] and (10.29), we get

$$\int_0^\pi \sin^{-1}(2x \sin \theta) \sin n\theta \, d\theta = \pi B_n(x)$$
$$= \pi \sum_{j=0}^\infty (-1)^j \binom{n+2j}{j} a_{n+2j} x^{n+2j}$$
$$= 0.$$

The above identity is also true for $n \in 2\mathbb{Z}$ because $\sin n\theta$ is an odd function.

[4.69] Applying Theorem 4.8 to $f(z) = \tan^{-1} z$ for $(x, n) \in \left[-\frac{1}{2}, \frac{1}{2}\right] \times (2\mathbb{N} - 1)$, and using (1.13), we get

$$\int_0^\pi \tan^{-1}(2x \cos \theta) \cos n\theta \, d\theta = \pi A_n(x)$$
$$= \pi \sum_{j=0}^\infty \binom{n+2j}{j} a_{n+2j} x^{n+2j}$$
$$= \pi \sum_{j=0}^\infty (-1)^{(n+2j-1)/2} \binom{n+2j}{j} \frac{x^{n+2j}}{n+2j}$$
$$= (-1)^{(n-1)/2} \pi \sum_{j=0}^\infty (-1)^j \binom{n+2j}{j} \frac{x^{n+2j}}{n+2j}.$$

10 Solutions to Special Integrals

[4.70] Applying Theorem 4.8 to $f(z) = \tan^{-1} z$ for $(x, n) \in \mathbb{R} \times (2\mathbb{N} - 2)$, and using (1.13), we get

$$\int_0^\pi \tan^{-1}(2x \cos\theta) \cos n\theta \, d\theta = \pi A_n(x)$$

$$= \pi \sum_{j=0}^\infty \binom{n+2j}{j} a_{n+2j} x^{n+2j}$$

$$= 0.$$

The above identity is also true for $n \in 2\mathbb{Z}$ because $\cos n\theta$ is an even function.

[4.71] Applying Theorem 4.13 to $f(z) = \tan^{-1} z$ for $(x, n) \in \left[-\frac{1}{2}, \frac{1}{2}\right] \times (2\mathbb{N} - 1)$, and using (1.13), we get

$$\int_0^\pi -i \tan^{-1}(2ix \sin\theta) \sin n\theta \, d\theta = \int_0^\pi \tanh^{-1}(2x \sin\theta) \sin n\theta \, d\theta$$

$$= \pi B_n(x)$$

$$= \pi \sum_{j=0}^\infty (-1)^j \binom{n+2j}{j} a_{n+2j} x^{n+2j}$$

$$= \pi \sum_{j=0}^\infty (-1)^j (-1)^{(n+2j-1)/2} \binom{n+2j}{j} \frac{x^{n+2j}}{n+2j}$$

$$= (-1)^{(n-1)/2} \pi \sum_{j=0}^\infty \binom{n+2j}{j} \frac{x^{n+2j}}{n+2j}.$$

(10.30)

[4.72] Applying Theorem 4.13 to $f(z) = \tan^{-1} z$ for $(x, n) \in \mathbb{R} \times (2\mathbb{N} - 2)$, and using (1.13) and (10.30), we get

$$\int_0^\pi \tanh^{-1}(2x \sin\theta) \sin n\theta \, d\theta = \pi B_n(x)$$

$$= \pi \sum_{j=0}^\infty (-1)^j \binom{n+2j}{j} a_{n+2j} x^{n+2j}$$

$$= 0.$$

The above identity is also true for $n \in 2\mathbb{Z}$ because $\sin n\theta$ is an odd function.

[4.73] Applying Theorem 4.8 to $f(z) = \tanh^{-1} z$ for $(x, n) \in \left[-\frac{1}{2}, \frac{1}{2}\right] \times (2\mathbb{N} - 1)$, and using (1.15), we get

$$\int_0^\pi \tanh^{-1}(2x \cos\theta) \cos n\theta \, d\theta = \pi A_n(x)$$

$$= \pi \sum_{j=0}^\infty \binom{n+2j}{j} a_{n+2j} x^{n+2j}$$

$$= \pi \sum_{j=0}^\infty \binom{n+2j}{j} \frac{x^{n+2j}}{n+2j}.$$

[4.74] Applying Theorem 4.8 to $f(z) = \tanh^{-1} z$ for $(x, n) \in \mathbb{R} \times (2\mathbb{N} - 2)$, and using (1.15), we get

$$\int_0^\pi \tanh^{-1}(2x \cos\theta) \cos n\theta \, d\theta = \pi A_n(x)$$

$$= \pi \sum_{j=0}^\infty \binom{n+2j}{j} a_{n+2j} x^{n+2j}$$

$$= 0.$$

The above identity is also true for $n \in 2\mathbb{Z}$ because $\cos n\theta$ is an even function.

[4.75] Applying Theorem 4.13 to $f(z) = \tanh^{-1} z$ for $(x, n) \in \left[-\frac{1}{2}, \frac{1}{2}\right] \times (2\mathbb{N} - 1)$, and using (1.15), we get

$$\int_0^\pi -i \tanh^{-1}(2ix \sin\theta) \sin n\theta \, d\theta = \int_0^\pi \tan^{-1}(2x \sin\theta) \sin n\theta \, d\theta$$

$$= \pi B_n(x)$$

$$= \pi \sum_{j=0}^\infty (-1)^j \binom{n+2j}{j} a_{n+2j} x^{n+2j}$$

$$= \pi \sum_{j=0}^\infty (-1)^j \binom{n+2j}{j} \frac{x^{n+2j}}{n+2j}.$$

(10.31)

[4.76] Applying Theorem 4.13 to $f(z) = \tanh^{-1} z$ for $(x, n) \in \mathbb{R} \times (2\mathbb{N} - 2)$, and using (1.15) and (10.31), we get

$$\int_0^\pi \tan^{-1}(2x \sin \theta) \sin n\theta \, d\theta = \pi B_n(x)$$

$$= \pi \sum_{j=0}^\infty (-1)^j \binom{n+2j}{j} a_{n+2j} x^{n+2j}$$

$$= 0.$$

The above identity is also true for $n \in 2\mathbb{Z}$ because $\sin n\theta$ is an odd function.

[4.77] Applying Theorem 4.8 to $f(z) = \sin z$ for $(x, n) \in \mathbb{R} \times (2\mathbb{N} - 1)$, and using (5.2) and (3.8), we get

$$\int_0^\pi \sin(2x \cos \theta) \cos n\theta \, d\theta = \pi A_n(x)$$

$$= \pi \sum_{j=0}^\infty \binom{n+2j}{j} a_{n+2j} x^{n+2j}$$

$$= \pi \sum_{j=0}^\infty (-1)^{(n+2j-1)/2} \binom{n+2j}{j} \frac{x^{n+2j}}{(n+2j)!}$$

$$= (-1)^{(n-1)/2} \pi x^n \sum_{j=0}^\infty (-1)^j \frac{x^{2j}}{j!(n+j)!}$$

$$= (-1)^{(n-1)/2} \pi J_n(2x).$$

[4.78] Applying Theorem 4.8 to $f(z) = \sin z$ for $(x, n) \in \mathbb{R} \times (2\mathbb{N} - 2)$, and using (5.2), we get

$$\int_0^\pi \sin(2x \cos \theta) \cos n\theta \, d\theta = \pi A_n(x)$$

$$= \pi \sum_{j=0}^\infty \binom{n+2j}{j} a_{n+2j} x^{n+2j}$$

$$= 0.$$

The above identity is also true for $n \in 2\mathbb{Z}$ because $\cos n\theta$ is an even function.

[4.79] Applying Theorem 4.13 to $f(z) = \sin z$ for $(x, n) \in \mathbb{R} \times (2\mathbb{N} - 1)$, and using (5.2) and (3.19), we get

$$\int_0^\pi -i \sin(2ix \sin\theta) \sin n\theta \, d\theta = \int_0^\pi \sinh(2x \sin\theta) \sin n\theta \, d\theta$$

$$= \pi \sum_{j=0}^\infty (-1)^j \binom{n+2j}{j} a_{n+2j} x^{n+2j}$$

$$= \pi B_n(x)$$

$$= \pi \sum_{j=0}^\infty (-1)^j (-1)^{(n+2j-1)/2} \binom{n+2j}{j} \frac{x^{n+2j}}{(n+2j)!} \qquad (10.32)$$

$$= (-1)^{(n-1)/2} x^n \pi \sum_{j=0}^\infty \frac{x^{2j}}{j!(n+j)!}$$

$$= (-1)^{(n-1)/2} \pi I_n(2x).$$

[4.80] Applying Theorem 4.13 to $f(z) = \sin z$ for $(x, n) \in \mathbb{R} \times (2\mathbb{N} - 2)$, and using (10.32) and (5.2), we get

$$\int_0^\pi \sinh(2x \sin\theta) \sin n\theta \, d\theta = \pi B_n(x)$$

$$= \pi \sum_{j=0}^\infty (-1)^j \binom{n+2j}{j} a_{n+2j} x^{n+2j}$$

$$= 0.$$

The above identity is also true for $n \in 2\mathbb{Z}$ because $\sin n\theta$ is an odd function.

[4.81] Applying Theorem 4.8 to $f(z) = \sinh z$ for $(x, n) \in \mathbb{R} \times (2\mathbb{N} - 1)$, and using (5.4) and (3.19), we get

$$\int_0^\pi \sinh(2x \cos\theta) \cos n\theta \, d\theta = \pi A_n(x)$$

$$= \pi \sum_{j=0}^\infty \binom{n+2j}{j} a_{n+2j} x^{n+2j}$$

$$= \pi \sum_{j=0}^\infty \binom{n+2j}{j} \frac{x^{n+2j}}{(n+2j)!}$$

$$= \pi x^n \sum_{j=0}^\infty \frac{x^{2j}}{j!(n+j)!}$$

$$= \pi I_n(2x).$$

[4.82] Applying Theorem 4.8 to $f(z) = \sinh z$ for $(x, n) \in \mathbb{R} \times (2\mathbb{N} - 2)$, and using (5.4), we get

$$\int_0^\pi \sinh(2x \cos\theta) \cos n\theta \, d\theta = \pi A_n(x)$$

$$= \pi \sum_{j=0}^\infty \binom{n+2j}{j} a_{n+2j} x^{n+2j}$$

$$= 0.$$

The above identity is also true for $n \in 2\mathbb{Z}$ because $\cos n\theta$ is an even function.

[4.83] Applying Theorem 4.13 to $f(z) = \sinh z$ for $(x, n) \in \mathbb{R} \times (2\mathbb{N} - 1)$, and using (5.4) and (3.8), we get

$$\int_0^\pi -i \sinh(2ix \sin\theta) \sin n\theta \, d\theta = \int_0^\pi \sin(2x \sin\theta) \sin n\theta \, d\theta$$

$$= \pi B_n(x)$$

$$= \pi \sum_{j=0}^\infty (-1)^j \binom{n+2j}{j} a_{n+2j} x^{n+2j}$$

$$= \pi \sum_{j=0}^\infty (-1)^j \binom{n+2j}{j} \frac{x^{n+2j}}{(n+2j)!} \quad (10.33)$$

$$= \pi x^n \sum_{j=0}^\infty (-1)^j \frac{x^{2j}}{j!(n+j)!}$$

$$= \pi J_n(2x).$$

[4.84] Applying Theorem 4.13 to $f(z) = \sinh z$ for $(x, n) \in \mathbb{R} \times (2\mathbb{N} - 2)$, and using (10.33) and (5.4), we get

$$\int_0^\pi \sin(2x \sin\theta) \sin n\theta \, d\theta = \pi B_n(x)$$

$$= \pi \sum_{j=0}^\infty (-1)^j \binom{n+2j}{j} a_{n+2j} x^{n+2j}$$

$$= 0.$$

The above identity is also true for $n \in 2\mathbb{Z}$ because $\sin n\theta$ is an odd function.

[4.85] Applying Theorem 4.8 to $f(z) = \cos z$ for $(x, n) \in \mathbb{R} \times (2\mathbb{N} - 2)$, and using (5.3) and (3.8), we get

$$\int_0^\pi \cos(2x\cos\theta)\cos n\theta\, d\theta = \pi A_n(x)$$

$$= \pi \sum_{j=0}^\infty \binom{n+2j}{j} a_{n+2j} x^{n+2j}$$

$$= \pi \sum_{j=0}^\infty (-1)^{(n+2j)/2} \binom{n+2j}{j} \frac{x^{n+2j}}{(n+2j)!}$$

$$= (-1)^{n/2} \pi x^n \sum_{j=0}^\infty \frac{(-1)^j x^{2j}}{j!(n+j)!}$$

$$= (-1)^{n/2} \pi J_n(2x).$$

[4.86] Applying Theorem 4.8 to $f(z) = \cos z$ for $(x, n) \in \mathbb{R} \times (2\mathbb{N} - 1)$, and using (5.3), we get

$$\int_0^\pi \cos(2x\cos\theta)\cos n\theta\, d\theta = \pi A_n(x)$$

$$= \pi \sum_{j=0}^\infty \binom{n+2j}{j} a_{n+2j} x^{n+2j}$$

$$= 0.$$

The above identity is also true for $n \in 2\mathbb{Z}+1$ because $\cos n\theta$ is an even function.

[4.87] Applying Theorem 4.11 to $f(z) = \cos z$ for $(x, n) \in \mathbb{R} \times (2\mathbb{N} - 2)$, and using (5.3) and (3.19), we get

$$\int_0^\pi \cos(2ix\sin\theta)\cos n\theta\, d\theta = \int_0^\pi \cosh(2x\sin\theta)\cos n\theta\, d\theta$$

$$= \pi B_n(x)$$

$$= \pi \sum_{j=0}^\infty (-1)^j \binom{n+2j}{j} a_{n+2j} x^{n+2j}$$

$$= \pi \sum_{j=0}^\infty (-1)^j (-1)^{(n+2j)/2} \binom{n+2j}{j} \frac{x^{n+2j}}{(n+2j)!}$$

$$= (-1)^{n/2} \pi x^n \sum_{j=0}^\infty \frac{x^{2j}}{j!(n+j)!}$$

$$= (-1)^{n/2} \pi I_n(2x). \tag{10.34}$$

10 Solutions to Special Integrals

[4.88] Applying Theorem 4.11 to $f(z) = \cos z$ for $(x, n) \in \mathbb{R} \times (2\mathbb{N} - 1)$, and using (10.34) and (5.3), we get

$$\int_0^\pi \cosh(2x \sin\theta) \cos n\theta \, d\theta = \pi B_n(x)$$

$$= \pi \sum_{j=0}^\infty (-1)^j \binom{n+2j}{j} a_{n+2j} x^{n+2j}$$

$$= 0.$$

The above identity is also true for $n \in 2\mathbb{Z} + 1$ because $\cos n\theta$ is an even function.

[4.89] Applying Theorem 4.8 to $f(z) = \cosh z$ for $(x, n) \in \mathbb{R} \times (2\mathbb{N} - 2)$, and using (5.5) and (3.19), we get

$$\int_0^\pi \cosh(2x \cos\theta) \cos n\theta \, d\theta = \pi A_n(x)$$

$$= \pi \sum_{j=0}^\infty \binom{n+2j}{j} a_{n+2j} x^{n+2j}$$

$$= \pi \sum_{j=0}^\infty \binom{n+2j}{j} \frac{x^{n+2j}}{(n+2j)!}$$

$$= \pi x^n \sum_{j=0}^\infty \frac{x^{2j}}{j!(n+j)!}$$

$$= \pi I_n(2x).$$

[4.90] Applying Theorem 4.8 to $f(z) = \cosh z$ for $(x, n) \in \mathbb{R} \times (2\mathbb{N} - 1)$, and using (5.5), we get

$$\int_0^\pi \cosh(2x \cos\theta) \cos n\theta \, d\theta = \pi A_n(x)$$

$$= \pi \sum_{j=0}^\infty \binom{n+2j}{j} a_{n+2j} x^{n+2j}$$

$$= 0.$$

The above identity is also true for $n \in 2\mathbb{Z} + 1$ because $\cos n\theta$ is an even function.

[4.91] Applying Theorem 4.11 to $f(z) = \cosh z$ for $(x, n) \in \mathbb{R} \times (2\mathbb{N} - 2)$, and using (5.5) and (3.8), we get

$$\int_0^\pi \cosh(2ix \sin\theta) \cos n\theta \, d\theta = \int_0^\pi \cos(2x \sin\theta) \cos n\theta \, d\theta$$

$$= \pi B_n(x)$$

$$= \pi \sum_{j=0}^\infty (-1)^j \binom{n+2j}{j} a_{n+2j} x^{n+2j}$$

$$= \pi \sum_{j=0}^\infty (-1)^j \binom{n+2j}{j} \frac{x^{n+2j}}{(n+2j)!} \quad (10.35)$$

$$= \pi x^n \sum_{j=0}^\infty (-1)^j \frac{x^{2j}}{j!(n+j)!}$$

$$= \pi J_n(2x).$$

[4.92] Applying Theorem 4.11 to $f(z) = \cosh z$ for $(x, n) \in \mathbb{R} \times (2\mathbb{N} - 1)$, and using (10.35) and (5.5), we get

$$\int_0^\pi \cos(2x \sin\theta) \cos n\theta \, d\theta = \pi B_n(x)$$

$$= \pi \sum_{j=0}^\infty (-1)^j \binom{n+2j}{j} a_{n+2j} x^{n+2j}$$

$$= 0.$$

The above identity is also true for $n \in 2\mathbb{Z} + 1$ because $\cos n\theta$ is an even function.

[4.93] Put $n = 0$ in [4.94].

[4.94] Applying Theorem 4.8 to $f(z) = (y+z)^m$ for $(x, y, m, n) \in \mathbb{R}^2 \times (\mathbb{N}-1) \times [0..m]$, and using the binomial theorem, we get

$$\int_0^\pi (y + 2x\cos\theta)^m \cos n\theta \, d\theta = \pi A_n(x)$$

$$= \pi \sum_{j=0}^\infty \binom{n+2j}{j} a_{n+2j} x^{n+2j}$$

$$= \pi \sum_{j=0}^{\lfloor (m-n)/2 \rfloor} \binom{n+2j}{j} \binom{m}{n+2j} y^{m-n-2j} x^{n+2j}.$$

10 Solutions to Special Integrals   317

[4.95] Applying Theorem 4.8 to $f(z) = (y+z)^m$ for $(x, y, m, n) \in \mathbb{R}^2 \times (\mathbb{N} - 1) \times ((\mathbb{N}-1) - [0..m])$, and using the binomial theorem, we get

$$\int_0^\pi (y + 2x\cos\theta)^m \cos n\theta \, d\theta = \pi A_n(x)$$

$$= \pi \sum_{j=0}^\infty \binom{n+2j}{j} a_{n+2j} x^{n+2j}$$

$$= 0.$$

The above identity is also true for $n \in 2\mathbb{Z} + 1$ because $\cos n\theta$ is an even function.

[4.96] *First Method:* Put $n = 0$ in [4.6].
Second Method: Applying Theorem 4.1 to $f(z) = e^{xz}$, and using (5.1), we get

$$\pi a_0 = \pi$$

$$= \int_0^\pi \frac{e^{xe^{i\theta}} + e^{xe^{-i\theta}}}{2} d\theta$$

$$= \int_0^\pi \frac{e^{x(\cos\theta + i\sin\theta)} + e^{x(\cos\theta - i\sin\theta)}}{2} d\theta \qquad (10.36)$$

$$= \int_0^\pi e^{x\cos\theta} \frac{e^{ix\sin\theta} + e^{-ix\sin\theta}}{2} d\theta$$

$$= \int_0^\pi e^{x\cos\theta} \cos(x\sin\theta) \, d\theta.$$

[4.97] Applying Theorem 4.4 to $f(z) = e^{xz}$ for $(x, n) \in \mathbb{R} \times \mathbb{N}$, and using (10.36) and (5.1), we get

$$\frac{\pi}{2} a_n = \frac{\pi}{2n!} x^n$$

$$= \int_0^\pi \frac{e^{xe^{i\theta}} + e^{xe^{-i\theta}}}{2} \cos n\theta \, d\theta \qquad (10.37)$$

$$= \int_0^\pi e^{x\cos\theta} \cos(x\sin\theta) \cos n\theta \, d\theta.$$

[4.98] Using [4.6] for $(x, n) \in \mathbb{R} \times \mathbb{N}$, and (10.37), we get

$$\frac{\pi}{n!}x^n = \int_0^\pi e^{x\cos\theta} \cos(n\theta - x\sin\theta)\, d\theta$$

$$= \int_0^\pi e^{x\cos\theta} \cos(x\sin\theta) \cos n\theta\, d\theta + \int_0^\pi e^{x\cos\theta} \sin(x\sin\theta) \sin n\theta\, d\theta$$

$$= \frac{\pi}{2n!}x^n + \int_0^\pi e^{x\cos\theta} \sin(x\sin\theta) \sin n\theta\, d\theta$$

$$\iff \int_0^\pi e^{x\cos\theta} \sin(x\sin\theta) \sin n\theta\, d\theta = \frac{\pi}{2n!}x^n.$$

[4.99] Replace n with $n - m$ in [4.6].

[4.100] Replace n with $n - m$ in [4.7].

[4.101] Put $x = 1$ in [4.102].

[4.102] Put $n = 0$ in [4.46].

[4.103] Put $x = 1$ in [4.104].

[4.104] Using [4.45] for $(x, n) \in (\mathbb{R} - (-1, 1)) \times \mathbb{N}$, we have

$$\int_0^\pi \left(\cos n\theta \log\left(\sqrt{1 + \frac{1}{x^2} + \frac{2}{x}\cos\theta}\right) + \sin n\theta \tan^{-1}\left(\frac{\sin\theta}{x + \cos\theta}\right) \right) d\theta$$

$$= (-1)^{n+1} \frac{\pi}{nx^n}. \tag{10.38}$$

Using [4.46] for $(x, n) \in (\mathbb{R} - (-1, 1)) \times \mathbb{N}$, we have

$$\int_0^\pi \left(\cos n\theta \log\left(\sqrt{1 + \frac{1}{x^2} + \frac{2}{x}\cos\theta}\right) - \sin n\theta \tan^{-1}\left(\frac{\sin\theta}{x + \cos\theta}\right) \right) d\theta = 0. \tag{10.39}$$

Taking the average of (10.38) and (10.39), we get

$$\int_0^\pi \cos n\theta \log\sqrt{1 + \frac{1}{x^2} + \frac{2}{x}\cos\theta}\, d\theta = (-1)^{n+1} \frac{\pi}{2nx^n}.$$

10 Solutions to Special Integrals   319

[4.105] Subtracting (10.39) from (10.38), and dividing the result by 2, we get

$$\int_0^\pi \sin n\theta \tan^{-1}\left(\frac{\sin\theta}{x+\cos\theta}\right) d\theta = (-1)^{n+1}\frac{\pi}{2nx^n}.$$

[4.106] Applying Theorem 4.7 to $f(z) = \log(1+z)$ for $x \in \left(-\frac{1}{2},\frac{1}{2}\right)$, and using the logarithmic series, we get

$$\int_0^\pi \log(1+2x\cos\theta)\,d\theta = \pi A_0(x)$$

$$= \pi \sum_{j=0}^\infty \binom{2j}{j} a_{2j} x^{2j}$$

$$= -\pi \sum_{j=1}^\infty \binom{2j}{j} \frac{x^{2j}}{2j}.$$

[4.107] Applying Theorem 4.8 to $f(z) = \log(1+z)$ for $(x,n) \in \left(-\frac{1}{2},\frac{1}{2}\right) \times \mathbb{N}$, and using the logarithmic series, we get

$$\int_0^\pi \log(1+2x\cos\theta)\cos n\theta\,d\theta = \pi A_n(x)$$

$$= \pi \sum_{j=0}^\infty \binom{n+2j}{j} a_{n+2j} x^{n+2j}$$

$$= \pi \sum_{j=0}^\infty (-1)^{n+1+2j} \binom{n+2j}{j} \frac{x^{n+2j}}{n+2j}$$

$$= (-1)^{n+1}\pi \sum_{j=0}^\infty \binom{n+2j}{j} \frac{x^{n+2j}}{n+2j}.$$

[4.108] Applying Theorem 4.7 to $f(z) = (1+z)^y$ for $(x,y) \in \left(-\frac{1}{2},\frac{1}{2}\right) \times \mathbb{R}$, and using the binomial theorem, we get

$$\int_0^\pi (1+2x\cos\theta)^y \, d\theta = \pi A_0(x)$$

$$= \pi \sum_{j=0}^\infty \binom{2j}{j} a_{2j} x^{2j}$$

$$= \pi \sum_{j=0}^\infty \binom{2j}{j} \frac{\prod_{k=0}^{2j-1}(y-k)}{(2j)!} x^{2j}.$$

[4.109] Applying Theorem 4.8 to $f(z) = (1+z)^y$ for $(x, y, n) \in \left(-\frac{1}{2}, \frac{1}{2}\right) \times \mathbb{R} \times \mathbb{N}$, and using the binomial theorem, we get

$$\int_0^\pi (1+2x\cos\theta)^y \cos n\theta \, d\theta = \pi A_n(x)$$

$$= \pi \sum_{j=0}^\infty \binom{n+2j}{j} a_{n+2j} x^{n+2j}$$

$$= \pi \sum_{j=0}^\infty \binom{n+2j}{j} \frac{\prod_{k=0}^{n+2j-1}(y-k)}{(n+2j)!} x^{n+2j}.$$

[4.110] Applying Theorem 4.17 to $f(z) = z^m$ for $(x, y, m) \in \mathbb{R} \times (\mathbb{R} - \{-x\}) \times (2\mathbb{N} - 2)$ and $n = 0$, we get

$$\frac{\pi}{2} T_0(x, y) = \pi \sum_{j=0}^\infty \binom{2j}{j} a_{2j} (xy)^j$$

$$= \pi \binom{m}{m/2} (xy)^{m/2}$$

$$= \int_0^\pi \frac{\left(xe^{i\theta} + ye^{-i\theta}\right)^m + \left(xe^{-i\theta} + ye^{i\theta}\right)^m}{2} \, d\theta$$

$$= \int_0^\pi \frac{1}{2} \left(((x+y)\cos\theta + i(x-y)\sin\theta)^m + ((x+y)\cos\theta - i(x-y)\sin\theta)^m\right) d\theta$$

$$= \int_0^\pi \frac{1}{2} \left(x^2 + y^2 + 2xy\cos 2\theta\right)^{m/2} \left(\left(\frac{(x+y)\cos\theta + i(x-y)\sin\theta}{\sqrt{x^2+y^2+2xy\cos 2\theta}}\right)^m\right.$$

$$+ \left(\frac{(x+y)\cos\theta - i(x-y)\sin\theta}{\sqrt{x^2+y^2+2xy\cos 2\theta}} \right)^m \right) d\theta$$

$$= \int_0^\pi (x^2+y^2+2xy\cos 2\theta)^{m/2} \frac{e^{im\phi(x,y,\theta)} + e^{-im\phi(x,y,\theta)}}{2} d\theta$$

$$= \int_0^\pi (x^2+y^2+2xy\cos 2\theta)^{m/2} \cos m\phi(x,y,\theta) \, d\theta, \tag{10.40}$$

where the function $\phi(x,y,\theta)$ is as defined in (4.48).

[4.111] Applying Theorem 4.17 to $f(z) = z^m$ for $(x, y, m) \in \mathbb{R} \times (\mathbb{R} - \{-x\}) \times (2\mathbb{N} - 1)$, $n = 0$, and using (10.40), we get

$$\frac{\pi}{2} T_0(x, y) = \pi \sum_{j=0}^\infty \binom{2j}{j} a_{2j}(xy)^j$$

$$= 0$$

$$= \int_0^\pi (x^2+y^2+2xy\cos 2\theta)^{m/2} \cos m\phi(x,y,\theta) \, d\theta,$$

where the function $\phi(x,y,\theta)$ is as defined in (4.48).

[4.112] Applying Theorem 4.17 to $f(z) = z^m$ for $(x, y, m, n) \in \mathbb{R} \times (\mathbb{R} - \{-x\}) \times (\mathbb{N} - 1)^2$ such that $n \leq m$ and $(m-n)/2$ is an integer, and using (10.40), we get

$$\frac{\pi}{2} T_n(x, y) = \frac{\pi}{2}(x^n + y^n) \sum_{j=0}^\infty \binom{n+2j}{j} a_{n+2j}(xy)^j$$

$$= \frac{\pi(x^n+y^n)}{2} \binom{m}{(m-n)/2}(xy)^{(m-n)/2}$$

$$= \int_0^\pi (x^2+y^2+2xy\cos 2\theta)^{m/2} \cos m\phi(x,y,\theta) \cos n\theta \, d\theta,$$

where the function $\phi(x,y,\theta)$ is as defined in (4.48).

[4.113] Applying Theorem 4.17 to $f(z) = z^m$ for $(x, y, m, n) \in \mathbb{R} \times (\mathbb{R} - \{-x\}) \times (\mathbb{N} - 1)^2$ such that $n > m$ or $(m-n)/2$ is not an integer, and using (10.40), we get

$$\frac{\pi}{2} T_n(x, y) = \frac{\pi}{2} (x^n + y^n) \sum_{j=0}^{\infty} \binom{n+2j}{j} a_{n+2j}(xy)^j$$

$$= 0$$

$$= \int_0^{\pi} \left(x^2 + y^2 + 2xy \cos 2\theta\right)^{m/2} \cos m\phi(x, y, \theta) \cos n\theta \, d\theta,$$

where the function $\phi(x, y, \theta)$ is as defined in (4.48).

[4.114] Applying Theorem 4.28 to $f(z) = z^m$ for $(x, y, m) \in \left(\mathbb{R}^2 - \{(0, 0)\}\right) \times (\mathbb{N} - 1)$, we get

$$\pi f(x) = \pi x^m = \int_0^{\pi} \frac{\left(x + y e^{-i\theta}\right)^m + \left(x + y e^{i\theta}\right)^m}{2} d\theta$$

$$= \int_0^{\pi} \frac{1}{2} \left((x + y\cos\theta + iy\sin\theta)^m + (x + y\cos\theta - iy\sin\theta)^m\right) d\theta$$

$$= \int_0^{\pi} \frac{1}{2} \left(x^2 + y^2 + 2xy\cos\theta\right)^{m/2} \left(\left(\frac{x + y\cos\theta + iy\sin\theta}{\sqrt{x^2 + y^2 + 2xy\cos\theta}}\right)^m \right.$$

$$\left. + \left(\frac{x + y\cos\theta - iy\sin\theta}{\sqrt{x^2 + y^2 + 2xy\cos\theta}}\right)^m\right) d\theta$$

$$= \int_0^{\pi} \left(x^2 + y^2 + 2xy\cos\theta\right)^{m/2} \frac{e^{im\phi(x,y,\theta)} + e^{-im\phi(x,y,\theta)}}{2} d\theta$$

$$= \int_0^{\pi} \left(x^2 + y^2 + 2xy\cos\theta\right)^{m/2} \cos m\phi(x, y, \theta) \, d\theta, \tag{10.41}$$

where the function $\phi(x, y, \theta)$ is as defined in (4.49).

[4.115] Applying Theorem 4.41 to $f(z) = z^m$ for $(x, y, m, n) \in \left(\mathbb{R}^2 - \{(0, 0)\}\right) \times \mathbb{N} \times [1..m]$ and using (10.41), we get

$$\frac{\pi}{2} S_n(x, y) = \frac{\pi y^n}{2} \sum_{j=0}^{\infty} \binom{n+j}{j} a_{n+j} x^j$$

$$= \frac{\pi}{2} \binom{m}{n} x^{m-n} y^n$$

$$= \int_0^{\pi} \left(x^2 + y^2 + 2xy\cos\theta\right)^{m/2} \cos m\phi(x, y, \theta) \cos n\theta \, d\theta,$$

where the function $\phi(x, y, \theta)$ is as defined in (4.49).

10 Solutions to Special Integrals   323

[4.116] Applying Theorem 4.41 to $f(z) = z^m$ for $(x, y, m, n) \in \left(\mathbb{R}^2 - \{(0, 0)\}\right) \times \mathbb{N} \times (\mathbb{N} - [1..m])$ and using (10.41), we get

$$\frac{\pi}{2} S_n(x, y) = \frac{\pi y^n}{2} \sum_{j=0}^{\infty} \binom{n+j}{j} a_{n+j} x^j$$

$$= 0$$

$$= \int_0^{\pi} \left(x^2 + y^2 + 2xy \cos\theta\right)^{m/2} \cos m\phi(x, y, \theta) \cos n\theta \, d\theta,$$

where the function $\phi(x, y, \theta)$ is as defined in (4.49).

[4.117] Put $n = 0$ in [4.118].

[4.118] Applying Theorem 4.17 to $f(z) = (1+z)^m$ for $(x, y, m, n) \in \mathbb{R}^2 \times (\mathbb{N} - 1)^2$, we get

$$\frac{\pi}{2} T_n(x, y) = \frac{\pi (x^n + y^n)}{2} \sum_{j=0}^{\infty} \binom{n+2j}{j} a_{n+2j} (xy)^j$$

$$= \frac{\pi (x^n + y^n)}{2} \sum_{j=0}^{\lfloor (m-n)/2 \rfloor} \binom{n+2j}{j} \binom{m}{n+2j} (xy)^j$$

$$= \int_0^{\pi} \frac{\left(1 + xe^{i\theta} + ye^{-i\theta}\right)^m + \left(1 + xe^{-i\theta} + ye^{i\theta}\right)^m}{2} \cos n\theta \, d\theta$$

$$= \int_0^{\pi} \frac{1}{2} \left((1 + (x+y)\cos\theta + i(x-y)\sin\theta)^m \right.$$
$$\left. + (1 + (x+y)\cos\theta - i(x-y)\sin\theta)^m\right) \cos n\theta \, d\theta$$

$$= \int_0^{\pi} \frac{1}{2} \left(1 + x^2 + y^2 + 2xy \cos 2\theta + 2(x+y)\cos\theta\right)^{m/2}$$

$$\left(\left(\frac{1 + (x+y)\cos\theta + i(x-y)\sin\theta}{\sqrt{1 + x^2 + y^2 + 2xy \cos 2\theta + 2(x+y)\cos\theta}}\right)^m \right.$$

$$\left. + \left(\frac{1 + (x+y)\cos\theta - i(x-y)\sin\theta}{\sqrt{1 + x^2 + y^2 + 2xy \cos 2\theta + 2(x+y)\cos\theta}}\right)^m\right) \cos n\theta \, d\theta$$

$$= \int_0^{\pi} \left(1 + x^2 + y^2 + 2xy \cos 2\theta + 2(x+y)\cos\theta\right)^{m/2}$$

$$\frac{e^{im\phi(x,y,\theta)} + e^{-im\phi(x,y,\theta)}}{2} \cos n\theta \, d\theta$$

$$= \int_0^{\pi} \left(1 + x^2 + y^2 + 2xy \cos 2\theta + 2(x+y)\cos\theta\right)^{m/2} \cos m\phi(x, y, \theta) \cos n\theta$$

$$d\theta,$$

where the function $\phi(x, y, \theta)$ is as defined in (4.50).

[4.119] Replace x with $1 + x$ in [4.114].

[4.120] Replace x with $1 + x$ in [1.115] and use the binomial theorem.

[4.121] Applying Theorem 4.17 to $f(z) = \log(1 + z)$ for $(x, y) \in \mathbb{R}^2$ such that $|x| + |y| \le 1$ and $n = 0$, we get

$$\pi T_0(x, y) = 2\pi \sum_{j=0}^{\infty} \binom{2j}{j} a_{2j}(xy)^j$$

$$= -\pi \sum_{j=1}^{\infty} \binom{2j}{j} \frac{(xy)^j}{j}$$

$$= 2 \int_0^{\pi} \frac{\log\left(1 + xe^{i\theta} + ye^{-i\theta}\right) + \log\left(1 + xe^{-i\theta} + ye^{i\theta}\right)}{2} d\theta$$

$$= \int_0^{\pi} (\log(1 + (x + y)\cos\theta + i(x - y)\sin\theta)$$
$$+ \log(1 + (x + y)\cos\theta - i(x - y)\sin\theta))\, d\theta$$

$$= \int_0^{\pi} \log\left(1 + x^2 + y^2 + 2(x + y)\cos\theta + 2xy\cos 2\theta\right) d\theta.$$
(10.42)

[4.122] Applying Theorem 4.17 to $f(z) = \log(1 + z)$ for $(x, y, n) \in \mathbb{R}^2 \times \mathbb{N}$ such that $|x| + |y| \le 1$ and using (10.42), we get

$$\pi T_n(x, y) = \pi (x^n + y^n) \sum_{j=0}^{\infty} \binom{n + 2j}{j} a_{n+2j}(xy)^j$$

$$= (-1)^{n+1}\pi (x^n + y^n) \sum_{j=0}^{\infty} \binom{n + 2j}{j} \frac{(xy)^j}{n + 2j}$$

$$= 2 \int_0^{\pi} \frac{\log\left(1 + xe^{i\theta} + ye^{-i\theta}\right) + \log\left(1 + xe^{-i\theta} + ye^{i\theta}\right)}{2} \cos n\theta\, d\theta$$

$$= \int_0^{\pi} \log\left(1 + x^2 + y^2 + 2(x + y)\cos\theta + 2xy\cos 2\theta\right) \cos n\theta\, d\theta.$$

10 Solutions to Special Integrals 325

[4.123] Put $y = 0$ in [4.121].

[4.124] Put $y = 0$ in [4.122].

[4.125] Applying Theorem 4.28 to $f(z) = \sin z$, we get

$$\pi f(x) = \pi \sin x = \int_0^\pi \frac{\sin\left(x + ye^{-i\theta}\right) + \sin\left(x + ye^{i\theta}\right)}{2}\, d\theta$$

$$= \int_0^\pi \sin\left(x + y\frac{e^{i\theta} + e^{-i\theta}}{2}\right) \cos\left(y\frac{e^{i\theta} - e^{-i\theta}}{2}\right) d\theta \quad (10.43)$$

$$= \int_0^\pi \sin(x + y\cos\theta) \cos(iy\sin\theta)\, d\theta$$

$$= \int_0^\pi \sin(x + y\cos\theta) \cosh(y\sin\theta)\, d\theta.$$

[4.126] Applying Theorem 4.28 to $f(z) = \sinh z$, we get

$$\pi f(x) = \pi \sinh x = \int_0^\pi \frac{\sinh\left(x + ye^{-i\theta}\right) + \sinh\left(x + ye^{i\theta}\right)}{2}\, d\theta$$

$$= \int_0^\pi \sinh\left(x + y\frac{e^{i\theta} + e^{-i\theta}}{2}\right) \cosh\left(y\frac{e^{i\theta} - e^{-i\theta}}{2}\right) d\theta \quad (10.44)$$

$$= \int_0^\pi \sinh(x + y\cos\theta) \cosh(iy\sin\theta)\, d\theta$$

$$= \int_0^\pi \sinh(x + y\cos\theta) \cos(y\sin\theta)\, d\theta.$$

[4.127] Applying Theorem 4.28 to $f(z) = \cos z$, we get

$$\pi f(x) = \pi \cos x = \int_0^\pi \frac{\cos\left(x + ye^{-i\theta}\right) + \cos\left(x + ye^{i\theta}\right)}{2}\, d\theta$$

$$= \int_0^\pi \cos\left(x + y\frac{e^{i\theta} + e^{-i\theta}}{2}\right) \cos\left(y\frac{e^{i\theta} - e^{-i\theta}}{2}\right) d\theta \quad (10.45)$$

$$= \int_0^\pi \cos(x + y\cos\theta) \cos(iy\sin\theta)\, d\theta$$

$$= \int_0^\pi \cos(x + y\cos\theta) \cosh(y\sin\theta)\, d\theta.$$

[4.128] Applying Theorem 4.28 to $f(z) = \cosh z$, we get

$$\pi f(x) = \pi \cosh x = \int_0^\pi \frac{\cosh\left(x + ye^{-i\theta}\right) + \cosh\left(x + ye^{i\theta}\right)}{2} d\theta$$

$$= \int_0^\pi \cosh\left(x + y\frac{e^{i\theta} + e^{-i\theta}}{2}\right) \cosh\left(y\frac{e^{i\theta} - e^{-i\theta}}{2}\right) d\theta \qquad (10.46)$$

$$= \int_0^\pi \cosh(x + y\cos\theta) \cosh(iy\sin\theta) \, d\theta$$

$$= \int_0^\pi \cosh(x + y\cos\theta) \cos(y\sin\theta) \, d\theta.$$

[4.129] Applying Theorem 4.41 to $f(z) = \sin z$ for $(x, y, n) \in \mathbb{R}^2 \times (2\mathbb{N} - 1)$ and using (10.43), (5.3), and (5.2), we get

$$\frac{\pi}{2} S_n(x, y) = \frac{\pi y^n}{2} \sum_{j=0}^\infty \binom{n+j}{j} a_{n+j} x^j$$

$$= \frac{\pi y^n}{2} \sum_{j=0}^\infty \binom{n+2j}{2j} (-1)^{(n+2j-1)/2} \frac{x^{2j}}{(n+2j)!}$$

$$= (-1)^{(n-1)/2} \frac{\pi y^n}{2n!} \sum_{j=0}^\infty (-1)^j \frac{x^{2j}}{(2j)!}$$

$$= (-1)^{(n-1)/2} \frac{\pi y^n}{2n!} \cos x$$

$$= \int_0^\pi \frac{\sin\left(x + ye^{-i\theta}\right) + \sin\left(x + ye^{i\theta}\right)}{2} \cos n\theta \, d\theta$$

$$= \int_0^\pi \sin(x + y\cos\theta) \cosh(y\sin\theta) \cos n\theta \, d\theta.$$

[4.130] Applying Theorem 4.41 to $f(z) = \sin z$ for $(x, y, n) \in \mathbb{R}^2 \times (2\mathbb{N} - 2)$ and using (10.43) and (5.2), we get

$$\frac{\pi}{2} S_n(x, y) = \frac{\pi y^n}{2} \sum_{j=0}^\infty \binom{n+j}{j} a_{n+j} x^j$$

$$= \frac{\pi y^n}{2} \sum_{j=0}^\infty \binom{n+2j+1}{2j+1} (-1)^{(n+2j)/2} \frac{x^{2j+1}}{(n+2j+1)!}$$

10 Solutions to Special Integrals

$$= (-1)^{n/2}\frac{\pi y^n}{2n!}\sum_{j=0}^{\infty}(-1)^j\frac{x^{2j+1}}{(2j+1)!}$$

$$= (-1)^{n/2}\frac{\pi y^n}{2n!}\sin x$$

$$= \int_0^{\pi}\frac{\sin\left(x+ye^{-i\theta}\right)+\sin\left(x+ye^{i\theta}\right)}{2}\cos n\theta\, d\theta$$

$$= \int_0^{\pi}\sin(x+y\cos\theta)\cosh(y\sin\theta)\cos n\theta\, d\theta.$$

[4.131] Applying Theorem 4.41 to $f(z) = \sinh z$ for $(x, y, n) \in \mathbb{R}^2 \times (2\mathbb{N} - 1)$ and using (10.44), (5.4), and (5.5), we get

$$\frac{\pi}{2}S_n(x, y) = \frac{\pi y^n}{2}\sum_{j=0}^{\infty}\binom{n+j}{j}a_{n+j}x^j$$

$$= \frac{\pi y^n}{2}\sum_{j=0}^{\infty}\binom{n+2j}{2j}\frac{x^{2j}}{(n+2j)!}$$

$$= \frac{\pi y^n}{2n!}\sum_{j=0}^{\infty}\frac{x^{2j}}{(2j)!}$$

$$= \frac{\pi y^n}{2n!}\cosh x$$

$$= \int_0^{\pi}\frac{\sinh\left(x+ye^{-i\theta}\right)+\sinh\left(x+ye^{i\theta}\right)}{2}\cos n\theta\, d\theta$$

$$= \int_0^{\pi}\sinh(x+y\cos\theta)\cos(y\sin\theta)\cos n\theta\, d\theta.$$

[4.132] Applying Theorem 4.41 to $f(z) = \sinh z$ for $(x, y, n) \in \mathbb{R}^2 \times (2\mathbb{N} - 2)$ and using (10.44) and (5.4), we get

$$\frac{\pi}{2}S_n(x, y) = \frac{\pi y^n}{2}\sum_{j=0}^{\infty}\binom{n+j}{j}a_{n+j}x^j$$

$$= \frac{\pi y^n}{2}\sum_{j=0}^{\infty}\binom{n+2j+1}{2j+1}\frac{x^{2j+1}}{(n+2j+1)!}$$

$$= \frac{\pi y^n}{2n!}\sum_{j=0}^{\infty}\frac{x^{2j+1}}{(2j+1)!}$$

$$= \frac{\pi y^n}{2n!} \sinh x$$

$$= \int_0^\pi \frac{\sinh\left(x + ye^{-i\theta}\right) + \sinh\left(x + ye^{i\theta}\right)}{2} \cos n\theta \, d\theta$$

$$= \int_0^\pi \sinh(x + y\cos\theta)\cos(y\sin\theta)\cos n\theta \, d\theta.$$

[4.133] Applying Theorem 4.41 to $f(z) = \cos z$ for $(x, y, n) \in \mathbb{R}^2 \times (2\mathbb{N} - 1)$ and using (10.45), (5.3), and (5.2), we get

$$\frac{\pi}{2} S_n(x, y) = \frac{\pi y^n}{2} \sum_{j=0}^\infty \binom{n+j}{j} a_{n+j} x^j$$

$$= \frac{\pi y^n}{2} \sum_{j=0}^\infty \binom{n+2j+1}{2j+1} (-1)^{(n+2j+1)/2} \frac{x^{2j+1}}{(n+2j+1)!}$$

$$= (-1)^{(n+1)/2} \frac{\pi y^n}{2n!} \sum_{j=0}^\infty (-1)^j \frac{x^{2j+1}}{(2j+1)!}$$

$$= (-1)^{(n+1)/2} \frac{\pi y^n}{2n!} \sin x$$

$$= \int_0^\pi \frac{\cos\left(x + ye^{-i\theta}\right) + \cos\left(x + ye^{i\theta}\right)}{2} \cos n\theta \, d\theta$$

$$= \int_0^\pi \cos(x + y\cos\theta)\cosh(y\sin\theta)\cos n\theta \, d\theta.$$

[4.134] Applying Theorem 4.41 to $f(z) = \cos z$ for $(x, y, n) \in \mathbb{R}^2 \times (2\mathbb{N} - 2)$ and using (10.45) and (5.3), we get

$$\frac{\pi}{2} S_n(x, y) = \frac{\pi y^n}{2} \sum_{j=0}^\infty \binom{n+j}{j} a_{n+j} x^j$$

$$= \frac{\pi y^n}{2} \sum_{j=0}^\infty \binom{n+2j}{2j} (-1)^{(n+2j)/2} \frac{x^{2j}}{(n+2j)!}$$

$$= (-1)^{n/2} \frac{\pi y^n}{2n!} \sum_{j=0}^\infty (-1)^j \frac{x^{2j}}{(2j)!}$$

$$= (-1)^{n/2} \frac{\pi y^n}{2n!} \cos x$$

$$= \int_0^\pi \frac{\cos\left(x + ye^{-i\theta}\right) + \cos\left(x + ye^{i\theta}\right)}{2} \cos n\theta \, d\theta$$

$$= \int_0^\pi \cos(x + y\cos\theta) \cosh(y\sin\theta) \cos n\theta \, d\theta.$$

[4.135] Applying Theorem 4.41 to $f(z) = \cosh z$ for $(x, y, n) \in \mathbb{R}^2 \times (2\mathbb{N} - 1)$ and using (10.46), (5.5), and (5.4), we get

$$\frac{\pi}{2} S_n(x, y) = \frac{\pi y^n}{2} \sum_{j=0}^\infty \binom{n+j}{j} a_{n+j} x^j$$

$$= \frac{\pi y^n}{2} \sum_{j=0}^\infty \binom{n+2j+1}{2j+1} \frac{x^{2j+1}}{(n+2j+1)!}$$

$$= \frac{\pi y^n}{2n!} \sum_{j=0}^\infty \frac{x^{2j+1}}{(2j+1)!}$$

$$= \frac{\pi y^n}{2n!} \sinh x$$

$$= \int_0^\pi \frac{\cosh\left(x + ye^{-i\theta}\right) + \cosh\left(x + ye^{i\theta}\right)}{2} \cos n\theta \, d\theta$$

$$= \int_0^\pi \cosh(x + y\cos\theta) \cos(y\sin\theta) \cos n\theta \, d\theta.$$

[4.136] Applying Theorem 4.41 to $f(z) = \cosh z$ for $(x, y, n) \in \mathbb{R}^2 \times (2\mathbb{N} - 2)$ and using (10.46) and (5.5), we get

$$\frac{\pi}{2} S_n(x, y) = \frac{\pi y^n}{2} \sum_{j=0}^\infty \binom{n+j}{j} a_{n+j} x^j$$

$$= \frac{\pi y^n}{2} \sum_{j=0}^\infty \binom{n+2j}{2j} \frac{x^{2j}}{(n+2j)!}$$

$$= \frac{\pi y^n}{2n!} \sum_{j=0}^\infty \frac{x^{2j}}{(2j)!}$$

$$= \frac{\pi y^n}{2n!} \cosh x$$

$$= \int_0^\pi \frac{\cosh\left(x + ye^{-i\theta}\right) + \cosh\left(x + ye^{i\theta}\right)}{2} \cos n\theta \, d\theta$$

$$= \int_0^\pi \cosh(x + y\cos\theta) \cos(y\sin\theta) \cos n\theta \, d\theta.$$

[4.137] Put $y = 1/x$ in [4.110].

[4.138] Put $y = 1/x$ in [4.111].

[4.139] Put $x = 1$ and replace y with x in [4.114].

[4.140] Put $y = 1/x$ in [4.114].

[4.141] Put $x = y = 1$ in [4.119].

[4.142] Put $x = y = 1$ in [4.120].

[4.143] Put $x = y = 1$ in [4.110].

[4.144] Put $x = y = 1$ in [4.111].

[4.145] Put $x = y = 1$ and $n = m$ in [4.112].

[4.146] Putting $x = 1/2$ in [4.123], we get

$$\int_0^\pi \log\left(1 + \frac{1}{4} + \cos\theta\right) d\theta = 0$$
$$\iff \int_0^\pi \log\left(\frac{5 + 4\cos\theta}{4}\right) d\theta = 0$$
$$\iff \int_0^\pi \log(5 + 4\cos\theta) \, d\theta = \int_0^\pi (\log 4) \, d\theta = 2\pi \log 2.$$

[4.147] Putting $x = -1$ in [4.123], we get

$$\int_0^\pi \log(2 - 2\cos\theta) \, d\theta = 0$$
$$\iff \int_0^\pi \log\left(4\sin^2\frac{\theta}{2}\right) d\theta = 0 \qquad (10.47)$$
$$\iff \int_0^\pi \log\left(2\sin\frac{\theta}{2}\right) d\theta = 0.$$

We also have

10   Solutions to Special Integrals

$$\int_0^\pi \log\left(2\sin\frac{\theta}{2}\right) d\theta = \int_0^\pi \log\left(2\sin\frac{\pi-\theta}{2}\right) d\theta$$
$$= \int_0^\pi \log\left(2\cos\frac{\theta}{2}\right) d\theta \qquad (10.48)$$
$$= 0.$$

Adding (10.47) and (10.48), we get

$$\int_0^\pi \log\left(4\sin\frac{\theta}{2}\cos\frac{\theta}{2}\right) d\theta = 0$$
$$\iff \int_0^\pi \log(2\sin\theta)\, d\theta = 0$$
$$\iff \int_0^\pi \log\sin\theta\, d\theta = -\int_0^\pi (\log 2)d\theta = -\pi\log 2.$$

[4.148]   Putting $x = 1$ in [4.124], we get

$$\int_0^\pi \log(2 + 2\cos\theta)\cos n\theta\, d\theta = (-1)^{n+1}\frac{\pi}{n}$$
$$\iff \int_0^\pi \log\left(4\cos^2\frac{\theta}{2}\right)\cos n\theta\, d\theta = (-1)^{n+1}\frac{\pi}{n}$$
$$\iff \int_0^\pi \log\left(2\cos\frac{\theta}{2}\right)\cos n\theta\, d\theta = (-1)^{n+1}\frac{\pi}{2n}$$
$$\iff \int_0^\pi \log\left(\cos\frac{\theta}{2}\right)\cos n\theta\, d\theta = (-1)^{n+1}\frac{\pi}{2n} - \int_0^\pi (\log 2)\cos n\theta\, d\theta$$
$$= (-1)^{n+1}\frac{\pi}{2n} - \log 2\left.\frac{\sin n\theta}{n}\right|_0^\pi$$
$$= (-1)^{n+1}\frac{\pi}{2n}.$$

[4.149]   Putting $x = 1/e^2$ in [4.123], we get

$$\int_0^\pi \log\left(1 + \frac{1}{e^4} + \frac{2}{e^2}\cos\theta\right) d\theta = 0$$
$$\iff \int_0^\pi \log\left(\frac{1}{e^2}\left(e^2 + \frac{1}{e^2} + 2\cos\theta\right)\right) d\theta = 0$$
$$\iff \int_0^\pi \log\left(e^2 + \frac{1}{e^2} + 2\cos\theta\right) d\theta = \int_0^\pi \log\left(e^2\right) d\theta = 2\pi.$$

[4.150] Put $y = -1/4x$ in [4.121] for $x \in \left[-1, -\frac{1}{4}\right] \cup \left[\frac{1}{4}, 1\right]$.

[4.151] Put $y = -1/4x$ in [4.122] for $(x, n) \in \left(\left[-1, -\frac{1}{4}\right] \cup \left[\frac{1}{4}, 1\right]\right) \times \mathbb{N}$.

[4.152] Put $x = 1/2$ in [4.150].

[4.153] Put $x = 1/2$ in [4.151] for $n \in 2\mathbb{N}$.

[4.154] Put $x = 1/2$ in [4.151] for $n \in 2\mathbb{N} - 1$.

[4.155] Put $y = 1/x$ in [4.117].

[4.156] Put $y = -1/x$ in [4.117].

[4.157] Put $x = y = 1$ in [4.117].

[4.158] Put $x = y = 1$ in [4.118].

[4.159] Put $x = 1$ and $y = -1$ in [4.117].

[4.160] Put $x = 1$ and $y = -1$ in [4.118] for $(m, n) \in (\mathbb{N} - 1) \times (2\mathbb{N} - 2)$.

[4.161] Put $x = 1$ and $y = -1$ in [4.118] for $(m, n) \in (\mathbb{N} - 1) \times (2\mathbb{N} - 1)$.

[4.162] Put $x = y = 1$ in [4.119].

[4.163] Put $x = y = 1$ in [4.120].

Chapter 11
Solutions to Generalized Bessel Functions

[5.1] Putting $\alpha = 2$ in (5.6), and using (5.5) and [3.6], we get

$$E_2(z^2) = \sum_{j=0}^{\infty} \frac{z^{2j}}{\Gamma(2j+1)}$$

$$= \sum_{j=0}^{\infty} \frac{z^{2j}}{(2j)!}$$

$$= \cosh z.$$

[5.2] From (5.7), we have

$$E_{\alpha,\beta}(z) = \sum_{j=0}^{\infty} \frac{z^j}{\Gamma(\alpha j + \beta)}$$

$$= \frac{1}{\Gamma(\beta)} + z \sum_{j=1}^{\infty} \frac{z^{j-1}}{\Gamma(\alpha(j-1)+\alpha+\beta)}$$

$$= \frac{1}{\Gamma(\beta)} + z \sum_{j=0}^{\infty} \frac{z^j}{\Gamma(\alpha j + \alpha + \beta)}$$

$$= \frac{1}{\Gamma(\beta)} + z E_{\alpha,\alpha+\beta}.$$

© The Author(s), under exclusive license to Springer Nature Singapore Pte Ltd. 2025
A. Mishra, *Special Integrals*, University Texts in the Mathematical Sciences,
https://doi.org/10.1007/978-981-97-7514-9_11

[5.3] From (5.8), we have

$$E_{\alpha,\beta-\alpha}^{\gamma}(z) - E_{\alpha,\beta-\alpha}^{\gamma-1}(z) = \sum_{j=0}^{\infty} \frac{\prod_{k=0}^{j-1}(\gamma+k)}{j!\Gamma(\alpha j+\beta-\alpha)}z^j - \sum_{j=0}^{\infty} \frac{\prod_{k=0}^{j-1}(\gamma+k-1)}{j!\Gamma(\alpha j+\beta-\alpha)}z^j$$

$$= \sum_{j=0}^{\infty} \frac{(\gamma+j-1-(\gamma-1))\prod_{k=0}^{j-2}(\gamma+k)}{j!\Gamma(\alpha j+\beta-\alpha)}z^j$$

$$= z\sum_{j=1}^{\infty} \frac{\prod_{k=0}^{j-2}(\gamma+k)}{(j-1)!\Gamma(\alpha(j-1)+\beta)}z^{j-1}$$

$$= z\sum_{j=0}^{\infty} \frac{\prod_{k=0}^{j-1}(\gamma+k)}{j!\Gamma(\alpha j+\beta)}z^j$$

$$= zE_{\alpha,\beta}^{\gamma}(z).$$

[5.4] From (5.9) and (3.6), we have

$$\beta E_{\alpha,\beta+1}^{\gamma,q}(z) + \alpha z\frac{d}{dz}\left(E_{\alpha,\beta+1}^{\gamma,q}(z)\right)$$

$$= \beta \sum_{j=0}^{\infty} \frac{\prod_{k=0}^{qj-1}(\gamma+k)}{j!\Gamma(\alpha j+\beta+1)}z^j + \alpha z\frac{d}{dz}\left(\sum_{j=0}^{\infty} \frac{\prod_{k=0}^{qj-1}(\gamma+k)}{j!\Gamma(\alpha j+\beta+1)}z^j\right)$$

$$= \beta \sum_{j=0}^{\infty} \frac{\prod_{k=0}^{qj-1}(\gamma+k)}{j!(\alpha j+\beta)\Gamma(\alpha j+\beta)}z^j + \alpha z\sum_{j=1}^{\infty} \frac{j\prod_{k=0}^{qj-1}(\gamma+k)}{j!(\alpha j+\beta)\Gamma(\alpha j+\beta)}z^{j-1}$$

$$= \sum_{j=0}^{\infty} \frac{(\alpha j+\beta)\prod_{k=0}^{qj-1}(\gamma+k)}{j!(\alpha j+\beta)\Gamma(\alpha j+\beta)}z^j$$

11 Solutions to Generalized Bessel Functions

$$= \sum_{j=0}^{\infty} \frac{\prod_{k=0}^{qj-1}(\gamma+k)}{j!\Gamma(\alpha j+\beta)} z^j$$

$$= E_{\alpha,\beta}^{\gamma,q}(z).$$

[5.5] From (5.10) and (3.6), we have

$$\phi(\alpha,\beta-1;z)+(1-\beta)\phi(\alpha,\beta;z)$$

$$=\sum_{j=0}^{\infty}\frac{z^j}{j!\Gamma(\alpha j+\beta-1)}+(1-\beta)\sum_{j=0}^{\infty}\frac{z^j}{j!\Gamma(\alpha j+\beta)}$$

$$=\sum_{j=-1}^{\infty}\frac{z^{j+1}}{(j+1)!\Gamma(\alpha(j+1)+\beta-1)}+\sum_{j=-1}^{\infty}\frac{(1-\beta)z^{j+1}}{(j+1)!\Gamma(\alpha(j+1)+\beta)}$$

$$=\frac{1}{\Gamma(\beta-1)}+\frac{1-\beta}{\Gamma(\beta)}+\sum_{j=0}^{\infty}\frac{z^{j+1}}{(j+1)!\Gamma(\alpha j+\alpha+\beta-1)}$$

$$+\sum_{j=0}^{\infty}\frac{(1-\beta)z^{j+1}}{(j+1)!\Gamma(\alpha j+\alpha+\beta)}$$

$$=\frac{\beta-1}{(\beta-1)\Gamma(\beta-1)}+\frac{1-\beta}{\Gamma(\beta)}+z\sum_{j=0}^{\infty}\left(\frac{1}{\Gamma(\alpha j+\alpha+\beta-1)}\right.$$

$$\left.+\frac{1-\beta}{\Gamma(\alpha j+\alpha+\beta)}\right)\frac{z^j}{(j+1)!}$$

$$=\frac{\beta-1}{\Gamma(\beta)}+\frac{1-\beta}{\Gamma(\beta)}+z\sum_{j=0}^{\infty}\left(\frac{\alpha j+\alpha+\beta-1}{(\alpha j+\alpha+\beta-1)\Gamma(\alpha j+\alpha+\beta-1)}\right.$$

$$\left.+\frac{1-\beta}{\Gamma(\alpha j+\alpha+\beta)}\right)\frac{z^j}{(j+1)!}$$

$$=z\sum_{j=0}^{\infty}\left(\frac{\alpha j+\alpha+\beta-1}{\Gamma(\alpha j+\alpha+\beta)}+\frac{1-\beta}{\Gamma(\alpha j+\alpha+\beta)}\right)\frac{z^j}{(j+1)!}$$

$$=z\sum_{j=0}^{\infty}\frac{\alpha(j+1)z^j}{(j+1)!\Gamma(\alpha j+\alpha+\beta)}$$

$$=\alpha z\sum_{j=0}^{\infty}\frac{z^j}{j!\Gamma(\alpha j+\alpha+\beta)}$$

$$=\alpha z\phi(\alpha,\alpha+\beta;z).$$

[5.6] From (5.10), we have

$$\frac{d\phi(\alpha,\beta;z)}{dz} = \frac{d}{dz}\left(\sum_{j=0}^{\infty}\frac{z^j}{j!\Gamma(\alpha j+\beta)}\right)$$

$$= \sum_{j=1}^{\infty}\frac{jz^{j-1}}{j!\Gamma(\alpha j+\beta)}$$

$$= \sum_{j=1}^{\infty}\frac{z^{j-1}}{(j-1)!\Gamma(\alpha j+\beta)}$$

$$= \sum_{j=0}^{\infty}\frac{z^j}{j!\Gamma(\alpha(j+1)+\beta)}$$

$$= \sum_{j=0}^{\infty}\frac{z^j}{j!\Gamma(\alpha j+\alpha+\beta)}$$

$$= \phi(\alpha,\alpha+\beta;z).$$

[5.7] Multiply [5.6] by az and use [5.5].

[5.8] Use (5.11) and [4.22].

[5.9] Use (5.11) and [4.23]. Proved in [4.23] for $n \in \mathbb{N} - 1$. For $n \in -\mathbb{N}$, using (5.12) and [4.23], we have

$$e(1,n) = e(1,-n)$$
$$= \frac{1}{\pi}\int_0^\pi e^{2\cos\theta}\cos(-n\theta)\,d\theta$$
$$= \frac{1}{\pi}\int_0^\pi e^{2\cos\theta}\cos n\theta\,d\theta.$$

[5.10] Use (5.11) and [4.24].

[5.11] Use (5.11) and [4.25]. Proved in [4.25] for $n \in \mathbb{N} - 1$. For $n \in -\mathbb{N}$, using (5.12) and [4.25], we have

$$e\left(x^2,n\right) = x^{-2n}e\left(x^2,-n\right)$$
$$= \frac{x^{-2n}}{\pi x^{-n}}\int_0^\pi e^{2x\cos\theta}\cos(-n\theta)\,d\theta$$
$$= \frac{1}{\pi x^n}\int_0^\pi e^{2x\cos\theta}\cos n\theta\,d\theta.$$

11 Solutions to Generalized Bessel Functions   337

[5.12] Use (5.11) and [4.77]. Proved in [4.77] for $n \in 2\mathbb{N} - 1$. For $n \in -(2\mathbb{N} - 1)$, using (5.12), [3.9], and [4.77], we have

$$e\left(-x^2, n\right) = (-1)^{-n} x^{-2n} e\left(-x^2, -n\right)$$
$$= \frac{(-1)^{(-n-1)/2}(-1)^{-n} x^{-2n}}{\pi x^{-n}} \int_0^\pi \sin(2x \cos \theta) \cos(-n\theta)\, d\theta$$
$$= \frac{(-1)^{(n-1)/2}}{\pi x^n} \int_0^\pi \sin(2x \cos \theta) \cos n\theta\, d\theta$$
$$= -\frac{J_{-n}(2x)}{x^n}$$
$$= \frac{J_n(2x)}{x^n}.$$

[5.13] Use (5.11) and [4.85]. Proved in [4.85] for $n \in 2\mathbb{N} - 2$. For $n \in -2\mathbb{N}$, using (5.12), [3.9], and [4.85], we have

$$e\left(-x^2, n\right) = (-1)^{-n} x^{-2n} e\left(-x^2, -n\right)$$
$$= \frac{(-1)^{-n/2}(-1)^{-n} x^{-2n}}{\pi x^{-n}} \int_0^\pi \cos(2x \cos \theta) \cos(-n\theta)\, d\theta$$
$$= \frac{(-1)^{n/2}}{\pi x^n} \int_0^\pi \cos(2x \cos \theta) \cos n\theta\, d\theta$$
$$= \frac{J_{-n}(2x)}{x^n}$$
$$= \frac{J_n(2x)}{x^n}.$$

[5.14] Use (5.11) and [4.79]. Proved in [4.79] for $n \in 2\mathbb{N} - 1$. For $n \in -(2\mathbb{N} - 1)$, using (5.12), [3.19], and [4.79], we have

$$e\left(x^2, n\right) = x^{-2n} e\left(x^2, -n\right)$$
$$= \frac{(-1)^{(-n-1)/2} x^{-2n}}{\pi x^{-n}} \int_0^\pi \sinh(2x \sin \theta) \sin(-n\theta)\, d\theta$$
$$= \frac{(-1)^{(n-1)/2}}{\pi x^n} \int_0^\pi \sinh(2x \sin \theta) \sin n\theta\, d\theta$$
$$= \frac{I_{-n}(2x)}{x^n}$$
$$= \frac{I_n(2x)}{x^n}.$$

[5.15] Use (5.11) and [4.87]. Proved in [4.87] for $n \in 2\mathbb{N} - 2$. For $n \in -2\mathbb{N}$, using (5.12), [3.19], and [4.87], we have

$$\begin{aligned} e\left(x^2, n\right) &= x^{-2n} e\left(x^2, -n\right) \\ &= \frac{(-1)^{-n/2} x^{-2n}}{\pi x^{-n}} \int_0^\pi \cosh(2x \sin\theta) \cos(-n\theta)\, d\theta \\ &= \frac{(-1)^{n/2}}{\pi x^n} \int_0^\pi \cosh(2x \sin\theta) \cos n\theta\, d\theta \\ &= \frac{I_{-n}(2x)}{x^n} \\ &= \frac{I_n(2x)}{x^n}. \end{aligned}$$

[5.16] Use (5.11) and [4.81]. Proved in [4.81] for $n \in 2\mathbb{N} - 1$. For $n \in -(2\mathbb{N} - 1)$, using (5.12), [3.19], and [4.81], we have

$$\begin{aligned} e\left(x^2, n\right) &= x^{-2n} e\left(x^2, -n\right) \\ &= \frac{x^{-2n}}{\pi x^{-n}} \int_0^\pi \sinh(2x \cos\theta) \cos(-n\theta)\, d\theta \\ &= \frac{1}{\pi x^n} \int_0^\pi \sinh(2x \cos\theta) \cos n\theta\, d\theta \\ &= \frac{I_{-n}(2x)}{x^n} \\ &= \frac{I_n(2x)}{x^n}. \end{aligned}$$

[5.17] Use (5.11) and [4.89]. Proved in [4.89] for $n \in 2\mathbb{N} - 2$. For $n \in -2\mathbb{N}$, using (5.12), [3.19], and [4.89], we have

$$\begin{aligned} e\left(x^2, n\right) &= x^{-2n} e\left(x^2, -n\right) \\ &= \frac{x^{-2n}}{\pi x^{-n}} \int_0^\pi \cosh(2x \cos\theta) \cos(-n\theta)\, d\theta \\ &= \frac{1}{\pi x^n} \int_0^\pi \cosh(2x \cos\theta) \cos n\theta\, d\theta \\ &= \frac{I_{-n}(2x)}{x^n} \\ &= \frac{I_n(2x)}{x^n}. \end{aligned}$$

11 Solutions to Generalized Bessel Functions   339

[5.18] Use (5.11) and [4.83]. Proved in [4.83] for $n \in 2\mathbb{N} - 1$. For $n \in -(2\mathbb{N} - 1)$, using (5.12), [3.9], and [4.83], we have

$$\begin{aligned}
e\left(-x^2, n\right) &= (-1)^{-n} x^{-2n} e\left(-x^2, -n\right) \\
&= \frac{(-1)^n x^{-2n}}{\pi x^{-n}} \int_0^\pi \sin(2x \sin\theta) \sin(-n\theta)\, d\theta \\
&= \frac{1}{\pi x^n} \int_0^\pi \sin(2x \sin\theta) \sin n\theta\, d\theta \\
&= -\frac{J_{-n}(2x)}{x^n} \\
&= \frac{J_n(2x)}{x^n}.
\end{aligned}$$

[5.19] Use (5.11) and [4.91]. Proved in [4.91] for $n \in 2\mathbb{N} - 2$. For $n \in -2\mathbb{N}$, using (5.12), [3.9], and [4.91], we have

$$\begin{aligned}
e\left(-x^2, n\right) &= (-1)^{-n} x^{-2n} e\left(-x^2, -n\right) \\
&= \frac{(-1)^n x^{-2n}}{\pi x^{-n}} \int_0^\pi \cos(2x \sin\theta) \cos(-n\theta)\, d\theta \\
&= \frac{1}{\pi x^n} \int_0^\pi \cos(2x \sin\theta) \cos n\theta\, d\theta \\
&= \frac{J_{-n}(2x)}{x^n} \\
&= \frac{J_n(2x)}{x^n}.
\end{aligned}$$

[5.20] Put $n = 0$ in [5.21].

[5.21] Put $y = 1$ in [5.47].

[5.22] Put $y = 1$ in [5.54].

[5.23] Put $y = 1$ in [5.52].

[5.24] From [5.26], we have

$$\pi = \int_0^\pi e^{(x+1)\cos\theta} \cos((x+1)\sin\theta)\, d\theta$$

$$\iff 1 = \frac{1}{\pi}\int_0^\pi e^{(x+1)\cos\theta} \cos(\sin\theta)\cos(x\sin\theta)\, d\theta \tag{11.1}$$

$$- \frac{1}{\pi}\int_0^\pi e^{(x+1)\cos\theta} \sin(\sin\theta)\sin(x\sin\theta)\, d\theta.$$

From [5.20], we have

$$e(x,0) = \frac{1}{\pi}\int_0^\pi e^{(x+1)\cos\theta} \cos((x-1)\sin\theta)\, d\theta$$

$$= \frac{1}{\pi}\int_0^\pi e^{(x+1)\cos\theta} \cos(\sin\theta)\cos(x\sin\theta)\, d\theta \tag{11.2}$$

$$+ \frac{1}{\pi}\int_0^\pi e^{(x+1)\cos\theta} \sin(\sin\theta)\sin(x\sin\theta)\, d\theta.$$

Adding (11.1) and (11.2), we get

$$e(x,0) + 1 = \frac{2}{\pi}\int_0^\pi e^{(x+1)\cos\theta} \cos(\sin\theta)\cos(x\sin\theta)\, d\theta$$

$$\iff e(x,0) = -1 + \frac{2}{\pi}\int_0^\pi e^{(x+1)\cos\theta} \cos(\sin\theta)\cos(x\sin\theta)\, d\theta.$$

[5.25] Subtracting (11.1) from (11.2), we get

$$e(x,0) - 1 = \frac{2}{\pi}\int_0^\pi e^{(x+1)\cos\theta} \sin(\sin\theta)\sin(x\sin\theta)\, d\theta$$

$$\iff e(x,0) = 1 + \frac{2}{\pi}\int_0^\pi e^{(x+1)\cos\theta} \sin(\sin\theta)\sin(x\sin\theta)\, d\theta.$$

[5.26] Replace x with $x+1$ in [4.96].

[5.27] Use (5.19) and [4.32].

[5.28] Use (5.19) and [4.31].

11 Solutions to Generalized Bessel Functions   341

[5.29] Use (5.19) and [4.33]. Proved in [4.33] for $n \in 2\mathbb{N} - 2$. For $n \in -2\mathbb{N}$, using (5.19) and [4.33], we have

$$\sinh(x^2, n) = x^{-2n} \sinh(x^2, -n)$$
$$= \frac{(-1)^{-n/2} x^{-2n}}{\pi x^{-n}} \int_0^\pi \left(\sin^2(x \cos\theta) \cosh^2(x \sin\theta) \right.$$
$$\left. + \cos^2(x \cos\theta) \sinh^2(x \sin\theta) \right) \cos(-n\theta) \, d\theta$$
$$= \frac{(-1)^{n/2}}{\pi x^n} \int_0^\pi \left(\sin^2(x \cos\theta) \cosh^2(x \sin\theta) \right.$$
$$\left. + \cos^2(x \cos\theta) \sinh^2(x \sin\theta) \right) \cos n\theta \, d\theta.$$

[5.30] Use (5.19) and [4.35]. Proved in [4.35] for $n \in 2\mathbb{N} - 2$. For $n \in -2\mathbb{N}$, using (5.19) and [4.35], we have

$$\sinh(x^2, n) = x^{-2n} \sinh(x^2, -n)$$
$$= \frac{x^{-2n}}{\pi x^{-n}} \int_0^\pi \left(\sinh^2(x \cos\theta) \cos^2(x \sin\theta) \right.$$
$$\left. + \cosh^2(x \cos\theta) \sin^2(x \sin\theta) \right) \cos(-n\theta) \, d\theta$$
$$= \frac{1}{\pi x^n} \int_0^\pi \left(\sinh^2(x \cos\theta) \cos^2(x \sin\theta) \right.$$
$$\left. + \cosh^2(x \cos\theta) \sin^2(x \sin\theta) \right) \cos n\theta \, d\theta.$$

[5.31] Use (5.20) and [4.37].

[5.32] Use (5.20) and [4.38].

[5.33] Use (5.20) and [4.39]. Proved in [4.39] for $n \in 2\mathbb{N} - 2$. For $n \in -2\mathbb{N}$, using (5.20) and [4.39], we have

$$\cosh(x^2, n) = x^{-2n} \cosh(x^2, -n)$$
$$= \frac{(-1)^{-n/2} x^{-2n}}{\pi x^{-n}} \int_0^\pi \left(\cos^2(x \cos\theta) \cosh^2(x \sin\theta) \right.$$
$$\left. + \sin^2(x \cos\theta) \sinh^2(x \sin\theta) \right) \cos(-n\theta) \, d\theta$$
$$= \frac{(-1)^{n/2}}{\pi x^n} \int_0^\pi \left(\cos^2(x \cos\theta) \cosh^2(x \sin\theta) \right.$$
$$\left. + \sin^2(x \cos\theta) \sinh^2(x \sin\theta) \right) \cos n\theta \, d\theta.$$

[5.34] Use (5.20) and [4.41]. Proved in [4.41] for $n \in 2\mathbb{N} - 2$. For $n \in -2\mathbb{N}$, using (5.20) and [4.41], we have

$$\cosh(x^2, n) = x^{-2n} \cosh(x^2, -n)$$
$$= \frac{x^{-2n}}{\pi x^{-n}} \int_0^\pi \left(\cosh^2(x\cos\theta)\cos^2(x\sin\theta)\right.$$
$$\left. + \sinh^2(x\cos\theta)\sin^2(x\sin\theta)\right)\cos(-n\theta)\, d\theta$$
$$= \frac{1}{\pi x^n} \int_0^\pi \left(\cosh^2(x\cos\theta)\cos^2(x\sin\theta)\right.$$
$$\left. + \sinh^2(x\cos\theta)\sin^2(x\sin\theta)\right)\cos n\theta\, d\theta.$$

[5.35] Replacing x with ix in [5.21] and using (5.14), (5.26), and (5.17), we get

$$e(ix, n) = \frac{2}{\pi(1 + i^n x^n)} \int_0^\pi e^{(ix+1)\cos\theta} \cos((ix-1)\sin\theta)\cos n\theta\, d\theta$$
$$\iff \cos(x, n) + i\sin(x, n)$$
$$= \frac{2}{\pi(1 + i^n x^n)} \int_0^\pi e^{ix\cos\theta} \cos(ix\sin\theta - \sin\theta) e^{\cos\theta} \cos n\theta\, d\theta$$
$$= \frac{2}{\pi(1 + i^n x^n)} \int_0^\pi (\cos(x\cos\theta) + i\sin(x\cos\theta))$$
$$(\cos(ix\sin\theta)\cos(\sin\theta) + \sin(ix\sin\theta)\sin(\sin\theta))e^{\cos\theta} \cos n\theta\, d\theta$$
$$= \frac{2}{\pi(1 + i^n x^n)} \int_0^\pi (\cos(x\cos\theta) + i\sin(x\cos\theta))$$
$$(\cosh(x\sin\theta)\cos(\sin\theta) + i\sinh(x\sin\theta)\sin(\sin\theta))e^{\cos\theta} \cos n\theta\, d\theta$$
$$= \frac{2}{\pi(1 + i^n x^n)} \int_0^\pi ((\cos(x\cos\theta)\cosh(x\sin\theta)\cos(\sin\theta)$$
$$- \sin(x\cos\theta)\sinh(x\sin\theta)\sin(\sin\theta))$$
$$+ i(\sin(x\cos\theta)\cosh(x\sin\theta)\cos(\sin\theta)$$
$$+ \cos(x\cos\theta)\sinh(x\sin\theta)\sin(\sin\theta))\, e^{\cos\theta} \cos n\theta\, d\theta$$
$$= \frac{2}{\pi(1 + i^n x^n)} \int_0^\pi (C_1(x, \theta) + iS_1(x, \theta))e^{\cos\theta} \cos n\theta\, d\theta.$$
(11.3)

For $(x, n) \in \mathbb{R} \times 2\mathbb{Z} - \{-1, 1\} \times 2(2\mathbb{Z} + 1)$, (11.3) becomes

$$\cos(x, n) + i\sin(x, n)$$
$$= \frac{2}{\pi(1 + (-1)^{n/2} x^n)} \int_0^\pi C_1(x, \theta) e^{\cos\theta} \cos n\theta\, d\theta \qquad (11.4)$$
$$+ \frac{2i}{\pi(1 + (-1)^{n/2} x^n)} \int_0^\pi S_1(x, \theta) e^{\cos\theta} \cos n\theta\, d\theta.$$

11 Solutions to Generalized Bessel Functions 343

Equating the imaginary part of (11.4) for $(x, n) \in \mathbb{R} \times 2\mathbb{Z} - \{-1, 1\} \times 2(2\mathbb{Z} + 1)$, we get

$$\sin(x, n) = \frac{2}{\pi \left(1 + (-1)^{n/2} x^n\right)} \int_0^\pi S_1(x, \theta) e^{\cos\theta} \cos n\theta \, d\theta. \tag{11.5}$$

[5.36] From (11.5), for $(x, n) \in \{-1, 1\} \times 2(2\mathbb{Z} + 1)$, we get

$$\int_0^\pi S_1(x, \theta) e^{\cos\theta} \cos n\theta \, d\theta = \frac{\pi}{2} \left(1 + (-1)^{n/2} x^n\right) \sin(x, n) = 0.$$

[5.37] For $(x, n) \in \mathbb{R} \times (2\mathbb{Z} + 1)$, (11.3) becomes

$$\cos(x, n) + i \sin(x, n)$$
$$= \frac{2(1 - i^n x^n)}{\pi (1 + i^n x^n)(1 - i^n x^n)} \int_0^\pi (C_1(x, \theta) + i S_1(x, \theta)) e^{\cos\theta} \cos n\theta \, d\theta$$
$$= \frac{2\left(1 + (-1)^{(n+1)/2} i x^n\right)}{\pi (1 + x^{2n})} \int_0^\pi (C_1(x, \theta) + i S_1(x, \theta)) e^{\cos\theta} \cos n\theta \, d\theta$$
$$= \left(\frac{2}{\pi(1 + x^{2n})} \int_0^\pi C_1(x, \theta) e^{\cos\theta} \cos n\theta \, d\theta \right. \tag{11.6}$$
$$+ \frac{2(-1)^{(n-1)/2} x^n}{\pi(1 + x^{2n})} \int_0^\pi S_1(x, \theta) e^{\cos\theta} \cos n\theta \, d\theta \bigg)$$
$$+ i \left(\frac{2}{\pi(1 + x^{2n})} \int_0^\pi S_1(x, \theta) e^{\cos\theta} \cos n\theta \, d\theta \right.$$
$$+ \frac{2(-1)^{(n+1)/2} x^n}{\pi(1 + x^{2n})} \int_0^\pi C_1(x, \theta) e^{\cos\theta} \cos n\theta \, d\theta \bigg).$$

Equating the imaginary part of (11.6) for $(x, n) \in \mathbb{R} \times (2\mathbb{Z} + 1)$, we get

$$\sin(x, n) = \frac{2}{\pi \left(1 + x^{2n}\right)} \int_0^\pi S_1(x, \theta) e^{\cos\theta} \cos n\theta \, d\theta$$
$$+ (-1)^{(n+1)/2} \frac{2 x^n}{\pi \left(1 + x^{2n}\right)} \int_0^\pi C_1(x, \theta) e^{\cos\theta} \cos n\theta \, d\theta.$$

[5.38] Equating the real part of (11.4) for $(x, n) \in \mathbb{R} \times 2\mathbb{Z} - \{-1, 1\} \times 2(2\mathbb{Z} + 1)$, we get

$$\cos(x, n) = \frac{2}{\pi \left(1 + (-1)^{n/2} x^n\right)} \int_0^\pi C_1(x, \theta) e^{\cos\theta} \cos n\theta \, d\theta. \tag{11.7}$$

[5.39] From (11.7) for $(x, n) \in \{-1, 1\} \times 2(2\mathbb{Z}+1)$, we get

$$\int_0^\pi C_1(x, \theta) e^{\cos\theta} \cos n\theta \, d\theta = \frac{\pi}{2}\left(1 + (-1)^{n/2} x^n\right) \cos(x, n) = 0.$$

[5.40] Equating the real part of (11.6) for $(x, n) \in \mathbb{R} \times (2\mathbb{Z}+1)$, we get

$$\cos(x, n) = \frac{2}{\pi\left(1+x^{2n}\right)} \int_0^\pi C_1(x, \theta) e^{\cos\theta} \cos n\theta \, d\theta$$

$$+ (-1)^{(n-1)/2} \frac{2x^n}{\pi\left(1+x^{2n}\right)} \int_0^\pi S_1(x, \theta) e^{\cos\theta} \cos n\theta \, d\theta.$$

[5.41] Replace x with ix in [5.35].

[5.42] Replace x with ix in [5.37].

[5.43] From [5.42] for $(x, n) \in \{-1, 1\} \times (2\mathbb{Z}+1)$, we get

$$\int_0^\pi S_2(x, \theta) e^{\cos\theta} \cos n\theta \, d\theta - x^n \int_0^\pi C_2(x, \theta) e^{\cos\theta} \cos n\theta \, d\theta$$
$$= \frac{\pi}{2}\left(1 - x^{2n}\right) \sinh(x, n) = 0$$
$$\iff \int_0^\pi S_2(x, \theta) e^{\cos\theta} \cos n\theta \, d\theta = x^n \int_0^\pi C_2(x, \theta) e^{\cos\theta} \cos n\theta \, d\theta.$$

[5.44] Replace x with ix in [5.38].

[5.45] Replace x with ix in [5.40].

[5.46] From [5.45] for $(x, n) \in \{-1, 1\} \times (2\mathbb{Z}+1)$, we get

$$\int_0^\pi C_2(x, \theta) e^{\cos\theta} \cos n\theta \, d\theta - x^n \int_0^\pi S_2(x, \theta) e^{\cos\theta} \cos n\theta \, d\theta$$
$$= \frac{\pi}{2}\left(1 - x^{2n}\right) \cosh(x, n) = 0$$
$$\iff \int_0^\pi C_2(x, \theta) e^{\cos\theta} \cos n\theta \, d\theta = x^n \int_0^\pi S_2(x, \theta) e^{\cos\theta} \cos n\theta \, d\theta.$$

11 Solutions to Generalized Bessel Functions   345

[5.47] Applying Theorem 4.17 to $f(z) = e^z$ for $n \in \mathbb{N} - 1$, and using (5.1) and (5.11), we get

$$\frac{\pi}{2} T_n(x, y) = \frac{\pi}{2} (x^n + y^n) \sum_{j=0}^{\infty} \binom{n+2j}{j} a_{n+2j}(xy)^j$$

$$= \frac{\pi}{2} (x^n + y^n) \sum_{j=0}^{\infty} \binom{n+2j}{j} \frac{(xy)^j}{(n+2j)!}$$

$$= \frac{\pi}{2} (x^n + y^n) \sum_{j=0}^{\infty} \frac{(xy)^j}{j!(n+j)!}$$

$$= \frac{\pi}{2} (x^n + y^n) e(xy, n)$$

$$= \int_0^{\pi} \frac{e^{xe^{i\theta} + ye^{-i\theta}} + e^{xe^{-i\theta} + ye^{i\theta}}}{2} \cos n\theta \, d\theta$$

$$= \int_0^{\pi} \frac{e^{(x+y)\cos\theta + i(x-y)\sin\theta} + e^{(x+y)\cos\theta - i(x-y)\sin\theta}}{2} \cos n\theta \, d\theta$$

$$= \int_0^{\pi} e^{(x+y)\cos\theta} \frac{e^{i(x-y)\sin\theta} + e^{-i(x-y)\sin\theta}}{2} \cos n\theta \, d\theta$$

$$= \int_0^{\pi} e^{(x+y)\cos\theta} \cos((x-y)\sin\theta) \cos n\theta \, d\theta. \quad (11.8)$$

For $n \in -\mathbb{N}$, using (5.12) and (11.8), we get

$$\int_0^{\pi} e^{(x+y)\cos\theta} \cos((x-y)\sin\theta) \cos n\theta \, d\theta$$

$$= \int_0^{\pi} e^{(x+y)\cos\theta} \cos((x-y)\sin\theta) \cos(-n\theta) \, d\theta$$

$$= \frac{\pi}{2} (x^{-n} + y^{-n}) e(xy, -n)$$

$$= \frac{\pi}{2} (x^{-n} + y^{-n}) (xy)^n e(xy, n)$$

$$= \frac{\pi}{2} (x^n + y^n) e(xy, n).$$

[5.48] Applying Theorem 4.17 to $f(z) = \sin z$ for $(x, y, n) \in \mathbb{R}^2 \times (2\mathbb{N} - 1)$, and using (5.2) and (5.11), we get

$$\frac{\pi}{2} T_n(x, y) = \frac{\pi}{2} (x^n + y^n) \sum_{j=0}^{\infty} \binom{n+2j}{j} a_{n+2j}(xy)^j$$

$$= \frac{\pi}{2}(x^n + y^n) \sum_{j=0}^{\infty} (-1)^{(n+2j-1)/2} \binom{n+2j}{j} \frac{(xy)^j}{(n+2j)!}$$

$$= (-1)^{(n-1)/2} \frac{\pi}{2}(x^n + y^n) \sum_{j=0}^{\infty} \frac{(-xy)^j}{j!(n+j)!}$$

$$= (-1)^{(n-1)/2} \frac{\pi}{2}(x^n + y^n) e(-xy, n)$$

$$= \int_0^{\pi} \frac{\sin(xe^{i\theta} + ye^{-i\theta}) + \sin(xe^{-i\theta} + ye^{i\theta})}{2} \cos n\theta \, d\theta$$

$$= \int_0^{\pi} \sin\left((x+y)\frac{e^{i\theta} + e^{-i\theta}}{2}\right) \cos\left((x-y)\frac{e^{i\theta} - e^{-i\theta}}{2}\right) \cos n\theta \, d\theta$$

$$= \int_0^{\pi} \sin((x+y)\cos\theta) \cos(i(x-y)\sin\theta) \cos n\theta \, d\theta$$

$$= \int_0^{\pi} \sin((x+y)\cos\theta) \cosh((x-y)\sin\theta) \cos n\theta \, d\theta. \quad (11.9)$$

For $n \in -(2\mathbb{N} - 1)$, using (5.12) and (11.9), we get

$$\int_0^{\pi} \sin((x+y)\cos\theta) \cosh((x-y)\sin\theta) \cos n\theta \, d\theta$$

$$= \int_0^{\pi} \sin((x+y)\cos\theta) \cosh((x-y)\sin\theta) \cos(-n\theta) \, d\theta$$

$$= (-1)^{(-n-1)/2} \frac{\pi}{2}(x^{-n} + y^{-n}) e(-xy, -n)$$

$$= (-1)^{(-n-1)/2} \frac{\pi}{2}(x^{-n} + y^{-n})(-1)^n (xy)^n e(-xy, n)$$

$$= (-1)^{(n-1)/2} \frac{\pi}{2}(x^n + y^n) e(-xy, n).$$

[5.49] Applying Theorem 4.17 to $f(z) = \sin z$ for $(x, y, n) \in \mathbb{R}^2 \times (2\mathbb{N} - 2)$, and using (5.2) and (11.9), we get

$$\frac{\pi}{2} T_n(x, y) = \frac{\pi}{2}(x^n + y^n) \sum_{j=0}^{\infty} \binom{n+2j}{j} a_{n+2j}(xy)^j$$

$$= 0$$

$$= \int_0^{\pi} \frac{\sin(xe^{i\theta} + ye^{-i\theta}) + \sin(xe^{-i\theta} + ye^{i\theta})}{2} \cos n\theta \, d\theta$$

$$= \int_0^{\pi} \sin((x+y)\cos\theta) \cosh((x-y)\sin\theta) \cos n\theta \, d\theta.$$

11 Solutions to Generalized Bessel Functions

The above identity is also true for $n \in -2\mathbb{N}$ because $\cos n\theta$ is an even function.

[5.50] Applying Theorem 4.17 to $f(z) = \cos z$ for $(x, y, n) \in \mathbb{R}^2 \times (2\mathbb{N} - 2)$, and using (5.3) and (5.11), we get

$$\frac{\pi}{2} T_n(x, y) = \frac{\pi}{2} (x^n + y^n) \sum_{j=0}^{\infty} \binom{n+2j}{j} a_{n+2j}(xy)^j$$

$$= \frac{\pi}{2} (x^n + y^n) \sum_{j=0}^{\infty} (-1)^{(n+2j)/2} \binom{n+2j}{j} \frac{(xy)^j}{(n+2j)!}$$

$$= (-1)^{n/2} \frac{\pi}{2} (x^n + y^n) \sum_{j=0}^{\infty} \frac{(-xy)^j}{j!(n+j)!}$$

$$= (-1)^{n/2} \frac{\pi}{2} (x^n + y^n) e(-xy, n)$$

$$= \int_0^\pi \frac{\cos\left(xe^{i\theta} + ye^{-i\theta}\right) + \cos\left(xe^{-i\theta} + ye^{i\theta}\right)}{2} \cos n\theta \, d\theta$$

$$= \int_0^\pi \cos\left((x+y)\frac{e^{i\theta} + e^{-i\theta}}{2}\right) \cos\left((x-y)\frac{e^{i\theta} - e^{-i\theta}}{2}\right) \cos n\theta \, d\theta$$

$$= \int_0^\pi \cos((x+y)\cos\theta) \cos(i(x-y)\sin\theta) \cos n\theta \, d\theta$$

$$= \int_0^\pi \cos((x+y)\cos\theta) \cosh((x-y)\sin\theta) \cos n\theta \, d\theta. \qquad (11.10)$$

For $n \in -2\mathbb{N}$, using (5.12) and (11.10), we get

$$\int_0^\pi \cos((x+y)\cos\theta) \cosh((x-y)\sin\theta) \cos n\theta \, d\theta$$

$$= \int_0^\pi \cos((x+y)\cos\theta) \cosh((x-y)\sin\theta) \cos(-n\theta) \, d\theta$$

$$= (-1)^{-n/2} \frac{\pi}{2} (x^{-n} + y^{-n}) e(-xy, -n)$$

$$= (-1)^{-n/2} \frac{\pi}{2} (x^{-n} + y^{-n}) (-1)^n (xy)^n e(-xy, n)$$

$$= (-1)^{n/2} \frac{\pi}{2} (x^n + y^n) e(-xy, n).$$

[5.51] Applying Theorem 4.17 to $f(z) = \cos z$ for $(x, y, n) \in \mathbb{R}^2 \times (2\mathbb{N} - 1)$, and using (5.3) and (11.10), we get

$$\frac{\pi}{2} T_n(x, y) = \frac{\pi}{2} (x^n + y^n) \sum_{j=0}^{\infty} \binom{n+2j}{j} a_{n+2j}(xy)^j$$

$$= 0$$

$$= \int_0^{\pi} \frac{\cos\left(xe^{i\theta} + ye^{-i\theta}\right) + \cos\left(xe^{-i\theta} + ye^{i\theta}\right)}{2} \cos n\theta \, d\theta$$

$$= \int_0^{\pi} \cos((x+y)\cos\theta) \cosh((x-y)\sin\theta) \cos n\theta \, d\theta.$$

The above identity is also true for $n \in -(2\mathbb{N} - 1)$ because $\cos n\theta$ is an even function.

[5.52] Applying Theorem 4.17 to $f(z) = \sinh z$ for $(x, y, n) \in \mathbb{R}^2 \times (2\mathbb{N} - 1)$, and using (5.4) and (5.11), we get

$$\frac{\pi}{2} T_n(x, y) = \frac{\pi}{2} (x^n + y^n) \sum_{j=0}^{\infty} \binom{n+2j}{j} a_{n+2j}(xy)^j$$

$$= \frac{\pi}{2} (x^n + y^n) \sum_{j=0}^{\infty} \binom{n+2j}{j} \frac{(xy)^j}{(n+2j)!}$$

$$= \frac{\pi}{2} (x^n + y^n) \sum_{j=0}^{\infty} \frac{(xy)^j}{j!(n+j)!}$$

$$= \frac{\pi}{2} (x^n + y^n) e(xy, n)$$

$$= \int_0^{\pi} \frac{\sinh\left(xe^{i\theta} + ye^{-i\theta}\right) + \sinh\left(xe^{-i\theta} + ye^{i\theta}\right)}{2} \cos n\theta \, d\theta$$

$$= \int_0^{\pi} \sinh\left((x+y)\frac{e^{i\theta} + e^{-i\theta}}{2}\right) \cosh\left((x-y)\frac{e^{i\theta} - e^{-i\theta}}{2}\right) \cos n\theta \, d\theta$$

$$= \int_0^{\pi} \sinh((x+y)\cos\theta) \cosh(i(x-y)\sin\theta) \cos n\theta \, d\theta$$

$$= \int_0^{\pi} \sinh((x+y)\cos\theta) \cos((x-y)\sin\theta) \cos n\theta \, d\theta. \tag{11.11}$$

For $n \in -(2\mathbb{N} - 1)$, using (5.12) and (11.11), we get

11 Solutions to Generalized Bessel Functions   349

$$\int_0^\pi \sinh\left((x+y)\cos\theta\right)\cos\left((x-y)\sin\theta\right)\cos n\theta\, d\theta$$
$$= \int_0^\pi \sinh\left((x+y)\cos\theta\right)\cos\left((x-y)\sin\theta\right)\cos(-n\theta)\, d\theta$$
$$= \frac{\pi}{2}\left(x^{-n}+y^{-n}\right)e(xy,-n)$$
$$= \frac{\pi}{2}\left(x^{-n}+y^{-n}\right)(xy)^n e(xy,n)$$
$$= \frac{\pi}{2}\left(x^n+y^n\right)e(xy,n).$$

[5.53] Applying Theorem 4.17 to $f(z) = \sinh z$ for $(x, y, n) \in \mathbb{R}^2 \times (2\mathbb{N} - 2)$, and using (5.4) and (11.11), we get

$$\frac{\pi}{2}T_n(x, y) = \frac{\pi}{2}\left(x^n+y^n\right)\sum_{j=0}^\infty \binom{n+2j}{j} a_{n+2j}(xy)^j$$
$$= 0$$
$$= \int_0^\pi \frac{\sinh\left(xe^{i\theta}+ye^{-i\theta}\right)+\sinh\left(xe^{-i\theta}+ye^{i\theta}\right)}{2}\cos n\theta\, d\theta$$
$$= \int_0^\pi \sinh\left((x+y)\cos\theta\right)\cos\left((x-y)\sin\theta\right)\cos n\theta\, d\theta.$$

The above identity is also true for $n \in -2\mathbb{N}$ because $\cos n\theta$ is an even function.

[5.54] Applying Theorem 4.17 to $f(z) = \cosh z$ for $(x, y, n) \in \mathbb{R}^2 \times (2\mathbb{N} - 2)$, and using (5.5) and (5.11), we get

$$\frac{\pi}{2}T_n(x, y) = \frac{\pi}{2}\left(x^n+y^n\right)\sum_{j=0}^\infty \binom{n+2j}{j} a_{n+2j}(xy)^j$$
$$= \frac{\pi}{2}\left(x^n+y^n\right)\sum_{j=0}^\infty \binom{n+2j}{j}\frac{(xy)^j}{(n+2j)!} \qquad (11.12)$$
$$= \frac{\pi}{2}\left(x^n+y^n\right)\sum_{j=0}^\infty \frac{(xy)^j}{j!(n+j)!}$$
$$= \frac{\pi}{2}\left(x^n+y^n\right)e(xy,n)$$

$$= \int_0^\pi \frac{\cosh\left(xe^{i\theta} + ye^{-i\theta}\right) + \cosh\left(xe^{-i\theta} + ye^{i\theta}\right)}{2} \cos n\theta \, d\theta$$

$$= \int_0^\pi \cosh\left((x+y)\frac{e^{i\theta} + e^{-i\theta}}{2}\right) \cosh\left((x-y)\frac{e^{i\theta} - e^{-i\theta}}{2}\right) \cos n\theta \, d\theta$$

$$= \int_0^\pi \cosh((x+y)\cos\theta) \cosh(i(x-y)\sin\theta) \cos n\theta \, d\theta$$

$$= \int_0^\pi \cosh((x+y)\cos\theta) \cos((x-y)\sin\theta) \cos n\theta \, d\theta.$$

For $n \in -2\mathbb{N}$, using (5.12) and (11.12), we get

$$\int_0^\pi \cosh((x+y)\cos\theta) \cos((x-y)\sin\theta) \cos n\theta \, d\theta$$

$$= \int_0^\pi \cosh((x+y)\cos\theta) \cos((x-y)\sin\theta) \cos(-n\theta) \, d\theta$$

$$= \frac{\pi}{2} \left(x^{-n} + y^{-n}\right) e(xy, -n)$$

$$= \frac{\pi}{2} \left(x^{-n} + y^{-n}\right) (xy)^n e(xy, n)$$

$$= \frac{\pi}{2} \left(x^n + y^n\right) e(xy, n).$$

[5.55] Applying Theorem 4.1 to $f(z) = \cosh z$ for $(x, y, n) \in \mathbb{R}^2 \times (2\mathbb{N} - 1)$, and using (5.5) and (11.12), we get

$$\frac{\pi}{2} T_n(x, y) = \frac{\pi}{2} \left(x^n + y^n\right) \sum_{j=0}^\infty \binom{n+2j}{j} a_{n+2j}(xy)^j$$

$$= 0$$

$$= \int_0^\pi \frac{\cosh\left(xe^{i\theta} + ye^{-i\theta}\right) + \cosh\left(xe^{-i\theta} + ye^{i\theta}\right)}{2} \cos n\theta \, d\theta$$

$$= \int_0^\pi \cosh((x+y)\cos\theta) \cos((x-y)\sin\theta) \cos n\theta \, d\theta.$$

The above identity is also true for $n \in -(2\mathbb{N} - 1)$ because $\cos n\theta$ is an even function.

[5.56] Put $y = 1/x$ in [5.47].

[5.57] Put $y = 1/x$ in [5.48].

[5.58] Put $y = 1/x$ in [5.49].

11 Solutions to Generalized Bessel Functions   351

[5.59] Put $y = 1/x$ in [5.50].

[5.60] Put $y = 1/x$ in [5.51].

[5.61] Put $y = 1/x$ in [5.52].

[5.62] Put $y = 1/x$ in [5.53].

[5.63] Put $y = 1/x$ in [5.54].

[5.64] Put $y = 1/x$ in [5.55].

[5.65] Put $n = 0$ in [5.56].

[5.66] Put $y = -1/x$ and $n = 0$ in [5.50].

[5.67] Put $n = 0$ in [5.63].

[5.68] Put $x = 1$ in [5.56].

[5.69] Put $x = 1$ in [5.57].

[5.70] Put $x = 1$ in [5.58].

[5.71] Put $x = 1$ in [5.59].

[5.72] Put $x = 1$ in [5.60].

[5.73] Put $x = 1$ in [5.61].

[5.74] Put $x = 1$ in [5.62].

[5.75] Put $x = 1$ in [5.63].

[5.76] Put $x = 1$ in [5.64].

[5.77] Put $x = 1$ in [5.29].

[5.78] Put $x = 1$ in [4.34].

[5.79] Put $x = 1$ in [5.30].

[5.80] Put $x = 1$ in [4.36].

[5.81] Put $x = 1$ in [5.33].

[5.82] Put $x = 1$ in [4.40].

[5.83] Put $x = 1$ in [5.34].

[5.84] Put $x = 1$ in [4.42].

[5.85] Put $x = 1$ in [4.96].

[5.86] Put $x = 1$ in [4.97].

[5.87] Putting $x = 1$ in [4.97], we get

$$\int_0^\pi e^{\cos\theta} \cos(\sin\theta) \cos n\theta \, d\theta = \frac{\pi}{2n!}. \tag{11.13}$$

Putting $x = -1$ in [4.97], we get

$$\int_0^\pi e^{-\cos\theta} \cos(\sin\theta) \cos n\theta \, d\theta = (-1)^n \frac{\pi}{2n!}. \tag{11.14}$$

Taking the average of (11.13) and (11.14), we get

$$\int_0^\pi \frac{e^{\cos\theta} + e^{-\cos\theta}}{2} \cos(\sin\theta) \cos n\theta \, d\theta = \int_0^\pi \cosh(\cos\theta) \cos(\sin\theta) \cos n\theta \, d\theta$$
$$= \left(\frac{1 + (-1)^n}{2}\right) \frac{\pi}{2n!}. \tag{11.15}$$

For $n \in 2\mathbb{N} - 2$, (11.15) becomes

$$\int_0^\pi \cosh(\cos\theta) \cos(\sin\theta) \cos n\theta \, d\theta = \frac{\pi}{2n!}.$$

[5.88] For $n \in 2\mathbb{N} - 1$, (11.15) becomes

$$\int_0^\pi \cosh(\cos\theta) \cos(\sin\theta) \cos n\theta \, d\theta = 0.$$

The above identity is also true for $n \in -(2\mathbb{N} - 1)$ because $\cos n\theta$ is an even function.

11 Solutions to Generalized Bessel Functions

[5.89] Subtracting (11.14) from (11.13) and dividing the result by 2, we get

$$\int_0^\pi \frac{e^{\cos\theta} - e^{-\cos\theta}}{2} \cos(\sin\theta) \cos n\theta \, d\theta = \int_0^\pi \sinh(\cos\theta) \cos(\sin\theta) \cos n\theta \, d\theta$$
$$= \left(\frac{1-(-1)^n}{2}\right) \frac{\pi}{2n!}. \tag{11.16}$$

For $n \in 2\mathbb{N} - 1$, (11.16) becomes

$$\int_0^\pi \sinh(\cos\theta) \cos(\sin\theta) \cos n\theta \, d\theta = \frac{\pi}{2n!}.$$

[5.90] For $n \in 2\mathbb{N} - 2$, (11.16) becomes

$$\int_0^\pi \sinh(\cos\theta) \cos(\sin\theta) \cos n\theta \, d\theta = 0.$$

The above identity is also true for $n \in -2\mathbb{N}$ because $\cos n\theta$ is an even function.

[5.91] Put $x = 1$ and $y = -1$ in [5.54].

[5.92] Put $x = 1$ and $y = -1$ in [5.55].

[5.93] From (5.14), (5.15) and (5.16), we have

$$\frac{e(iz,n) - e(-iz,n)}{2i}$$
$$= \frac{(\cos(z,n) + i\sin(z,n)) - (\cos(-z,n) + i\sin(-z,n))}{2i}$$
$$= \frac{(\cos(z,n) + i\sin(z,n)) - (\cos(z,n) - i\sin(z,n))}{2i}$$
$$= \frac{2i\sin(z,n)}{2i}$$
$$= \sin(z,n).$$

[5.94] From (5.14), (5.15), and (5.16), we have

$$\frac{e(iz, n) + e(-iz, n)}{2}$$
$$= \frac{(\cos(z, n) + i \sin(z, n)) + (\cos(-z, n) + i \sin(-z, n))}{2}$$
$$= \frac{(\cos(z, n) + i \sin(z, n)) + (\cos(z, n) - i \sin(z, n))}{2}$$
$$= \frac{2 \cos(z, n)}{2}$$
$$= \cos(z, n).$$

[5.95] From (5.14), (5.17), (5.18), (5.19), and (5.20), we have

$$\frac{e(z, n) - e(-z, n)}{2}$$
$$= \frac{(\cos(-iz, n) + i \sin(-iz, n)) - (\cos(iz, n) + i \sin(iz, n))}{2}$$
$$= \frac{(\cosh(-z, n) - \sinh(-z, n)) - (\cosh(z, n) - \sinh(z, n))}{2}$$
$$= \frac{(\cosh(z, n) + \sinh(z, n)) - (\cosh(z, n) - \sinh(z, n))}{2}$$
$$= \frac{2 \sinh(z, n)}{2}$$
$$= \sinh(z, n).$$

[5.96] From (5.14), (5.17), (5.18), (5.19), and (5.20), we have

$$\frac{e(z, n) + e(-z, n)}{2}$$
$$= \frac{(\cos(-iz, n) + i \sin(-iz, n)) + (\cos(iz, n) + i \sin(iz, n))}{2}$$
$$= \frac{(\cosh(-z, n) - \sinh(-z, n)) + (\cosh(z, n) - \sinh(z, n))}{2}$$
$$= \frac{(\cosh(z, n) + \sinh(z, n)) + (\cosh(z, n) - \sinh(z, n))}{2}$$
$$= \frac{2 \cosh(z, n)}{2}$$
$$= \cosh(z, n).$$

Chapter 12
Solutions to Series Sums Using Special Integrals

[6.1] Replacing x with $-1 + ix$ in [5.21], using (5.14), and $\phi(x) = \tan^{-1} x$, we get the LHS of [5.21] as

$$e(-1 + ix, n) = e\left(-\sqrt{1 + x^2}e^{-i\phi(x)}, n\right)$$

$$= \sum_{j=0}^{\infty} (-1)^j \frac{(1 + x^2)^{j/2} e^{-ij\phi(x)}}{j!(n+j)!}$$

$$= \sum_{j=0}^{\infty} \frac{(1+x^2)^j e^{-2ij\phi(x)}}{(2j)!(n+2j)!} - \sqrt{1+x^2} \sum_{j=0}^{\infty} \frac{(1+x^2)^j e^{-i(2j+1)\phi(x)}}{(2j+1)!(n+2j+1)!}$$

$$= \sum_{j=0}^{\infty} \frac{(1+x^2)^j (\cos 2j\phi(x) - i \sin 2j\phi(x))}{(2j)!(n+2j)!}$$

$$- \sqrt{1+x^2} \sum_{j=0}^{\infty} \frac{(1+x^2)^j (\cos(2j+1)\phi(x) - i \sin(2j+1)\phi(x))}{(2j+1)!(n+2j+1)!}$$

$$= \left(\sum_{j=0}^{\infty} \frac{(1+x^2)^j \cos 2j\phi(x)}{(2j)!(n+2j)!} - \sqrt{1+x^2} \sum_{j=0}^{\infty} \frac{(1+x^2)^j \cos(2j+1)\phi(x)}{(2j+1)!(n+2j+1)!}\right)$$

$$+ i\left(\sqrt{1+x^2} \sum_{j=0}^{\infty} \frac{(1+x^2)^j \sin(2j+1)\phi(x)}{(2j+1)!(n+2j+1)!} - \sum_{j=1}^{\infty} \frac{(1+x^2)^j \sin 2j\phi(x)}{(2j)!(n+2j)!}\right).$$

(12.1)

Replacing x with $-1 + ix$ in [5.21] and using $\phi(x) = \tan^{-1} x$, we simplify the factor of the integral on RHS of [5.21] as

$$\frac{2}{\pi} \cdot \frac{1}{1+(-1+ix)^n} = \frac{2}{\pi} \cdot \frac{1}{1+\left(1+x^2\right)^{n/2} e^{in(\pi-\phi(x))}}$$

$$= \frac{2}{\pi} \cdot \frac{1}{1+\left(1+x^2\right)^{n/2} \cos n(\pi-\phi(x)) + i\left(1+x^2\right)^{n/2} \sin n(\pi-\phi(x))}$$

$$= \frac{2}{\pi} \cdot \frac{1+\left(1+x^2\right)^{n/2} \cos n(\pi-\phi(x)) - i\left(1+x^2\right)^{n/2} \sin n(\pi-\phi(x))}{1+\left(1+x^2\right)^n + 2\left(1+x^2\right)^{n/2} \cos n(\pi-\phi(x))}.$$
(12.2)

Replacing x with $-1+ix$ in [5.21] and using (6.28) and (6.29), we simplify the integral on RHS of [5.21] as

$$\int_0^\pi e^{ix\cos\theta} \cos((ix-2)\sin\theta) \cos n\theta\, d\theta$$

$$= \int_0^\pi (\cos(x\cos\theta) + i\sin(x\cos\theta))$$

$$(\cos(ix\sin\theta)\cos(2\sin\theta) + \sin(ix\sin\theta)\sin(2\sin\theta))\cos n\theta\, d\theta$$

$$= \int_0^\pi (\cos(x\cos\theta) + i\sin(x\cos\theta))$$

$$(\cosh(x\sin\theta)\cos(2\sin\theta) + i\sinh(x\sin\theta)\sin(2\sin\theta))\cos n\theta\, d\theta$$
(12.3)

$$= \int_0^\pi (\cos(x\cos\theta)\cosh(x\sin\theta)\cos(2\sin\theta)$$

$$-\sin(x\cos\theta)\sinh(x\sin\theta)\sin(2\sin\theta))\cos n\theta\, d\theta$$

$$+ i\int_0^\pi (\cos(x\cos\theta)\sinh(x\sin\theta)\sin(2\sin\theta)$$

$$+\sin(x\cos\theta)\cosh(x\sin\theta)\cos(2\sin\theta))\cos n\theta\, d\theta$$

$$= \int_0^\pi C_3(x,\theta)\cos n\theta\, d\theta + i\int_0^\pi S_3(x,\theta)\cos n\theta\, d\theta.$$

Combining (12.1), (12.2), and (12.3), using [5.21], and $\phi(x) = \tan^{-1} x$, we get

$$\left(\sum_{j=0}^\infty \frac{(1+x^2)^j \cos 2j\phi(x)}{(2j)!(n+2j)!} - \sqrt{1+x^2} \sum_{j=0}^\infty \frac{(1+x^2)^j \cos(2j+1)\phi(x)}{(2j+1)!(n+2j+1)!}\right)$$

$$+ i\left(\sqrt{1+x^2}\sum_{j=0}^\infty \frac{(1+x^2)^j \sin(2j+1)\phi(x)}{(2j+1)!(n+2j+1)!} - \sum_{j=1}^\infty \frac{(1+x^2)^j \sin 2j\phi(x)}{(2j)!(n+2j)!}\right)$$

$$= \frac{2}{\pi} \cdot \frac{1+\left(1+x^2\right)^{n/2} \cos n(\pi-\phi(x)) - i\left(1+x^2\right)^{n/2} \sin n(\pi-\phi(x))}{1+\left(1+x^2\right)^n + 2\left(1+x^2\right)^{n/2} \cos n(\pi-\phi(x))}$$

$$\left(\int_0^\pi C_3(x,\theta)\cos n\theta\, d\theta + i\int_0^\pi S_3(x,\theta)\cos n\theta\, d\theta\right).$$
(12.4)

12 Solutions to Series Sums Using Special Integrals   357

We get the result by equating the real part of (12.4).

[6.2] Put $n = 0$ in [6.1].

[6.3] Equate the imaginary part of (12.4).

[6.4] Put $n = 0$ in [6.3].

[6.5] Put $x = 1$ in [6.4].

[6.6] Put $x = 1$ in [6.2].

[6.7] Replace x with ix in [6.1].

[6.8] Put $n = 0$ in [6.7].

[6.9] Replace x with ix in [6.3].

[6.10] Put $n = 0$ in [6.9].

[6.11] Replacing x with $1 + ix$ in [5.21], and using (5.14) and $\phi(x) = \tan^{-1} x$, we get the LHS of [5.21] as

$$e(1 + ix, n) = e\left(\sqrt{1 + x^2} e^{i\phi(x)}, n\right)$$

$$= \sum_{j=0}^{\infty} \frac{(1 + x^2)^{j/2} e^{ij\phi(x)}}{j!(n + j)!}$$

$$= \sum_{j=0}^{\infty} \frac{(1 + x^2)^j e^{2ij\phi(x)}}{(2j)!(n + 2j)!} + \sqrt{1 + x^2} \sum_{j=0}^{\infty} \frac{(1 + x^2)^j e^{i(2j+1)\phi(x)}}{(2j + 1)!(n + 2j + 1)!}$$

$$= \sum_{j=0}^{\infty} \frac{(1 + x^2)^j (\cos 2j\phi(x) + i \sin 2j\phi(x))}{(2j)!(n + 2j)!}$$

$$+ \sqrt{1 + x^2} \sum_{j=0}^{\infty} \frac{(1 + x^2)^j (\cos(2j + 1)\phi(x) + i \sin(2j + 1)\phi(x))}{(2j + 1)!(n + 2j + 1)!}$$

$$= \left(\sum_{j=0}^{\infty} \frac{(1 + x^2)^j \cos 2j\phi(x)}{(2j)!(n + 2j)!} + \sqrt{1 + x^2} \sum_{j=0}^{\infty} \frac{(1 + x^2)^j \cos(2j + 1)\phi(x)}{(2j + 1)!(n + 2j + 1)!} \right)$$

$$+ i \left(\sqrt{1 + x^2} \sum_{j=0}^{\infty} \frac{(1 + x^2)^j \sin(2j + 1)\phi(x)}{(2j + 1)!(n + 2j + 1)!} + \sum_{j=1}^{\infty} \frac{(1 + x^2)^j \sin 2j\phi(x)}{(2j)!(n + 2j)!} \right).$$

(12.5)

Replacing x with $1+ix$ in [5.21] and using $\phi(x) = \tan^{-1} x$, we simplify the factor of the integral on RHS of [5.21] as

$$\frac{2}{\pi} \cdot \frac{1}{1+(1+ix)^n} = \frac{2}{\pi} \cdot \frac{1}{1+\left(1+x^2\right)^{n/2} e^{in\phi(x)}}$$
$$= \frac{2}{\pi} \cdot \frac{1}{1+\left(1+x^2\right)^{n/2} \cos n\phi(x) + i\left(1+x^2\right)^{n/2} \sin n\phi(x)} \quad (12.6)$$
$$= \frac{2}{\pi} \cdot \frac{1+\left(1+x^2\right)^{n/2} \cos n\phi(x) - i\left(1+x^2\right)^{n/2} \sin n\phi(x)}{1+\left(1+x^2\right)^n + 2\left(1+x^2\right)^{n/2} \cos n\phi(x)}.$$

Replacing x with $1+ix$ in [5.21], we simplify the integral on RHS of [5.21] as

$$\int_0^\pi e^{(ix+2)\cos\theta} \cos(ix\sin\theta) \cos n\theta\, d\theta$$
$$= \int_0^\pi e^{2\cos\theta} e^{ix\cos\theta} \cosh(x\sin\theta) \cos n\theta\, d\theta$$
$$= \int_0^\pi e^{2\cos\theta} (\cos(x\cos\theta) + i\sin(x\cos\theta)) \cosh(x\sin\theta) \cos n\theta\, d\theta$$
$$= \int_0^\pi e^{2\cos\theta} \cos(x\cos\theta) \cosh(x\sin\theta) \cos n\theta\, d\theta$$
$$+ i \int_0^\pi e^{2\cos\theta} \sin(x\cos\theta) \cosh(x\sin\theta) \cos n\theta\, d\theta. \quad (12.7)$$

Combining (12.5), (12.6), and (12.7), using [5.21], and $\phi(x) = \tan^{-1} x$, we get

$$\left(\sum_{j=0}^\infty \frac{(1+x^2)^j \cos 2j\phi(x)}{(2j)!(n+2j)!} + \sqrt{1+x^2} \sum_{j=0}^\infty \frac{(1+x^2)^j \cos(2j+1)\phi(x)}{(2j+1)!(n+2j+1)!} \right)$$
$$+ i \left(\sqrt{1+x^2} \sum_{j=0}^\infty \frac{(1+x^2)^j \sin(2j+1)\phi(x)}{(2j+1)!(n+2j+1)!} + \sum_{j=1}^\infty \frac{(1+x^2)^j \sin 2j\phi(x)}{(2j)!(n+2j)!} \right)$$
$$= \frac{2}{\pi} \cdot \frac{1+\left(1+x^2\right)^{n/2} \cos n\phi(x) - i\left(1+x^2\right)^{n/2} \sin n\phi(x)}{1+\left(1+x^2\right)^n + 2\left(1+x^2\right)^{n/2} \cos n\phi(x)}$$
$$\left(\int_0^\pi e^{2\cos\theta} \cos(x\cos\theta) \cosh(x\sin\theta) \cos n\theta\, d\theta \right.$$
$$\left. + i \int_0^\pi e^{2\cos\theta} \sin(x\cos\theta) \cosh(x\sin\theta) \cos n\theta\, d\theta \right).$$
$$(12.8)$$

12 Solutions to Series Sums Using Special Integrals

We get the result by equating the real part of (12.8).

[6.12] Put $n = 0$ in [6.11].

[6.13] Equate the imaginary part of (12.8).

[6.14] Put $n = 0$ in [6.13].

[6.15] Put $x = 1$ in [6.12].

[6.16] Put $x = 1$ in [6.14].

[6.17] Replace x with ix in [6.11].

[6.18] Put $n = 0$ in [6.17].

[6.19] Replace x with ix in [6.13].

[6.20] Put $n = 0$ in [6.19].

[6.21] Replacing x with $xe^{i\phi}$ in [5.21] and using (5.14), we get the LHS of [5.21] as

$$e\left(xe^{i\phi}, n\right) = \sum_{j=0}^{\infty} \frac{x^j e^{ij\phi}}{j!(n+j)!}$$
$$= \sum_{j=0}^{\infty} \frac{x^j (\cos j\phi + i \sin j\phi)}{j!(n+j)!} \qquad (12.9)$$
$$= \sum_{j=0}^{\infty} \frac{x^j \cos j\phi}{j!(n+j)!} + i \sum_{j=1}^{\infty} \frac{x^j \sin j\phi}{j!(n+j)!}.$$

Replacing x with $xe^{i\phi}$ in [5.21], we simplify the factor of the integral on RHS of [5.21] as

$$\frac{2}{\pi} \cdot \frac{1}{1 + x^n e^{in\phi}} = \frac{2}{\pi} \cdot \frac{1}{1 + x^n \cos n\phi + ix^n \sin n\phi}$$
$$= \frac{2}{\pi} \cdot \frac{1 + x^n \cos n\phi - ix^n \sin n\phi}{1 + x^{2n} + 2x^n \cos n\phi}. \qquad (12.10)$$

Replacing x with $xe^{i\phi}$ in [5.21], and using (6.32) and (6.33), we simplify the integral on RHS of [5.21] as

$$\int_0^\pi e^{(xe^{i\phi}+1)\cos\theta} \cos\left((xe^{i\phi}-1)\sin\theta\right) \cos n\theta \, d\theta$$

$$= \int_0^\pi e^{(1+x\cos\phi+ix\sin\phi)\cos\theta} \cos((x\cos\phi-1)\sin\theta + ix\sin\phi\sin\theta) \cos n\theta \, d\theta$$

$$= \int_0^\pi e^{(1+x\cos\phi)\cos\theta}(\cos(x\sin\phi\cos\theta) + i\sin(x\sin\phi\cos\theta))$$
$$(\cos((x\cos\phi-1)\sin\theta)\cosh(x\sin\phi\sin\theta)$$
$$- i\sin((x\cos\phi-1)\sin\theta)\sinh(x\sin\phi\sin\theta))\cos n\theta \, d\theta$$

$$= \int_0^\pi (\cos(x\sin\phi\cos\theta)\cos((x\cos\phi-1)\sin\theta)\cosh(x\sin\phi\sin\theta)$$
$$+ \sin(x\sin\phi\cos\theta)\sin((x\cos\phi-1)\sin\theta)\sinh(x\sin\phi\sin\theta))$$
$$e^{(1+x\cos\phi)\cos\theta}\cos n\theta \, d\theta$$

$$+ i\int_0^\pi (\sin(x\sin\phi\cos\theta)\cos((x\cos\phi-1)\sin\theta)\cosh(x\sin\phi\sin\theta)$$
$$- \cos(x\sin\phi\cos\theta)\sin((x\cos\phi-1)\sin\theta)\sinh(x\sin\phi\sin\theta))$$
$$e^{(1+x\cos\phi)\cos\theta}\cos n\theta \, d\theta$$

$$= \int_0^\pi C_5(x,\phi,\theta)e^{(1+x\cos\phi)\cos\theta}\cos n\theta \, d\theta + i\int_0^\pi S_5(x,\phi,\theta)e^{(1+x\cos\phi)\cos\theta}$$
$$\cos n\theta \, d\theta. \tag{12.11}$$

Combining (12.9), (12.10), and (12.11), and using [5.21], we get

$$\sum_{j=0}^\infty \frac{x^j \cos j\phi}{j!(n+j)!} + i\sum_{j=1}^\infty \frac{x^j \sin j\phi}{j!(n+j)!}$$
$$= \frac{2}{\pi} \cdot \frac{1+x^n\cos n\phi - ix^n\sin n\phi}{1+x^{2n}+2x^n\cos n\phi} \tag{12.12}$$
$$\left(\int_0^\pi C_5(x,\phi,\theta)e^{(1+x\cos\phi)\cos\theta}\cos n\theta \, d\theta\right.$$
$$\left. + i\int_0^\pi S_5(x,\phi,\theta)e^{(1+x\cos\phi)\cos\theta}\cos n\theta \, d\theta\right).$$

We get the result by equating the real part of (12.12).

[6.22] Put $n = 0$ in [6.21].

[6.23] Equate the imaginary part of (12.12).

[6.24] Put $n = 0$ in [6.23].

12 Solutions to Series Sums Using Special Integrals

[6.25] Replace ϕ with $i\phi$ in [6.21].

[6.26] Put $n = 0$ in [6.25].

[6.27] Replace ϕ with $i\phi$ in [6.23].

[6.28] Put $n = 0$ in [6.27].

[6.29] Put $x = \sec\phi$ in [6.22].

[6.30] Put $x = -\sec\phi$ in [6.22].

[6.31] Put $x = \sec\phi$ in [6.24].

[6.32] Put $x = -\sec\phi$ in [6.24].

[6.33] Put $x = \text{sech}\,\phi$ in [6.26].

[6.34] Put $x = -\text{sech}\,\phi$ in [6.26].

[6.35] Put $x = \text{sech}\,\phi$ in [6.28].

[6.36] Put $x = -\text{sech}\,\phi$ in [6.28].

[6.37] Applying Theorem 6.13 to $g(z) = e^z$, we get

$$\begin{aligned}
f(1) \star e^x &= \sum_{j=0}^{\infty} \frac{a_j}{j!} x^j \\
&= -a_0 + \frac{1}{2\pi} \int_0^{\pi} \left(f\left(e^{i\theta}\right) + f\left(e^{-i\theta}\right)\right) \left(e^{xe^{i\theta}} + e^{xe^{-i\theta}}\right) d\theta \\
&= -a_0 + \frac{1}{2\pi} \int_0^{\pi} \left(f\left(e^{i\theta}\right) + f\left(e^{-i\theta}\right)\right) \left(e^{x(\cos\theta + i\sin\theta)} + e^{x(\cos\theta - i\sin\theta)}\right) d\theta \\
&= -a_0 + \frac{1}{2\pi} \int_0^{\pi} \left(f\left(e^{i\theta}\right) + f\left(e^{-i\theta}\right)\right) \left(e^{ix\sin\theta} + e^{-ix\sin\theta}\right) e^{x\cos\theta} d\theta \\
&= -f(0) + \frac{1}{\pi} \int_0^{\pi} \left(f\left(e^{i\theta}\right) + f\left(e^{-i\theta}\right)\right) \cos(x\sin\theta) e^{x\cos\theta} d\theta.
\end{aligned}$$

[6.38] Applying Theorem 6.13 to $g(z) = 1/(1-z) = \sum_{j=0}^{\infty} z^j$, we get

$$f(1) \star \frac{1}{1-x} = \sum_{j=0}^{\infty} a_j x^j$$

$$= -a_0 + \frac{1}{2\pi} \int_0^{\pi} \left(f\left(e^{i\theta}\right) + f\left(e^{-i\theta}\right)\right) \left(\frac{1}{1-xe^{i\theta}} + \frac{1}{1-xe^{-i\theta}}\right) d\theta$$

$$= -a_0 + \frac{1}{2\pi} \int_0^{\pi} \left(f\left(e^{i\theta}\right) + f\left(e^{-i\theta}\right)\right) \frac{2 - x\left(e^{i\theta} + e^{-i\theta}\right)}{1 + x^2 - x\left(e^{i\theta} + e^{-i\theta}\right)} d\theta$$

$$= -f(0) + \frac{1}{\pi} \int_0^{\pi} \left(f\left(e^{i\theta}\right) + f\left(e^{-i\theta}\right)\right) \frac{1 - x\cos\theta}{1 + x^2 - 2x\cos\theta} d\theta.$$

[6.39] Applying Theorem 6.13 to $g(z) = -\log(1-z) = \sum_{j=0}^{\infty} z^j/j$, we get

$$f(1) \star \frac{1}{1-x} = \sum_{j=1}^{\infty} \frac{a_j}{j} x^j$$

$$= -\frac{1}{2\pi} \int_0^{\pi} \left(f\left(e^{i\theta}\right) + f\left(e^{-i\theta}\right)\right) \left(\log\left(1 - xe^{i\theta}\right) + \log\left(1 - xe^{-i\theta}\right)\right) d\theta$$

$$= -\frac{1}{2\pi} \int_0^{\pi} \left(f\left(e^{i\theta}\right) + f\left(e^{-i\theta}\right)\right) \log\left(1 + x^2 - x\left(e^{i\theta} + e^{-i\theta}\right)\right) d\theta$$

$$= -\frac{1}{2\pi} \int_0^{\pi} \left(f\left(e^{i\theta}\right) + f\left(e^{-i\theta}\right)\right) \log\left(1 + x^2 - 2x\cos\theta\right) d\theta.$$

[6.40] Applying Theorem 6.13 to $g(z) = z/(1-z)^2 = \sum_{j=1}^{\infty} jz^j$, we get

$$f(1) \star \frac{x}{(1-x)^2} = \sum_{j=1}^{\infty} ja_j x^j$$

$$= \frac{1}{2\pi} \int_0^{\pi} \left(f\left(e^{i\theta}\right) + f\left(e^{-i\theta}\right)\right) \left(\frac{xe^{i\theta}}{\left(1-xe^{i\theta}\right)^2} + \frac{xe^{-i\theta}}{\left(1-xe^{-i\theta}\right)^2}\right) d\theta$$

$$= \frac{1}{2\pi} \int_0^{\pi} x\left(f\left(e^{i\theta}\right) + f\left(e^{-i\theta}\right)\right)$$
$$\frac{e^{i\theta}\left(1 + x^2 e^{-2i\theta} - 2xe^{-i\theta}\right) + e^{-i\theta}\left(1 + x^2 e^{2i\theta} - 2xe^{i\theta}\right)}{\left(1 + x^2 - x\left(e^{i\theta} + e^{-i\theta}\right)\right)^2} d\theta$$

$$= \frac{1}{2\pi} \int_0^{\pi} x\left(f\left(e^{i\theta}\right) + f\left(e^{-i\theta}\right)\right) \frac{e^{i\theta} + e^{-i\theta} + x^2\left(e^{i\theta} + e^{-i\theta}\right) - 4x}{\left(1 + x^2 - 2x\cos\theta\right)^2} d\theta$$

$$= \frac{1}{\pi} \int_0^{\pi} x\left(f\left(e^{i\theta}\right) + f\left(e^{-i\theta}\right)\right) \frac{(1+x^2)\cos\theta - 2x}{\left(1 + x^2 - 2x\cos\theta\right)^2} d\theta.$$

12 Solutions to Series Sums Using Special Integrals   363

[6.41] *First Method:* Proved in [5.24].
  Second Method: Applying Theorem 6.13 to $f(z) = e^z$ and $g(z) = e^z$, we get

$$e^1 \star e^x = \sum_{j=0}^{\infty} \frac{x^j}{(j!)^2} = e(x, 0)$$

$$= -1 + \frac{1}{2\pi} \int_0^{\pi} \left(e^{e^{i\theta}} + e^{e^{-i\theta}}\right)\left(e^{xe^{i\theta}} + e^{xe^{-i\theta}}\right) d\theta$$

$$= -1 + \frac{1}{2\pi} \int_0^{\pi} \left(e^{\cos\theta + i\sin\theta} + e^{\cos\theta - i\sin\theta}\right)\left(e^{x\cos\theta + ix\sin\theta} + e^{x\cos\theta - ix\sin\theta}\right) d\theta$$

$$= -1 + \frac{1}{2\pi} \int_0^{\pi} e^{(x+1)\cos\theta} \left(e^{i\sin\theta} + e^{-i\sin\theta}\right)\left(e^{ix\sin\theta} + e^{-ix\sin\theta}\right) d\theta$$

$$= -1 + \frac{2}{\pi} \int_0^{\pi} e^{(x+1)\cos\theta} \cos(\sin\theta)\cos(x\sin\theta)\, d\theta.$$

[6.42] *First Method:* Proved in [5.25].
  Second Method: Applying Theorem 6.14 to $f(z) = e^z$ and $g(z) = e^z$, we get

$$e^1 \star e^x = \sum_{j=0}^{\infty} \frac{x^j}{(j!)^2} = e(x, 0)$$

$$= 1 - \frac{1}{2\pi} \int_0^{\pi} \left(e^{e^{i\theta}} - e^{e^{-i\theta}}\right)\left(e^{xe^{i\theta}} - e^{xe^{-i\theta}}\right) d\theta$$

$$= 1 - \frac{1}{2\pi} \int_0^{\pi} \left(e^{\cos\theta + i\sin\theta} - e^{\cos\theta - i\sin\theta}\right)\left(e^{x\cos\theta + ix\sin\theta} - e^{x\cos\theta - ix\sin\theta}\right) d\theta$$

$$= 1 - \frac{1}{2\pi} \int_0^{\pi} e^{(x+1)\cos\theta} \left(e^{i\sin\theta} - e^{-i\sin\theta}\right)\left(e^{ix\sin\theta} - e^{-ix\sin\theta}\right) d\theta$$

$$= 1 + \frac{2}{\pi} \int_0^{\pi} e^{(x+1)\cos\theta} \sin(\sin\theta)\sin(x\sin\theta)\, d\theta.$$

[6.43] Subtract [6.42] from [6.41].

[6.44] Applying Theorem 6.13 to $g(z) = f(z)$, we get

$$f(1) \star f(1) = \sum_{j=0}^{\infty} a_j^2$$
$$= -f(0)^2 + \frac{1}{2\pi} \int_0^{\pi} \left(f\left(e^{i\theta}\right) + f\left(e^{-i\theta}\right) \right)^2 d\theta.$$

[6.45] Applying Theorem 6.13 to $f(z) = e^z = \sum_{j=0}^{\infty} z^j/j!$ and $g(z) = 1/(1-z) = \sum_{j=0}^{\infty} z^j$, we get

$$e^x \star \frac{1}{1-y} = \sum_{j=0}^{\infty} \frac{(xy)^j}{j!} = e^{xy}$$

$$= -1 + \frac{1}{2\pi} \int_0^{\pi} \left(e^{xe^{i\theta}} + e^{xe^{-i\theta}} \right) \left(\frac{1}{1-ye^{i\theta}} + \frac{1}{1-ye^{-i\theta}} \right) d\theta$$

$$= -1 + \frac{1}{2\pi} \int_0^{\pi} \left(e^{x\cos\theta + ix\sin\theta} + e^{x\cos\theta - ix\sin\theta} \right) \frac{2 - y\left(e^{i\theta} + e^{-i\theta}\right)}{1 + y^2 - y\left(e^{i\theta} + e^{-i\theta}\right)} d\theta$$

$$= -1 + \frac{1}{2\pi} \int_0^{\pi} e^{x\cos\theta} \left(e^{ix\sin\theta} + e^{-ix\sin\theta} \right) \frac{2 - 2y\cos\theta}{1 + y^2 - 2y\cos\theta} d\theta$$

$$= -1 + \frac{2}{\pi} \int_0^{\pi} e^{x\cos\theta} \cos(x\sin\theta) \frac{1 - y\cos\theta}{1 + y^2 - 2y\cos\theta} d\theta$$

$$\iff \int_0^{\pi} \frac{1 - y\cos\theta}{1 + y^2 - 2y\cos\theta} e^{x\cos\theta} \cos(x\sin\theta) d\theta = \frac{\pi}{2} \left(e^{xy} + 1 \right).$$

[6.46] Use [6.39].

[6.47] Putting $f(x) = \log(1+x)$ in [6.46], we get

$$\int_0^x \frac{\log(1+x)}{x} dx$$

$$= -\frac{1}{2\pi} \int_0^{\pi} \left(\log\left(1 + e^{i\theta}\right) + \log\left(1 + e^{-i\theta}\right) \right) \log\left(1 + x^2 - 2x\cos\theta\right) d\theta$$

$$= -\frac{1}{2\pi} \int_0^{\pi} \log\left(2 + e^{i\theta} + e^{-i\theta}\right) \log\left(1 + x^2 - 2x\cos\theta\right) d\theta$$

$$= -\frac{1}{2\pi} \int_0^{\pi} \log\left(2 + 2\cos\theta\right) \log\left(1 + x^2 - 2x\cos\theta\right) d\theta$$

$$= -\frac{1}{2\pi} \int_0^{\pi} \log\left(4\cos^2 \frac{\theta}{2}\right) \log\left(1 + x^2 - 2x\cos\theta\right) d\theta$$

$$= -\frac{1}{\pi} \int_0^{\pi} \log\left(2\cos \frac{\theta}{2}\right) \log\left(1 + x^2 - 2x\cos\theta\right) d\theta.$$

[6.48] Putting $f(x) = \sin x$ in [6.46], we get

$$\int_0^x \frac{\sin x}{x} dx = -\frac{1}{2\pi} \int_0^\pi \left(\sin\left(e^{i\theta}\right) + \sin\left(e^{-i\theta}\right)\right) \log\left(1 + x^2 - 2x\cos\theta\right) d\theta$$

$$= -\frac{1}{\pi} \int_0^\pi \sin\left(\frac{e^{i\theta} + e^{-i\theta}}{2}\right) \cos\left(\frac{e^{i\theta} - e^{-i\theta}}{2}\right) \log\left(1 + x^2 - 2x\cos\theta\right) d\theta$$

$$= -\frac{1}{\pi} \int_0^\pi \sin(\cos\theta) \cos(i\sin\theta) \log\left(1 + x^2 - 2x\cos\theta\right) d\theta$$

$$= -\frac{1}{\pi} \int_0^\pi \sin(\cos\theta) \cosh(\sin\theta) \log\left(1 + x^2 - 2x\cos\theta\right) d\theta.$$

[6.49] Putting $f(x) = \sinh x$ in [6.46], we get

$$\int_0^x \frac{\sinh x}{x} dx = -\frac{1}{2\pi} \int_0^\pi \left(\sinh\left(e^{i\theta}\right) + \sinh\left(e^{-i\theta}\right)\right) \log\left(1 + x^2 - 2x\cos\theta\right) d\theta$$

$$= -\frac{1}{\pi} \int_0^\pi \sinh\left(\frac{e^{i\theta} + e^{-i\theta}}{2}\right) \cosh\left(\frac{e^{i\theta} - e^{-i\theta}}{2}\right) \log\left(1 + x^2 - 2x\cos\theta\right) d\theta$$

$$= -\frac{1}{\pi} \int_0^\pi \sinh(\cos\theta) \cosh(i\sin\theta) \log\left(1 + x^2 - 2x\cos\theta\right) d\theta$$

$$= -\frac{1}{\pi} \int_0^\pi \sinh(\cos\theta) \cos(\sin\theta) \log\left(1 + x^2 - 2x\cos\theta\right) d\theta.$$

[6.50] Applying Theorem 6.13 to $f(z) = \log(1 - z) = -\sum_{j=1}^\infty z^j/j$ and $g(z) = 1/(1-z) = \sum_{j=0}^\infty z^j$, we get

$$\log(1-x) \star \frac{1}{1-y} = -\sum_{j=1}^\infty \frac{(xy)^j}{j} = \log(1-xy)$$

$$= \frac{1}{2\pi} \int_0^\pi \left(\frac{1}{1-ye^{i\theta}} + \frac{1}{1-ye^{-i\theta}}\right) \left(\log\left(1 - xe^{i\theta}\right) + \log\left(1 - xe^{-i\theta}\right)\right) d\theta$$

$$= \frac{1}{2\pi} \int_0^\pi \frac{2 - y\left(e^{i\theta} + e^{-i\theta}\right)}{1 + y^2 - y\left(e^{i\theta} + e^{-i\theta}\right)} \log\left(1 + x^2 - x\left(e^{i\theta} + e^{-i\theta}\right)\right) d\theta$$

$$= \frac{1}{\pi} \int_0^\pi \frac{1 - y\cos\theta}{1 + y^2 - 2y\cos\theta} \log\left(1 + x^2 - 2x\cos\theta\right) d\theta.$$

[6.51] Putting $f(z) = \log(1+z)$ in [6.38], we get

$$\log(1+x) = \frac{1}{\pi}\int_0^\pi \left(\log\left(1+e^{i\theta}\right) + \log\left(1+e^{-i\theta}\right)\right)\frac{1-x\cos\theta}{1+x^2-2x\cos\theta}\,d\theta$$

$$= \frac{1}{\pi}\int_0^\pi \log\left(2+e^{i\theta}+e^{-i\theta}\right)\frac{1-x\cos\theta}{1+x^2-2x\cos\theta}\,d\theta$$

$$= \frac{1}{\pi}\int_0^\pi \log\left(2+2\cos\theta\right)\frac{1-x\cos\theta}{1+x^2-2x\cos\theta}\,d\theta$$

$$= \frac{1}{\pi}\int_0^\pi \log\left(4\cos^2\frac{\theta}{2}\right)\frac{1-x\cos\theta}{1+x^2-2x\cos\theta}\,d\theta$$

$$= \frac{2}{\pi}\int_0^\pi \log\left(2\cos\frac{\theta}{2}\right)\frac{1-x\cos\theta}{1+x^2-2x\cos\theta}\,d\theta.$$

[6.52] Putting $f(z) = \sin z$ in [6.38], we get

$$\sin x = \frac{1}{\pi}\int_0^\pi \left(\sin\left(e^{i\theta}\right) + \sin\left(e^{-i\theta}\right)\right)\frac{1-x\cos\theta}{1+x^2-2x\cos\theta}\,d\theta$$

$$= \frac{2}{\pi}\int_0^\pi \sin\left(\frac{e^{i\theta}+e^{-i\theta}}{2}\right)\cos\left(\frac{e^{i\theta}-e^{-i\theta}}{2}\right)\frac{1-x\cos\theta}{1+x^2-2x\cos\theta}\,d\theta$$

$$= \frac{2}{\pi}\int_0^\pi \sin(\cos\theta)\cos(i\sin\theta)\frac{1-x\cos\theta}{1+x^2-2x\cos\theta}\,d\theta$$

$$= \frac{2}{\pi}\int_0^\pi \sin(\cos\theta)\cosh(\sin\theta)\frac{1-x\cos\theta}{1+x^2-2x\cos\theta}\,d\theta.$$

[6.53] Putting $f(z) = \sinh z$ in [6.38], we get

$$\sinh x = \frac{1}{\pi}\int_0^\pi \left(\sinh\left(e^{i\theta}\right) + \sinh\left(e^{-i\theta}\right)\right)\frac{1-x\cos\theta}{1+x^2-2x\cos\theta}\,d\theta$$

$$= \frac{2}{\pi}\int_0^\pi \sinh\left(\frac{e^{i\theta}+e^{-i\theta}}{2}\right)\cosh\left(\frac{e^{i\theta}-e^{-i\theta}}{2}\right)\frac{1-x\cos\theta}{1+x^2-2x\cos\theta}\,d\theta$$

$$= \frac{2}{\pi}\int_0^\pi \sinh(\cos\theta)\cosh(i\sin\theta)\frac{1-x\cos\theta}{1+x^2-2x\cos\theta}\,d\theta$$

$$= \frac{2}{\pi}\int_0^\pi \sinh(\cos\theta)\cos(\sin\theta)\frac{1-x\cos\theta}{1+x^2-2x\cos\theta}\,d\theta.$$

12 Solutions to Series Sums Using Special Integrals   367

[6.54] Applying Theorem 6.13 to $f(z) = g(z) = 1/(1-z) = \sum_{j=0}^{\infty} z^j$, we get

$$\frac{1}{1-x} \star \frac{1}{1-y} = \sum_{j=0}^{\infty} (xy)^j = \frac{1}{1-xy}$$

$$= -1 + \frac{1}{2\pi} \int_0^\pi \left(\frac{1}{1-xe^{i\theta}} + \frac{1}{1-xe^{-i\theta}} \right) \left(\frac{1}{1-ye^{i\theta}} + \frac{1}{1-ye^{-i\theta}} \right) d\theta$$

$$= -1 + \frac{1}{2\pi} \int_0^\pi \frac{2 - x\left(e^{i\theta} + e^{-i\theta}\right)}{1+x^2 - x\left(e^{i\theta} + e^{-i\theta}\right)} \cdot \frac{2 - y\left(e^{i\theta} + e^{-i\theta}\right)}{1+y^2 - y\left(e^{i\theta} + e^{-i\theta}\right)} d\theta$$

$$= -1 + \frac{1}{2\pi} \int_0^\pi \frac{2 - 2x\cos\theta}{1+x^2 - 2x\cos\theta} \cdot \frac{2 - 2y\cos\theta}{1+y^2 - 2y\cos\theta} d\theta$$

$$= -1 + \frac{2}{\pi} \int_0^\pi \frac{1 - x\cos\theta}{1+x^2 - 2x\cos\theta} \cdot \frac{1 - y\cos\theta}{1+y^2 - 2y\cos\theta} d\theta$$

$$\iff \int_0^\pi \frac{1 - x\cos\theta}{1+x^2 - 2x\cos\theta} \cdot \frac{1 - y\cos\theta}{1+y^2 - 2y\cos\theta} d\theta = \frac{\pi}{2} \cdot \frac{2-xy}{1-xy}.$$

[6.55] Applying Theorem 6.13 to $f(z) = z/(1-z)^2 = \sum_{j=1}^{\infty} jz^j$ and $g(z) = \log(1-z) = -\sum_{j=1}^{\infty} z^j/j$, we get

$$\frac{x}{(1-x)^2} \star \log(1-y) = -\sum_{j=1}^{\infty} (xy)^j = -\frac{xy}{1-xy}$$

$$= \frac{1}{2\pi} \int_0^\pi \left(\frac{xe^{i\theta}}{(1-xe^{i\theta})^2} + \frac{xe^{-i\theta}}{(1-xe^{-i\theta})^2} \right) \left(\log\left(1-ye^{i\theta}\right) + \log\left(1-ye^{-i\theta}\right) \right) d\theta$$

$$= \frac{1}{2\pi} \int_0^\pi x \frac{e^{i\theta}\left(1+x^2e^{-2i\theta} - 2xe^{-i\theta}\right) + e^{-i\theta}\left(1+x^2e^{2i\theta} - 2xe^{i\theta}\right)}{\left(1+x^2 - x\left(e^{i\theta} + e^{-i\theta}\right)\right)^2}$$
$$\log\left(1+y^2 - y\left(e^{i\theta} + e^{-i\theta}\right)\right) d\theta$$

$$= \frac{1}{2\pi} \int_0^\pi x \frac{e^{i\theta} + e^{-i\theta} + x^2\left(e^{i\theta} + e^{-i\theta}\right) - 4x}{\left(1+x^2 - 2x\cos\theta\right)^2} \log\left(1+y^2 - 2y\cos\theta\right) d\theta$$

$$= \frac{1}{\pi} \int_0^\pi x \log\left(1+y^2 - 2y\cos\theta\right) \frac{(1+x^2)\cos\theta - 2x}{\left(1+x^2 - 2x\cos\theta\right)^2} d\theta$$

$$\iff \int_0^\pi x \log\left(1+y^2 - 2y\cos\theta\right) \frac{(1+x^2)\cos\theta - 2x}{\left(1+x^2 - 2x\cos\theta\right)^2} d\theta = -\frac{\pi xy}{1-xy}.$$

[6.56] Putting $f(x) = \sum_{j=0}^{n} a_j x^j$ in [6.38], we get

$$2a_0 + \sum_{j=1}^{n} a_j x^j = a_0 + f(x)$$

$$= \frac{1}{\pi} \int_0^{\pi} \left(f\left(e^{i\theta}\right) + f\left(e^{-i\theta}\right) \right) \frac{1 - x\cos\theta}{1 + x^2 - 2x\cos\theta} \, d\theta$$

$$= \frac{1}{\pi} \int_0^{\pi} \left(\sum_{j=0}^{n} a_j \left(e^{ji\theta} + e^{-ji\theta}\right) \right) \frac{1 - x\cos\theta}{1 + x^2 - 2x\cos\theta} \, d\theta$$

$$= \frac{2}{\pi} \int_0^{\pi} \left(\sum_{j=0}^{n} a_j \cos j\theta \right) \frac{1 - x\cos\theta}{1 + x^2 - 2x\cos\theta} \, d\theta.$$

[6.57] Applying Theorem 6.13 to $f(z) = g(z) = 1/\sqrt{1-z} = \sum_{j=0}^{\infty} \binom{2j}{j} z^j / 2^{2j}$, we get

$$\frac{1}{\sqrt{1-x}} \star \frac{1}{\sqrt{1-x}} = \sum_{j=0}^{\infty} \binom{2j}{j}^2 \frac{x^{2j}}{2^{4j}}$$

$$= -1 + \frac{1}{2\pi} \int_0^{\pi} \left(\frac{1}{\sqrt{1 - xe^{i\theta}}} + \frac{1}{\sqrt{1 - xe^{-i\theta}}} \right)^2 d\theta. \tag{12.13}$$

Putting $x = 1/4$ in (12.13), we get

$$\sum_{j=0}^{\infty} \frac{1}{2^{8j}} \binom{2j}{j}^2 = -1 + \frac{1}{2\pi} \int_0^{\pi} \left(\frac{1}{\sqrt{1 - \frac{1}{4}e^{i\theta}}} + \frac{1}{\sqrt{1 - \frac{1}{4}e^{-i\theta}}} \right)^2 d\theta$$

$$= -1 + \frac{1}{2\pi} \int_0^{\pi} \frac{2 - \frac{1}{4}\left(e^{i\theta} + e^{i\theta}\right) + 2\sqrt{\frac{17}{16} - \frac{1}{4}\left(e^{i\theta} + e^{i\theta}\right)}}{\frac{17}{16} - \frac{1}{4}\left(e^{i\theta} + e^{i\theta}\right)} \, d\theta$$

$$= -1 + \frac{1}{2\pi} \int_0^{\pi} \frac{2 - \frac{1}{2}\cos\theta + 2\sqrt{\frac{17}{16} - \frac{1}{2}\cos\theta}}{\frac{17}{16} - \frac{1}{2}\cos\theta} \, d\theta$$

$$= -1 + \frac{1}{2\pi} \int_0^\pi \frac{32 - 8\cos\theta + 8\sqrt{17 - 8\cos\theta}}{17 - 8\cos\theta} d\theta$$

$$= -1 + \frac{1}{2\pi} \int_0^\pi \left(1 + \frac{15 + 8\sqrt{9 + 16\sin^2\frac{\theta}{2}}}{9 + 16\sin^2\frac{\theta}{2}}\right) d\theta$$

$$= -\frac{1}{2} + \frac{1}{2\pi} \int_0^\pi \frac{15 + 8\sqrt{9 + 16\sin^2\frac{\theta}{2}}}{9 + 16\sin^2\frac{\theta}{2}} d\theta.$$

[6.58] Put $x = 1$ in [6.41].

[6.59] Put $x = 1$ in [6.42].

Appendix A
Mapping from the Manuscript to the Book

Manuscript page#	Manuscript entry#	Book entry#
1	(1)	[1.1]
1	(2)	[1.2]
1	(3)	[1.3]
1	(4) *Incorrect Identity*	[1.4] *Corrected Identity*
1	(5)	[1.5]
1	(6) *Incorrect Identity*	[1.6] *Corrected Identity*
1	(7)	[1.7]
1	(8)	[1.8]
2	(9)	[1.9]
2	(10)	[1.10]
2	(11)	[1.11]
2	(12)	[1.12]
2	(13)	[1.13]
3	(14)	[1.14]
3	(15)	[1.15]
3	(16)	[1.16]
3	(17)	[1.17]
3	(18)	[1.18]
4	(19)	[1.20]
4	(20)	[1.19]
4	(21)	[1.21]
4	(22)	[1.22]
4	(23)	[1.23]
4	(24)	[1.24]
4	(25)	[1.25]
5	(26)	[1.26]
5	(27)	[1.27]
5	(28)	[1.28]
5	(29)	[1.29]
5	(30)	[1.30]
5	(31)	[1.31]
5	(32)	[1.32]

Appendix A: Mapping from the Manuscript to the Book

Manuscript page#	Manuscript entry#	Book entry#
6	(33)	[1.33]
6	(34)	[4.22]
6	(35)	[1.34]
6	(36)	[1.35]
6	(37)	[1.36]
6	(38)	[1.37]
6	(39)	[1.38]
7	(40)	[1.39]
7	(41)	[1.40]
7	(42)	[1.41]
7	(43)	[1.42]
7	(44)	[1.43]
7	(45)	[1.44]
7	(46)	[1.45]
7	(47)	[1.46]
8	(48)	[1.47]
8	(49)	[1.48]
8	(50)	[1.49]
8	(51)	[1.50]
8	(52)	[1.51]
8	(53)	[1.52]
8	(54)	[1.53]
9	(55)	[1.54]
9	(56)	[1.55]
9	(57)	[1.56]
9	(58)	[1.57]
9	(59)	[1.58]
9	(60)	[1.59]
9	(61)	[1.60]
9	(62)	[1.61]
10	(63)	[1.62]
10	(64)	[1.63]
10	(65)	[1.64]
10	(66)	[1.65]
10	(67)	[1.66]
10	(68)	[1.67]
11	(69)	[1.68]
11	(70)	[1.69]
11	(71)	[1.73]
11	(72)	[1.83]
12	(1) *Incorrect Identity*	[2.1] *Corrected Identity*
12	(2)	[2.2]
12	(3)	[2.3]
13	(4)	[2.4]
13	(5)	[2.5]
13	(6)	[2.6]
13	(7)	[2.7]
13	(8)	[2.8]
13	(9)	[2.9]

Appendix A: Mapping from the Manuscript to the Book 373

Manuscript page#	Manuscript entry#	Book entry#
14	(10)	[2.10]
14	(11)	[2.11]
14	(12)	[2.12]
14	(13)	[2.13]
14	(14)	[2.14]
15	(15)	[2.15]
15	(16)	[2.16]
15	(17)	[2.17]
15	(18)	[2.18]
15	(19)	[2.19]
16	(20)	[2.20]
16	(21)	[2.21]
16	(22)	[2.22]
16	(23)	[2.23]
16	(24)	[2.24]
16	(25)	[2.25]
16	(26)	[2.26]
17	(1)	[2.27]
17	(2)	[2.28]
17	(3)	[2.29]
17	(4)	[2.30]
18	(5)	[2.31]
18	(6)	[2.32]
18	(7)	[2.33]
18	(8)	[2.34]
18	(9)	[2.35]
18	(10)	[2.36]
19	(11)	[2.37]
19	(12)	[2.38]
19	(13)	[2.39]
19	(14)	[2.40]
20	(15)	[2.41]
20	(16)	[2.42]
20	(17)	[2.43]
20	(18)	[2.44]
21	(19)	[2.45]
21	(20)	[2.46]
21	(21)	[2.47]
21	(22)	[2.48]
21	(23)	[2.49]
21	(24)	[2.50]
21	(25)	[2.51]
22	(26)	[2.52]
23	A Binomial Identity	[1.70]
24–27	Fibonacci Series	
28–29	A Functional Equation	
30–34	Rational Approximations to $\sqrt{2}$	
34–35	Series Expansions of $\sin x$ and $\cos x$	
36 and 38	Approximation to $n!$	
37	Series for $\tan^{-1} x$	Example 1.8

Appendix A: Mapping from the Manuscript to the Book

Manuscript page#	Manuscript entry#	Book entry#
37	Series for $\sin^{-1} x$	[2.8]
39	(1)	[4.1]
39	(2)	[4.2–4.3]
39	(3)	[4.4–4.5]
39	(4)	[4.6–4.7]
39	(5)	[4.8–4.9]
40	(6) *Incorrect Identity*	[4.10–4.11] *Corrected Identities*
40	(7)	[4.12–4.13]
40	(8)	[4.14–4.15]
40	(9)	[4.16–4.17]
41	(10)	[4.18–4.19]
41	(11)	[4.20–4.21]
41	(12)	[4.22]
41	(13)	[4.23]
41	(14)	[4.24]
42	(15)	[4.25]
42	(16)	[4.26–4.27]
43	(17)	[4.28]
43	(18) *Incorrect Identity*	[4.29–4.30] *Corrected Identities*
43	(19)	[4.31–4.32]
43	(20)	[4.33–4.34]
44	(21)	[4.35–4.36]
44	(22)	[4.37–4.38]
44	(23)	[4.39–4.40]
44	(24)	[4.41–4.42]
45	(25)	[4.43–4.44]
45	(26)	[4.45–4.46]
46	(27)	[4.47–4.48]
47	(28)	[4.49–4.50]
47	(29)	[4.51]
48	(30)	[4.52–4.53]
49	(31)	[4.54]
49	Theorem 5 *Incorrect Theorem*	Theorem 4.16 *Corrected Theorem*
50	(32)	[4.55–4.56]
50	Theorem 4 *Incorrect Theorem*	Theorem 4.15 *Corrected Theorem*
51–52	(33)	[4.57–4.58]
52	(34–35)	[4.59–4.60]
53	(36)	[4.61–4.62]
53	(37)	[4.63–4.64]
53	(38)	[4.65–4.66]
53	(39) *Incorrect Identity*	[4.67–4.68] *Corrected Identities*
54	(40)	[4.69–4.70]
54	(41)	[4.71–4.72]
54	(42)	[4.73–4.74]
54	(43)	[4.75–4.76]
55	(44)	[4.77–4.78]
55	(45)	[4.79–4.80]
55	(46)	[4.81–4.82]
55	(47)	[4.83–4.84]

Appendix A: Mapping from the Manuscript to the Book 375

Manuscript page#	Manuscript entry#	Book entry#
56	(48)	[4.85–4.86]
56	(49)	[4.87–4.88]
56	(50)	[4.89–4.90]
56	(51)	[4.91–4.92]
57	(52)	[4.94–4.95]
57	(53)	[4.93]
57	(54) *Incorrect Identity*	[4.96–4.97] *Corrected Identities*
57	(55)	[4.98]
58	(56)	[4.99–4.100]
58	(57)	[4.96]
58	(58)	[4.101]
58	(59)	[4.102]
59	(60)	[4.103]
59	(61)	[4.104–4.105]
59	(62)	[4.106]
60	(63)	[4.107]
60	(64)	[4.108]
60	(65)	[4.109]
61	Theorem 1	Theorem 4.1
61	Last two lines	Corollary 4.3
62	Generalization of Theorem 1	Theorem 4.2
62–63	A Special Case	Theorem 4.4 *A Generalization of the "A Special Case"*
64	Theorem 2	Theorems 4.5–4.6
65–66	Theorem 3	Theorems 4.7–4.14
67	Theorem 6	Theorem 4.17
68	(1)	Corollary 4.18
68	(2)	Corollary 4.19
69	(3)	Corollary 4.20
69	(4)	Corollary 4.21
70	(5)	Corollary 4.22
70	(6)	Corollary 4.23
71	(7)	Corollary 4.24
71	(8)	Corollary 4.25
72	(9)	Corollary 4.26
72	(10)	Corollary 4.27
73	Theorem 7	Theorem 4.41
73	Note	Theorem 4.28
74	(1)	Corollary 4.42
74	(2)	Corollary 4.43
75	(3)	Corollary 4.44
75	(4)	Corollary 4.45
76	(5)	Corollary 4.46
76	(6)	Corollary 4.47
77	(7)	Corollary 4.48
77	(8) *Incorrect Corollary*	Corollary 4.49 *Corrected Corollary*
78	(9)	Corollary 4.50
78	(10)	Corollary 4.51

Appendix A: Mapping from the Manuscript to the Book

Manuscript page#	Manuscript entry#	Book entry#
79	(11) (First Part)	Corollary 4.52
79	(11) (Second Part)	Corollary 4.53
79	(11) Note	Corollaries 4.29 – 4.40
80	Theorem 8	Theorem 4.28
81	(1) *Incorrect Identity*	[4.96] *Corrected Identity*
81	(2) *Incorrect Identity*	[4.96–4.97] *Corrected Identities*
81	(3) *Incorrect Identity*	[4.110–4.111] *Corrected Identities*
81	(4) *Incorrect Identity*	[4.112–4.113] *Corrected Identities*
82	(5)	[4.114]
82	(6)	[4.115–4.116]
82	(7)	[4.117]
83	(8)	[4.118–4.119]
83	(9)	[4.120]
84	(10)	[4.121]
84	(11)	[4.122]
84	(12)	[4.123]
84	(13)	[4.124]
85	(14)	[4.125]
85	(15)	[4.126]
85	(16)	[4.127]
85	(17)	[4.128]
85	(18)	[4.129–4.130]
86	(19)	[4.131–4.132]
86	(20)	[4.133–4.134]
87	(21)	[4.135–4.136]
87	(22)	[4.137–4.138]
87	(23)	[4.139]
88	(24)	[4.140]
88	(25)	[4.141]
88	(27)	[4.143–4.144]
88	(28)	[4.145]
88	(29)	[4.146]
88	(30)	[4.147]
89	(31)	[4.148]
89	(32)	[4.149]
89	(33)	[4.150]
89	(34)	[4.151]
90	(35)	[4.152]
90	(36)	[4.153–4.154]
90	(37)	[4.155]
91	(38)	[4.156]
91	(39)	[4.157]
91	(40)	[4.159]
91	(41)	[4.158]
92	(42)	[4.160–4.161]
92	(43)	[4.163]
92	(43)	[4.142]
92	(44)	[1.71–1.72]

Appendix A: Mapping from the Manuscript to the Book 377

Manuscript page#	Manuscript entry#	Book entry#
93–94	The Z–Functions	Definition 5.8
95	(4)	[5.21–5.23]
96	(5)	[5.20] and [5.24–5.25]
96	(6)	[5.26]
97	(1)	[5.8]
97	(2)	[5.9]
97	(3)	[5.10]
97	(4)	[5.11]
97	(5)	[5.12–5.13]
97	(6)	[5.14–5.15]
98	(7)	[5.16–5.17]
98	(8)	[5.18–5.19]
99	Circulo–Z Functions	(5.14)
100	Circulo–Z Functions	Definition 5.11
100	Hypo–Z Functions	(5.17) and (5.18)
101	Hypo–Z Functions	Definition 5.12
101	(1) *Incorrect Identity*	(5.14) *Corrected Identity*
102	(2) *Incorrect Identities*	[5.93–5.94] *Corrected Identities*
102	(3) *Incorrect Identities*	[5.95–5.96] *Corrected Identities*
102	(4) *Incorrect Identities*	(5.17) and (5.18) *Corrected Identities*
103	$C_1(\theta)$	(5.27)
103	$C_2(\theta)$	(5.29)
103	$S_1(\theta)$	(5.26)
103	$S_2(\theta)$	(5.28)
103–104	(1)	[5.35–5.37]
104	(2)	[5.38–5.40]
105	(3)	[5.41–5.43]
105–106	(4) *Incorrect Identities*	[5.44–5.46] *Corrected Identities*
107	(1)	[5.27–5.28]
107	(2)	[4.33–4.34] and [5.29]
107	(3)	[4.35–4.36] and [5.30]
108	(4)	[5.31–5.32]
108	(5)	[4.39–4.40] and [5.33]
108	(6)	[4.41–4.42] and [5.34]
109	(1)	[5.47]
109	(2)	[5.48–5.49]
109	(3)	[5.50–5.51]
109	(4)	[5.52–5.53]
110	(5) *Incorrect Identities*	[5.54–5.55] *Corrected Identities*
110	(6)	[5.56]
110	(7)	[5.57–5.58]
110	(8)	[5.59–5.60]
111	(9)	[5.61–5.62]
111	(10)	[5.63–5.64]
111	(11)	[5.65–5.67]
112	(12)	[5.68]
112	(13)	[5.69–5.70]
112	(14)	[5.71–5.72]
112	(15)	[5.73–5.74]
112	(16)	[5.75–5.76]

378 Appendix A: Mapping from the Manuscript to the Book

Manuscript page#	Manuscript entry#	Book entry#
113	(17)	[5.77–5.78]
113	(18)	[5.79–5.80]
113	(19)	[5.81–5.82]
113	(20)	[5.83–5.84]
114	(21)	[5.75]
114	(22)	[5.75–5.76]
114	(23)	[5.73–5.74]
114	(24)	[5.85]
114	(25) *Incorrect Identity*	[4.1] *Corrected Identity*
114	(26)	[5.86]
115	(27)	[5.87–5.88]
115	(28)	[5.89–5.90]
115	(29)	[5.91]
115	(30) *Incorrect Identity*	[5.91–5.92] *Corrected Identities*
115	(31) *Incorrect Identity*	[5.8] *Corrected Identity*
115	(32) *Incorrect Identity*	[5.9] *Corrected Identity*
115	(33) *Incorrect Identity*	
116	(34) *Incorrect Identity*	
116	(35) *Incorrect Identity*	
116	(36) *Incorrect Identity*	
117	$C_1(\theta)$	(6.28)
117	$C_2(\theta)$	(6.30)
117	$S_1(\theta)$	(6.29)
117	$S_2(\theta)$	(6.31)
118	(1)	[6.1]
118	(2)	[6.2]
119	(3)	[6.3]
119	(4)	[6.4]
120	(5) *Incorrect Identity*	[6.5] *Corrected Identity*
120	(6) *Incorrect Identity*	[6.6] *Corrected Identity*
121	(7) *Incorrect Identity*	[6.7] *Corrected Identity*
121	(8)	[6.8]
122	(9) *Incorrect Identity*	[6.9] *Corrected Identity*
122	(10)	[6.10]
123	(11) *Incorrect Identity*	[6.11] *Corrected Identity*
123	(12) *Incorrect Identity*	[6.12] *Corrected Identity*
124	(13) *Incorrect Identity*	[6.13] *Corrected Identity*
124	(14) *Incorrect Identity*	[6.14] *Corrected Identity*
125	(15) *Incorrect Identity*	[6.15] *Corrected Identity*
125	(16) *Incorrect Identity*	[6.16] *Corrected Identity*
126	(17) *Incorrect Identity*	[6.17] *Corrected Identity*
126	(18) *Incorrect Identity*	[6.18] *Corrected Identity*
127	(19) *Incorrect Identity*	[6.19] *Corrected Identity*
127	(20) *Incorrect Identity*	[6.20] *Corrected Identity*
128	$C_1(\theta)$	(6.32)
128	$S_1(\theta)$	(6.33)
128	$C_2(\theta)$	(6.34)
128	$S_2(\theta)$	(6.35)
129	(21)	[6.21]
129	(22)	[6.22]

Appendix A: Mapping from the Manuscript to the Book 379

Manuscript page#	Manuscript entry#	Book entry#
129	(23)	[6.23]
130	(24)	[6.24]
130	(25)	[6.25]
130	(26)	[6.26]
131	(27)	[6.27]
131	(28)	[6.28]
132	(29)	[6.29]
132	(30)	[6.30]
132	(31)	[6.31]
132	(32)	[6.32]
133	(33)	[6.33]
133	(34)	[6.34]
133	(35)	[6.35]
133	(36)	[6.36]
134	Note *Incorrect Identities*	
135	Definition	Definition 6.12
136	General Formula for the Functional Transformation	Theorem 6.13
137	(1)	[6.37]
137	(2)	[6.38]
137	(3)	[6.39]
138	(4) *Incorrect Identity*	[6.40] *Corrected Identity*
138	(5)	[6.41]
138	(6)	[6.42]
138	(7)	[6.43]
138	(8)	[6.44]
139	(9)	[6.45] *Generalized Identity*
139	(10)	[6.46]
139	(11)	[6.47]
139	(12)	[6.48]
140	(13)	[6.49]
140	(14)	[6.50] *Generalized Identity*
140	(15)	[6.51]
140	(16)	[6.52]
141	(17)	[6.53]
141	(18)	[6.54] *Generalized Identity*
141	(19) *Incorrect Identity*	[6.55] *Corrected Identity*
141	(20)	[6.56]
142	(21), Line 1 *Incorrect Identity*	[1.19] *Corrected Identity*
142	(21), Lines 2–4 *Incorrect Identity*	[6.57] *Corrected Identity*
142	(22)	[6.58]
142	(23)	[6.59]
143	(1)	[1.73]
143	(2)	[1.74]
143	(3)	[1.75]
144	(4)	[1.76]
144	(5)	[1.77]
144	(6)	[1.78]

Appendix A: Mapping from the Manuscript to the Book

Manuscript page#	Manuscript entry#	Book entry#
145	(7)	[1.79]
145	(8)	[1.80]
145	(9)	[1.81]
146	(10) *Incorrect Identity*	[1.82] *Corrected Identity*
146	(20)	[1.92]
147	(11)	[1.83]
147	(12)	[1.84]
147–148	(13)	[1.85]
148	(14)	[1.86]
148	(15)	[1.87]
149	(16)	[1.88]
149	(17)	[1.89]
150	(18)	[1.90]
150	(19)	[1.91]

References

Ahlfors, L. V. (2017). *Complex analysis* (3rd ed.). McGraw Hill.
Apostol, T. M. (2017). *Calculus volume 1* (2nd ed.). Wiley.
Berndt, B. C. (1994). *Ramanujan's notebooks*. Part IV: Springer-Verlag.
Boros, G., & Moll, V. H. (2006). *Irresistible integrals*. Cambridge University Press.
Borwein, J. M., & Borwein, P. B. (1987). *Pi and the AGM: A study in analytic number theory and computational complexity*. Wiley Interscience Publication.
Bromwich, T. J. I. (2018). *Elementary integrals: A short table*. Forgotten Books.
Brown, J. W., & Churchill, R. V. (2017a). *Complex variables and applications* (9th ed.). McGraw Hill.
Brown, J. W., & Churchill, R. V. (2017b). *Fourier series and boundary value problems* (8th ed.). McGraw Hill.
Carslaw, H. S. (2018). *Introduction to the theory of Fourier's series and integrals*. Forgotten Books.
Convay, J. B. (1995). *Functions of one complex variable I* (2nd ed.). Springer-Verlag.
Edwards, J. (1896). *Integral calculus for beginners*. Macmillan and Co.
Erdélyi, A. (1955). *Higher transcendental functions (Vol.* III): McGraw Hill.
Hardy, G. H. (1921). *A course of pure mathematics* (3rd ed.). Cambridge University Press.
Hoffman, K., & Kunje, R. (2015). *Linear algebra* (2nd ed.). Pearson.
Lang, S. (2013). *Complex analysis* (4th ed.). Springer.
Lehmer, D. H. (1985). Interesting series involving the central binomial coefficient. *The American Mathematical Monthly, 92*(7), 449–457.
Loney, S. L. (1928). *Plane trigonometry*. Part II: Cambridge University Press.
Maron, I. A. (1973). *Problems in calculus of one variable*. Mir Publishers.
Mishra, A. (1997). Unpublished manuscript.
Mittag-Leffler, G. (1903). Sur la nouvelle fonction $E_\alpha(x)$. *C. R. Acad. Sci. Paris, 137*, 554–558.
Piskunov, N. (1996a). *Differential and integral calculus volume 1*. Mir Publishers.
Piskunov, N. (1996b). *Differential and integral calculus volume 2*. Mir Publishers.
Prabhakar, T. R. (1971). A singular integral equation with a generalized Mittag Leffler function in the kernel. *Yokohama Mathematical Journal, 19*, 7–15.
Shukla, A. K., & Prajapati, J. C. (2007). On a generalization of Mittag-Leffler function and its properties. *Journal of Mathematical Analysis and Applications, 336*, 797–811.

Thomas, G. B., & Finney, R. L. (1995). *Calculus and analytic geometry* (6th ed.). Narosa Publishing House.
Thomas, G. B., Hass, J., Heil, C., & Weir, M. D. (2018). *Thomas' calculus* (14th ed.). Pearson.
Tolstov, G. P. (1977). *Fourier series*. Dover Publications, Inc.
Watson, G. N. (2011). *A treatise on the theory of Bessel functions* (2nd ed.). Cambridge University Press.
Whittaker, E. T., & Watson, G. N. (2016). *A course of modern analysis* (4th ed.). Cambridge University Press.
Wiman, A. (1905). Über den fundamentalsatz in der teorie der funktionen $E_\alpha(x)$. *Acta Mathematica, 29*, 191–201.
Wright, E. M. (1933). On the coefficients of power series having exponential singularities. *Journal of the London Mathematical Society,* s1-8:71–79.

The manufacturer's authorised representative in the EU is Springer Nature Customer Service Centre GmbH, Europaplatz 3, 69115 Heidelberg, Germany. If you have any concerns regarding our products, please contact ProductSafety@springernature.com

Printed and bound by CPI Group (UK) Ltd, Croydon, CR0 4YY

26/03/2026

02078939-0009